Mechanics of Motor Proteins and the Cytoskeleton

Mechanics of Motor Proteins and the Cytoskeleton

Jonathon Howard
Department of Physiology and Biophysics
University of Washington
and
Max Planck Institute for
Molecular Cell Biology and Genetics

Sinauer Associates, Inc. • Publishers
Sunderland, Massachusetts

About the Cover

A rat kinesin molecule taking 8 nm steps along a microtubule. (Data courtesy of Nick Carter and Rob Cross.)

Mechanics of Motor Proteins and the Cytoskeleton

Copyright © 2001 by Sinauer Associates, Inc.

All rights reserved. This book may not be published in whole or in part without permission from the publisher. For information, address Sinauer Associates, 23 Plumtree Road, Sunderland, MA 01375 U.S.A.

FAX: 413-549-1118

Email: publish@sinauer.com; Internet: www.sinauer.com

Library of Congress Cataloging-in-Publication Data

Howard, Jonathon, 1957-
　Mechanics of motor proteins and the cytoskeleton / Jonathon Howard
　　p. cm.
　Includes bibliographical references and index.
　ISBN 978-0-87893-333-4
　　1. Cytoskeletal proteins. I. Title.

QP552.C96 H69 2001
572'.6—dc21

2001017017

To Libby Howard and the memory of Harry Stanton Howard

Brief Contents

1 Introduction 1

PART I Physical Principles 7

    2 Mechanical Forces 9
    3 Mass, Stiffness, and Damping of Proteins 29
    4 Thermal Forces and Diffusion 49
    5 Chemical Forces 75
    6 Polymer Mechanics 99

PART II Cytoskeleton 117

    7 Structures of Cytoskeletal Filaments 119
    8 Mechanics of the Cytoskeleton 135
    9 Polymerization of Cytoskeletal Filaments 151
    10 Force Generation by Cytoskeletal Filaments 165
    11 Active Polymerization 179

PART III Motor Proteins 195

    12 Structures of Motor Proteins 197
    13 Speeds of Motors 213
    14 ATP Hydrolysis 229
    15 Steps and Forces 245
    16 Motility Models: From Crossbridges to Motion 263

Afterword 285

Appendix 287

Bibliography 339

Index 361

Preface

This book is for biology, physics, and engineering students who want to learn about the mechanics of molecules and how it applies to the morphology and motility of cells. The book has three parts—mechanical principles, the cytoskeleton, and motor proteins. The first part, especially, has a large number of equations, but the nonmathematical reader should not be put off: All the key concepts are explained in words, and there are lots of boxed examples and figures—think of the equations as definitions and summaries of main results. For the mathematically inclined, detailed proofs of all the important results are included in the Appendix. To keep track of the symbols, a table of parameters is provided on the front endpapers, and tables of physical constants and units are provided on the rear endpapers. For biologists interested in motor proteins, a shorter course could comprise Chapters 1–3, the first two sections of Chapters 4 and 5, Chapter 7, and Chapters 12–16. For biophysicists and bioengineers more interested in molecular mechanics, a shorter course could comprise Chapters 1–6 (including the Appendix and Problem sets), followed by selected sections of Parts II and III, or readings from the structural biology and protein machine literature.

The book assumes a rudimentary knowledge of cell biology that can be found in textbooks such as *The Cell*, Second Edition (Cooper, 2000), *Molecular Biology of the Cell*, Third Edition (Alberts, 1994), or *Molecular Cell Biology*, Fourth Edition (Lodish, 2000). One of these books would serve as a useful reference about proteins and to get a broader picture of the structure and function of different cells. The book also assumes familiarity with freshman physics, though all concepts are introduced from first principles. A standard elementary physics text is *Physics* (Resnick et al., 1992); *The Feynman Lectures on Physics* (Feynman et al., 1963) is a wonderful text, but it is difficult.

A comment on units. In keeping with the molecular theme of the book, all energies are expressed per molecule, rather than per mole, as is common in chemistry and biochemistry textbooks. SI units are used throughout, with the exception of micron, which is used instead of micrometer. Following the SI guidelines, angstroms and calories are avoided. The advantage of using SI units and thinking in terms of molecules rather than moles is that energies, forces, and displacements can be readily calculated—for example, thermal energy equals 4×10^{-21} J = 4 pN $\times$ 1 nm.

This book is not a comprehensive survey of motor proteins and the cytoskeleton. The references are by no means exhaustive, and I apologize to all my colleagues whose work has not been cited. Instead I refer the reader to the following. There are a number of histories: They include *Machina Carnis: The Biochemistry of Muscular Contraction in Its Historical Development* (Needham,

1971), *Reflections on Muscle* (Huxley, 1980), and *The Cytoskeleton: An Introductory Survey* (Schliwa, 1986). Review articles by H. E. Huxley (1996) and I. R. Gibbons (1981) are also highly recommended. There are also a number of excellent books on the cytoskeleton and motor proteins. They include *The Structural Basis of Muscular Contraction* (Squire, 1981), *Molecules of the Cytoskeleton* (Amos and Amos, 1991), *Muscle Contraction* (Bagshaw, 1993), *Guidebook to Cytoskeletal and Motor Proteins* (Kreis and Vale, 1999), and *Cell Movements: From Molecules to Motility*, Second Edition (Bray, 2001).

Acknowledgments

I would like to thank the many students and colleagues who read various chapters in various drafts. Special thanks go to Taylor Allen and his students at Oberlin College, Clive Bagshaw, Dennis Bray, Harold Erickson, Peter Gillespie, Al Gordon, Freddie Gutfreund, Bertil Hille, Anthony Hyman, Tony Hyman, Andy Hunter, Tom Milac, Bill Parson, Ravi Sawhney, and Bill Schief. Many colleagues have contributed figures and I am indebted to them. Importantly, I thank the staff of Sinauer Associates, and Kerry Falvey in particular, who have been a pleasure to work with, and who have made this book a reality. There are no doubt many errors in it, but they are of my own making.

I also thank members of my lab for their patience for my neglect over the last five years as I have been writing this book. My research would not be possible without them and the generous support of the National Institutes of Health, the Human Frontier Science Program, the University of Washington, the Pew Charitable Trusts, and the Max Planck Gesellschaft. I'd especially like to thank Wayne Crill for recruiting me to Seattle, and providing a wonderful and nurturing environment in which to begin as an independent scientist. This book was begun in 1996 while I was a John Simon Guggenheim fellow on sabbatical leave in Heidelberg at the Max Planck Institute for Medical Research and the European Molecular Biology Laboratory; special thanks to my hosts Wolf Almers and Tony Hyman. This book is the culmination of twenty years of doing science; during these years I have been very fortunate to have many wonderful teachers, most notably Simon Laughlin, Jim Hudspeth, and Bertil Hille. Finally, I would like to thank my wife Karla Neugebauer, without whose support this book could not have been written.

CHAPTER

1

Introduction

Motor proteins are molecular machines that convert the chemical energy derived from the hydrolysis of ATP into mechanical work used to power cellular motility. The focus of this book is on how they operate. How do they move? How do they generate force? What paths do they follow on their cytoskeletal filaments? How much fuel do they consume, and with what efficiency? What makes these questions especially fascinating is that motor proteins are unusual machines that do what no man-made machines do—they convert chemical energy to mechanical energy directly, rather than via an intermediate such as heat or electrical energy. Over the last ten years, rapid progress has been made in understanding biological motors; now seems like an opportune time to bring these new findings together and to set out the physical principles behind the operation of molecular machines.

The study of motor proteins begins with myosin, which drives the contraction of muscle (Figure 1.1). Myosin was first isolated as a complex with actin filaments by Kühne and coworkers (1864), though it was not until the 1940s that the complex was dissociated into the separate proteins, myosin and actin (Straub, 1941–1942; Szent-Györgyi, 1941–1942). Dynein, which drives the beating of sperm and cilia, was identified in 1963 (Gibbons, 1963). Kinesin, which moves organelles along microtubules, was purified in 1985 (Brady, 1985; Vale et al., 1985). Since then, the pace of research has accelerated, and the last decade has been an age of discovery of motor proteins. Thanks in part to the genome projects, the founding members of the three families of motor proteins—myosin, dynein, and kinesin—have been joined by hundreds of newfound relatives whose amino acid sequences contain regions that are similar to the motor domains of the original proteins (see, for example, http://www.mrc-lmb.cam.ac.uk/myosin/myosin.html; http://www.blocks.fhcrc.org/~kinesin/index.html). These new motors participate in a wide range of processes that occur in all cells, not just

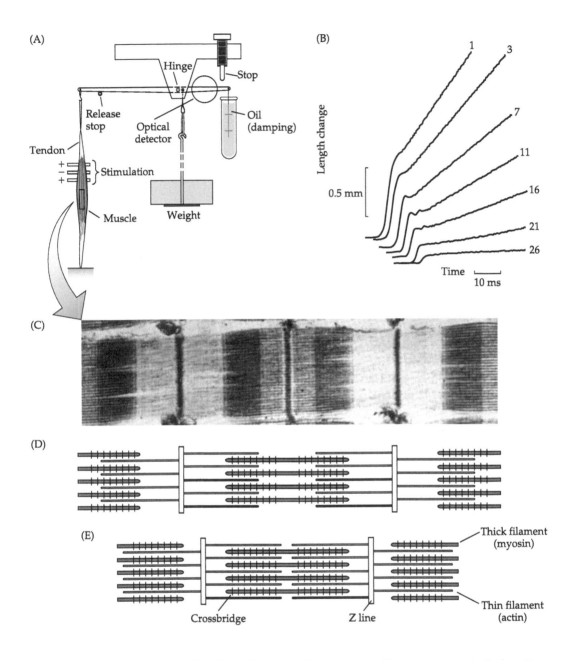

specialized motile cells such as muscle and sperm. These processes include mitosis, cell division, organelle transport, and organelle synthesis (Bloom and Endow, 1995; Sellers and Goodson, 1995; Hirokawa, 1998). Because different motors have evolved for different functions, there is a great diversity of motor properties; even within one family, different motors go in different directions at vastly different speeds. Thus understanding how motor proteins work requires finding general principles that encompass the whole spectrum of motors, not just muscle

Figure 1.1 Muscle contraction
(A) An apparatus for measuring the contraction of skeletal muscle under various loads. The muscle is activated by electrical stimulation, but is unable to shorten due to the releasable stop located to the right of the muscle–lever arm attachment. When the stop is released, the muscle shortens and the lever rotates counterclockwise about the hinge. The position of the lever, and hence the length of the muscle, is detected optically. The load is varied by placing different weights just to the right of the hinge. The oil damps out the oscillations of the apparatus. (B) Individual traces showing the shortening at different loads. After the initial jump following release of the stop, the muscle shortens at an approximately constant speed that depends on the load (in grams, numbers next to each trace). (A plot of the speed as a function of the load is shown in Figure 16.5A.) The traces are offset for clarity. The initial jumps are due to the rapid relaxation of the strain in the tendons, which become highly stretched following activation of the muscle prior to the release. (C) Electron micrograph of the sarcomere, the structural repeat of striated muscle. (D) The interdigitating thick filaments (myosin) and thin filaments (actin). (E) Contraction is due to the sliding of the filaments, which, in turn, is driven by the myosin crossbridges (depicted as protrusions from the thick filament). The thin filaments in C, D, and E are 1 μm long. (A and B after Jewell and Wilkie, 1958; C courtesy of Hugh Huxley.)

myosin or sperm dynein. In fact, the diversity of motors proves to be useful. Much has been learned about how motors work by correlating the performance of a motor with its structure, and making comparisons to other motors.

Over the last decade, as all these new motors were being discovered, a new appreciation for the roles of the cytoskeletal filaments in cell motility was being gained. It is now firmly established that the polymerization and depolymerization of actin filaments (Cortese et al., 1989; Loisel et al., 1999) and microtubules (Hotani and Miyamoto, 1990; Dogterom and Yurke, 1997; Fygenson et al., 1997) can lead to motility and force generation, even in the absence of motor proteins. For example, the crawling of cells across surfaces is driven by actin polymerization (Figure 1.2). Although polymerization in vitro requires no other proteins (just actin or tubulin), polymerization in cells is tightly regulated by protein complexes—such as the Arp2/3 complex, the kinetichore, and the centromere—that bind to the ends of the filaments. The last ten years have seen a concerted effort to identify the protein components of these complexes, and an excellent review can be found in Kreis and Vale (1999). Thus biological force generation is a general property of the cytoskeleton and is not restricted to motor proteins.

Meanwhile, the structural and physical basis for motility has been placed on a firm foundation. The atomic structures of myosin (Rayment et al., 1993b) and kinesin (Kull et al., 1996) have been solved. So too have the structures of their tracks, the actin filament (Holmes et al., 1990; Kabsch et al., 1990) and the microtubule (Lowe and Amos, 1998; Nogales et al., 1998; Nogales et al., 1999; Amos, 2000). Concurrently, techniques have been developed to study the movement of individual motor molecules (Howard et al., 1989), and to measure their steps (Svoboda et al., 1993; Finer et al., 1994; Molloy et al., 1995a), forces (Hunt et al., 1994; Svoboda and Block, 1994; Meyhöfer and Howard, 1995), and chemistry (Funatsu et al., 1995; Ishijima et al., 1998) with molecular precision.

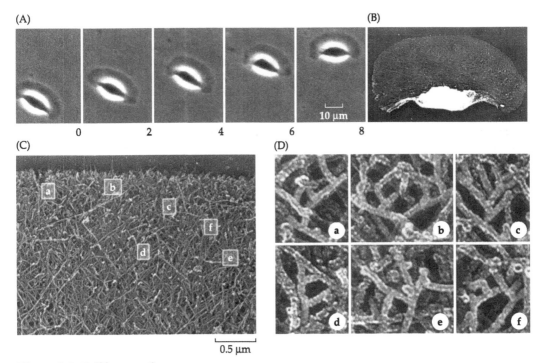

Figure 1.2 Cell locomotion
(A) Time-lapse images of a keratocyte, taken from the skin of the African clawed toad *Xenopus*, moving across a glass surface. Time is shown in minutes. Images were taken using a phase-contrast microscope. (B) Electron micrograph of the keratocyte shown in (A). (C) The leading edge of a keratocyte, showing the meshwork of crisscrossed actin filaments. (D) Higher magnification of selected regions of (C) showing that the actin filaments are branched. Forward motion is driven by the polymerization of the actin filaments adjacent to the membrane of the leading edge and simultaneous retraction of the rear edge of the cell. (From Svitkina and Borisy, 1999.)

The structural and single-molecule results reinforce the concept of proteins as machines (Alberts, 1998): According to this view, a molecular motor is an assembly of mechanical parts—springs, levers, swivels, and latches—that move in a coordinated fashion as ATP is hydrolyzed and directed motion produced (Figure 1.3). Single-molecule mechanical and optical techniques are now being applied to many biochemical processes mediated by other molecular machines; these include ATP synthesis (Noji et al., 1997), DNA transcription (Wang et al., 1998), and DNA replication (Wuite et al., 2000). The techniques can even be used to record from molecules on the surfaces of intact cells (Benoit et al., 2000; Sako et al., 2000; Schutz et al., 2000), and recordings from inside cells should soon be possible. Thus single-molecule techniques are becoming to cell biology what the patch-clamp technique and single ion-channel recordings are to neurobiology (Neher and Sakmann, 1976; Sakmann and Neher, 1995).

To understand how molecular motors operate requires mechanical concepts such as force, elasticity, damping, and work. Although these concepts can be

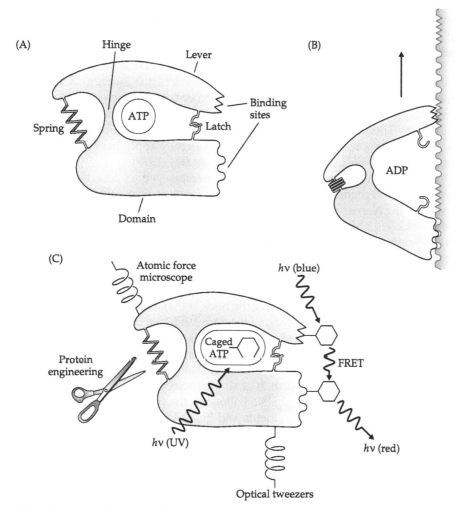

Figure 1.3 Protein as machine
(A) A protein as an assembly of mechanical components including a lever, a hinge, a spring, a latch, and various binding sites. The constituent amino acids give the protein a rigidity similar to that of hard plastic. If a 5-nm-diameter protein were enlarged 10 million times, it would be 50 mm across and similar to a plastic toy in size and rigidity. However, in order to mimic the highly damped motion characteristic of a real protein, the plastic toy would have to be immersed in a very viscous liquid, such as treacle (with viscosity 10 million times that of water). If this were done, the protein machine and the toy machine would change shape at vastly different rates: If the protein relaxed with a time constant of 100 ns in response to an applied force, the toy would relax with a time constant of 1 second, ten million times slower. (B) The opening and closing of a cleft, together with the coordinated binding and unbinding from a surface, can produce directed motion. (C) Single-molecule techniques. Optical techniques can be used to release a caged ligand such as ATP and measure its binding to the protein. The relative movement of protein domains can be monitored by fluorescence resonance energy transfer (FRET). Optical tweezers and atomic force microscopes can exert forces on the protein and therefore can be used to measure stiffness. Site-directed mutagenesis can be used to make specific alterations to a protein in order to identify the mechanical components and to facilitate the binding of chemical probes.

found in lower-level college physics textbooks (e.g., Resnick et al., 1992), there is a problem. These textbooks are primarily concerned with macroscopic systems such as billiard balls and cannonballs, whose motions are qualitatively different to the highly damped, diffusive motion of isolated molecules. It may be obvious how heating a gas creates pressure in a piston, but how does a single molecule generate force? The lack of elementary textbooks on small-system mechanics for cell biologists is in stark contrast to the situation for neurobiologists. There are several excellent textbooks that explain electrical concepts such as voltage, resistance, and capacitance as they apply to cells and membranes (Katz, 1966; Hille, 1992; Junge, 1992; Aidley, 1998).

The problem is even more serious when it comes to the interaction between mechanics and chemistry. Electrochemistry is a mature field, and excellent treatments of the Nernst equation and the likes can be found in physical chemistry textbooks (e.g., Moore, 1972; Atkins, 1986). This provides a solid theoretical framework for understanding ion pumps, ion channels, and even electron transport. But "mechanochemistry" is not mentioned in the physical chemistry textbooks. Biomechanics has no molecular foundation.

A major goal of this book is to provide a physical foundation for molecular mechanics. An important feature of the book is that all the results are derived from first principles, though many of the derivations are relegated to an Appendix so as not to obscure the underlying concepts. Part I explains how small particles like proteins respond to mechanical, thermal, and chemical forces. It is Mechanics 101 for biologists. But it should also be interesting to physics students because the frenzied motion of single molecules is very different from the ballistic motion of cannonballs. Following the physical principles, Part II focuses on the cytoskeletal filaments and Part III focuses on motor proteins. The treatments are unified in the sense that they are organized around principles rather than proteins. The chapters are not centered on a particular filament or a particular motor; instead, they focus on topics such as structure, chemistry, and mechanics, and the different filaments or motors are discussed together. I hope that the highly reductionist approach is vindicated in the last chapter, which synthesizes the molecular data into models that account quantitatively for the contraction of muscle and motility of kinesin.

Physical Principles

PART I

The motion of a football is very familiar to us, and it is well described in undergraduate physics textbooks: The foot provides a force to get the ball moving, inertia keeps it going, and gravity pulls it to the ground. But the motion of a molecule, such as a protein, is very different. The higgledy-piggledy thermal motion of microscopic particles is beyond our everyday experience and is usually only described in advanced textbooks on thermal or statistical physics, which are read by very few biology students (and few physics majors, for that matter). This is unfortunate because to understand life inside the cell, it is essential to understand how proteins and other molecules move and change shape in response to mechanical and thermal forces. Furthermore, forces play a central role in biochemistry: It is through the thermal agitation of molecular structure that proteins and other molecules reach the high-energy transition states that are essential intermediates in chemical reactions.

The ambitious goal of the first part of this book is to connect these concepts—mechanical forces, thermal forces, and chemical forces—and develop molecular mechanics from first principles in a way that is accessible to biology, physics, and chemistry students alike. Chapter 2 introduces the mechanical concepts of force, elasticity, damping, and energy. Chapter 3 describes the material properties of proteins and shows that their overdamped motion is fundamentally different from the motion of macroscopic objects. Chapter 4 shows how molecules diffuse under the action of thermal forces, and Chapter 5 shows how mechanical and thermal forces influence chemical reactions. Chapter 6 applies these concepts to the mechanical properties of polymers.

CHAPTER

2

Mechanical Forces

The central concept of this book is force. Motor proteins and other molecular machines are able to move and do work because they generate force, for it is force that drives change and motion. But what is force? Where does it come from? And what effect does it have on proteins and cells? These questions will be answered in the following five chapters that constitute Part I of this text.

This chapter is an introduction to Newtonian mechanics. It begins with the definition of force and the calculation of the magnitudes of the various forces that act on molecules. The three fundamental mechanical elements are then introduced. These elements—the spring, the dashpot, and the mass—are the building blocks of complex mechanical devices such as protein machines. I describe how these elements move individually in response to forces, and how different combinations of elements respond to forces in different ways, some combinations moving monotonically, and other combinations undergoing oscillatory motion. The chapter ends with the definitions of work and energy.

Although much of this material is contained in standard undergraduate physics textbooks such as Resnick et al. (1992) and Feynman et al. (1963), we will discover that most of the mechanics in the textbooks is irrelevant to molecular and cellular biology. The reason is that proteins and other biomolecules are so small that the inertial forces are comparatively small and can usually be ignored, whereas the viscous forces from the surrounding fluid are usually large and dominate the mechanical responses. Consequently, gravity is negligible, and the oscillatory motions characteristic of inertial systems such as planets and pendulums, systems that occupy so much of the mechanics textbooks, simply do not occur at the single-molecule level.

Figure 2.1 Deformation of an elastic object
A force, F, is applied at one end while the other end is held fixed.

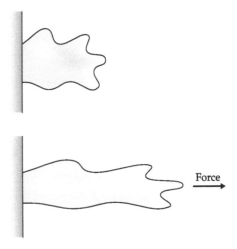

Force

A **force** is an influence—a push or a pull in everyday experience—that causes a free particle to accelerate or that causes a constrained object, such as that shown in Figure 2.1, to become deformed. Forces arise from many different physical processes. Several of these are summarized in Table 2.1, together with their approximate magnitudes. Force is often confused with work, which is the product of the force with the distance over which the force acts. If there is no motion, then there is no work, even though a force has been exerted. This is where the confusion arises: Try to tell a weightlifter who has just tried and failed to lift the heaviest weight that no work has been done!

An object can be subject to several forces simultaneously. The **net force** is the sum of all the individual forces. This seems obvious. However, such a reckoning of forces relies on an important concept, namely that the effect of a force is independent of its physical origin. For example, an elastic force can be exactly counteracted by a viscous force. As we will see, the elastic and viscous forces discussed in this chapter can, in turn, be counteracted by the thermal and chemical forces described in Chapters 4 and 5.

The SI unit of force is the **newton**. (Other SI units can be found in the table on the rear endpapers of this book.) One newton, written 1 N, corresponds approximately to a weight of 100 grams, or about 4 ounces. Although 1 N is a modest force in our everyday experience, it is billions of times larger than the forces that operate at the molecular level. For single molecules, forces are more appropriately measured in piconewtons, where 1 pN equals 10^{-12} N (see the table on the back endpapers for other SI prefixes). How small is a piconewton? It is equal to the weight of one red blood cell. It is also equal to the optical pressure exerted by a laser pointer on a screen. And it is approximately equal to the maximum force generated by a muscle divided by the number of myosin molecules acting in parallel in the muscle.

Table 2.1 Examples of forces acting on molecules

Type of force	Diagram	Approximate magnitude
Elastic		1–100 pN
Covalent		10,000 pN
Viscous		1–1000 pN
Collisional		10^{-12} to 10^{-9} pN for 1 collision/s
Thermal		100–1000 pN
Gravity		10^{-9} pN
Centrifugal		$< 10^{-3}$ pN
Electrostatic and van der Waals		1–1000 pN
Magnetic		$\ll 10^{-6}$ pN

Newton's **first law of motion** states that if an object has no net force acting on it, then it will remain at rest or, if it is moving, it will continue to move at constant velocity. Newton's **second law** states that if an object is subject to a net force, F, then it will accelerate according to the equation

$$F = ma \qquad (2.1)$$

where m is the mass. This equation says that the larger the mass, m, the slower its acceleration, a (if the force is constant). Many of the parameters used in this book are listed in the Table of Parameters on the endpapers. The unit of mass is the kilogram, kg, and the unit of acceleration is m/s^2. Thus 1 N = 1 kg·m/s^2. An important consequence of Newton's second law is that if an object is stationary or moving at constant velocity, then there is no net force acting on it (the forces are all balanced).

Several examples of mechanical forces on molecules follow. Forces can be transferred to proteins either by direct contact with the atoms of other molecules, or by the interaction of the protein with a field, such as a gravitational field or an electric or magnetic field.

Example 2.1 Physical forces and their magnitudes at the single-molecule level

ELASTIC FORCES. If an object is connected to a spring of stiffness κ that is stretched a distance x beyond its resting length, then the object will experience a force of $F = \kappa x$. For a motor protein, the stiffness might be about 1 mN/m = 1 pN/nm. If the spring is strained through a distance of 1 nm = 10^{-9} m, a distance appropriate to the size of proteins, then the force exerted on the object is 1 pN.

VISCOUS FORCES. If an object is held fixed in a moving liquid or is moving through a stationary fluid, then it will experience a viscous, or drag, force from the liquid. The force is proportional to the relative velocity, v, between the object and the fluid according to $F = \gamma v$. The constant of proportionality, γ, is called the drag coefficient. The drag coefficient is related to the size and the shape of the object as well as the viscosity. For example, for a sphere of radius r moving through a liquid of viscosity η, the drag coefficient is $6\pi\eta r$ (Stokes' law, Chapter 3). The viscous forces on proteins are large. For a globular protein of diameter 6 nm, corresponding to a molecular mass of ~100 kDa (see Table 2.2), the drag coefficient measured by centrifugation studies at 20°C is ~ 60 pN·s/m (Creighton, 1993), in good agreement with Stokes' law. The average instantaneous thermal speed of such a protein in solution at standard temperatures is ~8 m/s (this is a consequence of thermally driven collisions from the surrounding solvent molecules, Chapter 4). The corresponding viscous force is therefore ~480 pN.

COLLISIONAL AND THERMAL FORCES. If an object is struck by another, it experiences a force equal to the rate of change in momentum (mv) of the striking particle, $F = d(mv)/dt$. For example, the mass of a water molecule is ~30 × 10^{-27} kg, the average speed associated with its kinetic energy is ~600 m/s (Chapter 4), and therefore its momentum is ~18 × 10^{-24} kg·m/s. If a protein were struck head-on every second by a water molecule that bounced straight back, then the average force would be equal to 36 × 10^{-12} pN (twice the momentum for an elastic collision). This is a very small force. However, in solution a huge number of collisions take place per second. The collisions come from all directions, and the resulting randomly directed force, called the thermal force, drives diffusion. The average instantaneous thermal force acting on a 100 kDa protein is on the order of the viscous force, or ~500 pN (Chapter 4).

OPTICAL FORCES. Another example of a collisional force is optical pressure. Because photons have momentum, they exert a force when they are diffracted by an object. The momentum of a photon is $h\nu/c = h/n\lambda$, where h is Planck's constant, ν is the frequency of the light, c is the speed of light, n is the refractive index, and λ is the wavelength (in a vacuum). If an object in water ($n = 1.33$) absorbs one green photon ($\lambda = 500$ nm) per second, the corresponding optical force on it is 1.0×10^{-15} pN (the values for the physical constants can be found in the table on the endpapers). This is a very small force. Even if a molecule adsorbs 10^9 photons per second, which would require very bright laser illumination, the optical force would still be only 10^{-6} pN.

GRAVITY. An object of mass m experiences a gravitational force of magnitude mg, where g is the acceleration due to gravity, equal to ~9.8 m/s^2 at the Earth's surface. With a mass of only 166×10^{-24} kg, a 100 kDa protein experiences a gravitational force of only 1.6×10^{-9} pN. At the single-molecule level, gravitational forces are very small and can be ignored.

CENTRIFUGAL FORCES. An object spinning in a centrifuge experiences a centrifugal force equal to ma_c. Ultracentrifuges are capable of generating centrifugal accelerations, a_c, in excess of 100,000 times that of gravity. The associated centrifugal forces on molecules are still quite modest, ~160×10^{-18} N = ~160×10^{-6} pN for our 100 kDa protein, but this is large enough to cause the protein to drift at an average speed of ~3 μm/s (using the drag coefficient from Table 2.2). The slow drift is superimposed on the rapid, randomly directed thermal motion. At this speed the protein will sediment through a distance of 100 mm, a typical length of a centrifuge tube, in about 10 hours.

ELECTROSTATIC FORCES. A particle with charge q, in an electric field of strength E, will experience a force $F = qE$. An ion such as sodium experiences an electrostatic force when it moves through an ion channel in the plasma membrane. The charge on the ion is 160×10^{-21} coulombs (see the table of physical constants on the rear endpapers), and the electric field across a typical plasma membrane is 15×10^6 V/m (60 mV potential across the 4-nm-thick membrane). The corresponding force is 2.4 pN. A similar-sized force exists between two monovalent ions in water that are separated by 1 nm (Problem 2.7): The force will be smaller in a salt solution due to charge screening, but will be larger in the interior of proteins where the dielectric constant is low.

Van der Waals forces are also electrostatic: They arise from the charge separation induced by nearby atoms. Van der Waals forces can be as high as 100 pN per nm^2 of protein–protein interface (Appendix 3.1).

MAGNETIC FORCES. Magnetic forces are very small at the molecular level because molecules interact only very weakly with magnetic fields. For example, the maximum force on a proton, the nucleus with the largest magnetic moment, in the strongest nuclear magnetic resonance (NMR) machines is only on the order of 10^{-12} pN. Thus even for a huge protein with 3000 amino acids and 60,000 atoms, subject to a very strong magnetic field, the magnetic force is less than 10^{-6} pN.

Table 2.2 *Physical properties of a globular protein of molecular mass 100 kDa*

Property	Value	Comment
Mass	166×10^{-24} kg	Mass of 1 mole/Avogadro constant
Density	1.38×10^3 kg/m^3	1.38 times the density of water
Volume	120 nm^3	Mass/density
Radius	3 nm	Assuming it is spherical
Drag coefficient[a]	60 pN·s/m	From Stokes' law (Chapter 3)
Diffusion coefficient[a]	67 µm^2/s	From the Einstein relation (Chapter 4)
Average speed[b]	8.6 m/s	From the Equipartition principle (Chapter 4)

Note: 1 nm = 10^{-9} m, but 1 nm^3 = (1 nm)^3 = 10^{-27} m^3.
[a]In water at 20°C
[b]Root-mean-square (the square root of the average value of the square of the velocity)

Motion of Springs, Dashpots, and Masses Induced by Applied Forces

All mechanical devices can be built with three fundamental mechanical elements—the spring, the dashpot, and the mass. A protein or other molecule can be thought of as a mechanical device composed of atoms that have mass, connected by bonds that have elasticity, like springs. A wind-up toy can be thought of as a mechanical device composed of a spring, some bars and levers that have mass, and some hinges that contribute a little friction and damping. In this section we consider how individual mechanical elements move under the influence of an applied force. In the next section we will consider combinations of elements. The individual motions are summarized in Figure 2.2 and can be described as follows:

MASS. According to Newton's second law, a force causes a mass to undergo a constant acceleration equal to F/m (Equation 2.1). The greater the mass—that is, the greater the inertia—the smaller the acceleration. Because acceleration is the rate of change of velocity ($a = dv/dt$), a constant acceleration means that the velocity increases linearly with time. If the initial velocity is zero, then the speed at time t will be given by $v(t) = at$. Because the velocity is the rate of change in displacement, a linearly increasing velocity means that the position will increase parabolically with time. If the initial displacement is zero, then the displacement at time t will be given by $x(t) = ½ at^2$. This equation describes the motion of a free-falling ball.

DASHPOT. A dashpot is an idealized mechanical element that is fixed at one end and responds to a force applied at the other end by elongating at a constant velocity. The velocity of elongation of a dashpot is equal to F/γ, where γ

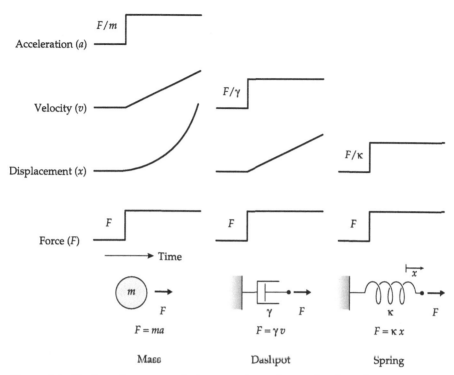

Figure 2.2 Motion of a mass, a dashpot, and a spring under the influence of a constant external force

is the drag coefficient (Equation 2.2). Because the velocity is constant, the length of a dashpot increases linearly with time as shown in Figure 2.2. A dashpot is analogous to a spoon in a jar of honey. When the jar is held fixed in one hand, it is easy to pull the spoon out slowly, but to pull it out quickly requires a large force. Indeed, if one pulls fast enough it is even possible to pick up the jar. The higher the viscosity of the honey, the higher the drag coefficient, and the greater the force needed to attain a certain speed. The dashpot is used as a model to describe how an object moves in a fluid. We can think of a submerged object as being connected to an (imaginary) dashpot whose drag coefficient is proportional to the viscosity according to Stokes' law (Chapter 3).

In Figure 2.2, the right-hand end of the dashpot moves at constant speed. This is a reflection of the fact that there is no net force acting at this point: The external force F is exactly balanced by the internal drag force, F_d. In other words, $F + F_d = 0$, in accordance with Newton's second law (if the net force were not zero, then there would be acceleration). This means that the drag force is

$$F_d = -\gamma v \qquad (2.2)$$

where the minus sign represents the fact that the drag force opposes the movement. The drag coefficient, γ, has units N·s/m.

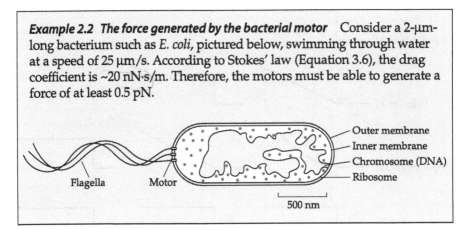

Example 2.2 The force generated by the bacterial motor Consider a 2-μm-long bacterium such as *E. coli*, pictured below, swimming through water at a speed of 25 μm/s. According to Stokes' law (Equation 3.6), the drag coefficient is ~20 nN·s/m. Therefore, the motors must be able to generate a force of at least 0.5 pN.

SPRING. A spring is a mechanical element whose length increases in response to an applied force. Like the dashpot, one end of the spring is held fixed while the force is applied to the other end. The increase in length of the spring above its resting length equals F/κ, where κ is the spring constant. The greater the spring constant—that is, the stiffer or less compliant the spring—the smaller the extension for a given force.

After the onset of the force, the right-hand end of the spring is stationary. There is no acceleration because there is no net force acting on this point. The external force F is exactly balanced by the internal, elastic force F_s. In other words, $F + F_s = 0$, again in accordance with Newton's second law. This means that the force exerted by the spring is

$$F_s = -\kappa x \qquad (2.3)$$

where the minus sign represents the fact that the elastic force is a restoring one that opposes the movement.

If a spring has constant stiffness, meaning that the stiffness is independent of the force or extension, then we say that it obeys **Hooke's law**. As we will see in the next chapter, Hooke's law is a good approximation for the stretching of many materials, provided that the forces and resulting extensions are not too large.

Motion of Combinations of Mechanical Elements

When different mechanical elements are put together, their response to applied forces becomes more complex and interesting. In this section we consider how pairs of mechanical elements move. There are two qualitatively different behaviors: monotonic "creeping" motion and oscillatory "ringing" motion.

MASS AND DASHPOT. Consider a mass and dashpot connected in series as shown in Figure 2.3A (left panel). This is a model for the movement of a cell or a protein through a liquid (Figure 2.3A, middle). The mass experiences an applied force (F) pulling it to the right. The motion is opposed by a drag force,

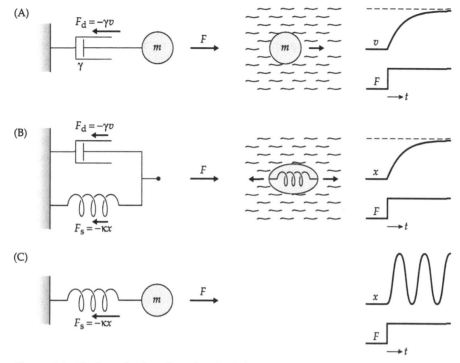

Figure 2.3 Motion of pairs of mechanical elements
(A) The mass and dashpot model represents an object that is damped by a viscous fluid. (B) A spring and dashpot model represents a low-mass object (like a protein) that is deformed in a viscous fluid. (C) The mass and spring model represents an undamped system.

F_d, whose value is $-\gamma v$. The net force on the mass is therefore $F + F_d = F - \gamma v$. The net force acting on the mass causes it to accelerate according to $ma = F - \gamma v$. Because acceleration is the rate of change of velocity, $a = dv/dt$, we can rewrite this equation as

$$m\frac{dv}{dt} + \gamma v = F \qquad (2.4)$$

Now suppose that initially the mass is stationary and that at time zero a constant force F is applied. At first, when the speed is low, the drag force is small, and the mass will undergo a constant acceleration ($a = F/m$), leading to a linear increase in velocity. However, as the velocity increases, the drag force becomes significant, causing a decrease in the net force acting on the mass and therefore a decrease in the acceleration. As a result, the velocity begins to level off. Finally, the drag force approaches the applied force, the acceleration drops to zero, and the velocity approaches the **terminal velocity** equal to F/γ. This motion is described by the equation

$$v(t) = \frac{F}{\gamma}\left[1 - \exp\left(-\frac{t}{\tau}\right)\right] \qquad \tau = \frac{m}{\gamma} \qquad (2.5)$$

which is plotted in Figure 2.3A (right). That this equation is a solution to the previous equation can be verified by differentiating Equation 2.5 and substituting the derivative into Equation 2.4. The time constant τ at which the velocity approaches the terminal velocity depends on the mass and damping. The higher the mass, the slower the acceleration and the longer the time before the drag force becomes significant. Conversely, the higher the drag coefficient, the greater the drag force for a given speed, and so the shorter the time before the drag force becomes limiting. Smaller objects have smaller time constants: The mass is proportional to (length)^3, whereas the damping coefficient is proportional to (length)^1; therefore the time constant scales with (length)^2, becoming very small as the dimension gets smaller. This relationship is illustrated in the next two examples.

Example 2.3 The inertia of a bacterium Consider a bacterium swimming through water at 25 μm/s. How long will the bacterium continue to coast after its motors have stopped? We model the bacterium as a mass and dashpot. The corresponding equation of motion is $m\frac{dv}{dt} + \gamma v = 0$. After the motors stop and the flagella cease beating, the speed will decrease exponentially according to $v(t) = v(0)\exp(-t/\tau)$, with a time constant, τ, equal to m/γ. The mass is approximately equal to $\frac{4}{3}\pi r^3 \rho$, where r is the radius (~1 μm) and ρ is the density (~ 1000 kg/m^3), or ~4 × 10^{-15} kg. From Example 2.2, the drag coefficient is 20 nN·s/m. The time constant is therefore 0.2 μs.

The total distance that the bacterium coasts is

$$x = \int_0^\infty v(t) \cdot dt = \int_0^\infty v(0)\exp-(t/\tau) \cdot dt = v(0) \cdot \tau$$

For an initial speed, $v(0) = 25$ μm/s, this distance is only ~5 pm = 0.05 Å, less than the diameter of a water molecule! Thus a bacterium has very little inertia to keep it moving forward (Berg, 1993).

Example 2.4 The persistence of protein movements The time constant, m/γ, of a globular, 100 kDa protein is ~2.8 ps, using the mass and drag coefficients from Table 2.2. In other words, after the protein gains speed due to molecular collisions with solvent molecules, the velocity persists for only a very short time as other collisions rapidly randomize the protein's direction of travel. Given that the average instantaneous speed of such a protein is 8.6 m/s (see Table 2.2), the average distance that the protein moves before its speed is randomized by molecular collisions is only 24 pm, or 0.24 Å.

SPRING AND DASHPOT. A spring in parallel with a dashpot (see Figure 2.3B, left panel) is a model for a compliant object that is deformed in a liquid, such as a protein that undergoes a global (i.e., large-scale) conformational change. It can also be used to model a viscoelastic material, such as skin, that takes a finite time to adopt a new shape (see Figure 2.3B, middle). In this case, the applied force is opposed by the sum of the viscous and elastic forces: $F = \gamma v + \kappa x$. Because the velocity is the derivative of displacement, this equation becomes

$$\gamma \frac{dx}{dt} + \kappa x = F \tag{2.6}$$

If the spring is unstrained at time zero, then an applied force will initially only be opposed by the dashpot (since the elastic force is initially zero). Thus the spring and dashpot together begin to move at a constant speed equal to F/γ. However, as the spring becomes more elongated, the elastic force increases, the speed begins to decrease, and the displacement begins to level off. Finally, the elastic force approaches the applied force, the velocity drops to zero, and the spring approaches its final extension equal to F/κ. This motion is described by the equation

$$x(t) = \frac{F}{\kappa}\left[1 - \exp\left(-\frac{t}{\tau}\right)\right] \qquad \tau = \frac{\gamma}{\kappa} \tag{2.7}$$

which is plotted in Figure 2.3B (right). This equation is analogous to that describing a mass and dashpot. However, in this case the time constant depends on the damping and stiffness: The higher the damping, the smaller the velocity and the longer the time before the elastic force becomes significant. Conversely, the higher the spring constant, the greater the elastic force for a given elongation, and so the shorter the time before the elastic force becomes limiting.

Example 2.5 The timescale of protein conformational changes Consider a protein that is initially held in a strained conformation, perhaps due to an internal strut (Figure A on page 18, left panel). A mechanical model for this arrangement is a spring in parallel with a dashpot and a latch (Figure B, left panel). Now suppose that the constraint is suddenly relieved (Figure A, middle panel). This is equivalent to releasing the latch (Figure B, middle panel). We expect that the protein will change shape and relax into its unstrained conformation (Figures A and B, right panel) with a time constant on the order of γ/κ, the drag coefficient divided by the spring constant. Figure C shows an energy diagram in which the stiff-latched state moves into the relaxed conformation of the more compliant unlatched state. For a roughly globular protein with a molecular mass of ~100 kDa, the drag coefficient is ~60 pN·s/m (see Table 2.2 and Appendix 3.3). If the elastic element has a stiffness of ~4 pN/nm, comparable to that of the myosin crossbridge (Chapter 16), then the relaxation time constant will be 15 ns. This model provides a general picture for the timescale of protein conformational changes: Local chemical changes,

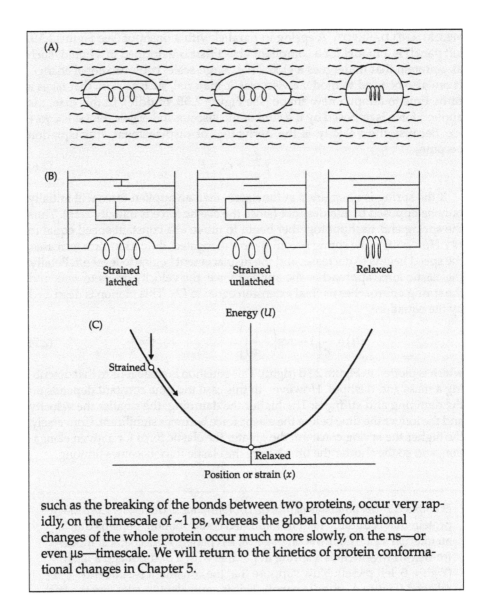

such as the breaking of the bonds between two proteins, occur very rapidly, on the timescale of ~1 ps, whereas the global conformational changes of the whole protein occur much more slowly, on the ns—or even μs—timescale. We will return to the kinetics of protein conformational changes in Chapter 5.

MASS AND SPRING. The mass on a spring (see Figure 2.3C, left panel) is a familiar mechanical system found in physics and chemistry textbooks, where it is used to describe the vibrations of tuning forks and atomic bonds. The net force acting on the mass is the applied force minus the elastic force, and therefore $ma = F + F_s = F - \kappa x$. Because the acceleration is the second derivative of the displacement, we can rearrange this equation to obtain

$$m\frac{d^2x}{dt^2} + \kappa x = F \tag{2.8}$$

Suppose that the mass is initially stationary and that at time zero a constant force is applied. At first, when the displacement is small and the elastic force is also small, the particle undergoes constant acceleration and the displacement increases parabolically. As the displacement increases, the elastic force increases, and, when the displacement reaches F/κ, the net force acting on the particle is zero. At this point the acceleration is also zero. However, at this time the particle has reached its maximum velocity and its inertia keeps it moving. As it overshoots what will be the new average displacement of F/κ, the restoring force in the spring increases and the particle gradually slows down and finally stops when the displacement reaches $2F/\kappa$. Now the elastic restoring force causes the particle to accelerate back to its initial position. It overshoots the average displacement as it returns to the initial position, whereupon it begins another cycle of oscillation. The resulting sinusoidal oscillation, called **harmonic motion**, is described by the equation

$$x(t) = \frac{F}{\kappa}[1 - \cos(\omega t)] \qquad \omega = \sqrt{\frac{\kappa}{m}} \tag{2.9}$$

This equation is plotted in Figure 2.3C. ω is the frequency of the oscillations expressed in units of radians per second. However, it is more familiar to express frequency in terms of cycles per second, or hertz (Hz); the relationship between the two frequencies is $f = \omega/2\pi$. If the mass is increased, the acceleration will decrease, and the longer the time until the elastic force becomes significant. A higher mass is therefore associated with a low frequency of oscillation. Conversely, when the stiffness is increased, the elastic force rises more quickly, and frequency of oscillation increases.

Example 2.6 Vibration of chemical bonds Covalent chemical bonds can be thought of as having stiffness. The manifestation of this stiffness is that they vibrate with a frequency given by Equation 2.9. The vibration can be detected spectroscopically when the molecule absorbs light of the same frequency as the molecular vibration. For example, the fundamental vibration frequency of the H–Cl bond in HCl is $\upsilon = 89.6 \times 10^{12}$ Hz (2990 cm^{-1}), corresponding to a wavelength of $c/89.6 \times 10^{12} = 3.53$ μm (in the infrared) (Atkins, 1986). The appropriate mass is ~1.63×10^{-27} kg (approximately the mass of the hydrogen nucleus), and so the stiffness is $m\omega^2 = 4\pi^2\nu^2 m = 517$ N/m.

> **Example 2.7 Protein vibrations** Consider the motor protein myosin again. The motor domain has a mass of ~160×10^{-24} kg and the stiffness is 4 pN/nm (4 mN/m). The vibration frequency is calculated to be ~10^9 Hz. This corresponds to a period of oscillation of 1 ns. By contrast, the relaxation time calculated in Example 2.5 is 15 ns. Does the protein oscillate when it detaches from the actin filament, or does it creep exponentially into its relaxed state? The answer requires solution of the full model, with mass, spring, and dashpot. The solution, provided in the next section, shows that the protein creeps rather than rings.

Motion of a Mass and Spring with Damping

In this section, we consider how a system comprising all three elements—the mass on a spring subject to damping (Figure 2.4A)—moves in response to an applied force. This is a simple mechanical model for a protein undergoing a large-scale conformational change that is damped by the surrounding fluid, and possibly by internal viscosity (Figure 2.4B). This three-element model captures the main qualitative features of more complex models in that it can display oscillatory or monotonic motions depending on the strength of the damping.

The equation of motion is

$$m\frac{d^2x}{dt^2} + \gamma\frac{dx}{dt} + \kappa x = F \quad (2.10)$$

The solution depends on whether the damping is small or large (Appendix 2.1). When the damping is small, $\gamma^2 < 4m\kappa$, the motion is oscillatory, like that of a mass on a spring except that the amplitude of the oscillation gradually decreases as the vibration dies out. This is shown in Figure 2.4C. We say that the motion is **underdamped**. As the damping decreases, the oscillation dies out more slowly and the motion more closely approximates harmonic motion. On the other hand, when the damping is large, $\gamma^2 > 4m\kappa$, the motion is monotonic, like that of a damped spring, as shown in Figure 2.4D. In this case, we say that the motion is **overdamped**. The overdamped motion is actually associated with two time constants, though only one time constant is apparent in Figure 2.4B. The fast time constant corresponds to the time that it takes for the mass to accelerate to a velocity of ~F/γ, whereas the slow time constant corresponds to the relaxation of the spring and dashpot. The reason why the fast component is not seen in the figure is that its amplitude is much smaller than that of the slow component: Even when the motion is moderately overdamped ($\gamma^2/4m\kappa = 3$), the amplitude of the fast component is only ~10% that of the slow component. When the system is even more strongly damped, the fast component can be ignored altogether; under these conditions, we can ignore the mass and just consider the spring and dashpot. In the highly over-

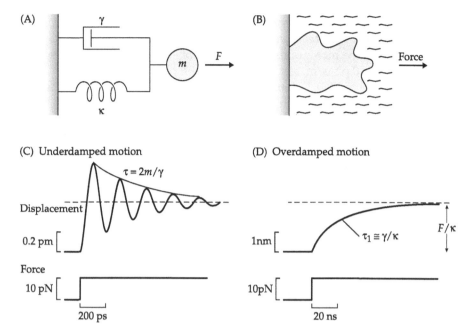

Figure 2.4 Motion of a mass and spring with damping
The mechanical model (A) is used to describe the motion of an elastic solid in a fluid (B). (C) Underdamped motion. In this example, $\gamma^2/4m\kappa \cong 0.007$, and the motion corresponds to that of a hypothetical globular protein that is both very large (16 MDa) and very rigid (stiffness 30 N/m) and experiences unrealistically little damping from the fluid ($\gamma = 150$ pN·s/m). (D) Overdamped motion. In this example, $\gamma^2/4m\kappa \cong 1400$, and the motion corresponds to that of a more realistic protein of molecular mass 100 kDa and stiffness 4 pN/nm.

damped case, the inertial term can be dropped, and we can describe the motion without invoking Newton's second law!

The condition for overdamping is that $\gamma^2 > 4m\kappa$. This makes intuitive sense because the system should become more damped as the drag coefficient increases: When the relaxation time-constant of the spring (γ/κ) becomes greater than that of the mass (m/γ), the kinetic energy of the mass is unable to sustain the oscillation. There is another way of thinking about the motion. If the system is overdamped, then the inertial force (ma) is always smaller than the viscous force (γv) (Appendix 2.1). Conversely, if the inertial force is always smaller than the viscous force, then the motion is overdamped (Appendix 2.1). This is an important observation: The quality of the motion—whether it is oscillatory or monotonic—depends on the relative contribution of the inertial forces (that tend to produce oscillations) and the viscous forces (that tend to damp the oscillations out). It turns out that inertial forces are usually very small at the microscopic and molecular levels, so that the overdamped case usually applies.

> **Example 2.8 Motor proteins are overdamped** Consider the motor protein myosin again. The motor domain has a mass of ~160 × 10^{-24} kg, the drag coefficient is ~60 pN·s/m, and the stiffness is ~4 pN/nm. In this case $\gamma^2/4m\kappa$ is equal to 1400. Because this is much greater than 1, it follows that the motion is highly overdamped. Thus, when the force exerted by a motor protein abruptly changes—for example, if it enters a new chemical state as described in Example 2.5—then the protein will relax monotonically into its new equilibrium conformation without undergoing oscillations, as shown in Figure 2.4D. The time constant of the relaxation will be given by the damped spring, namely ~15 ns.

Work, Energy, and Heat

It is important not to confuse force with work or energy. If a force, F, is applied to a mechanical system and the system moves through a distance x_0, then work has been done on it. The **work**, w, equals the force times the distance. If the force is a constant (independent of position), then the work done is Fx_0. More generally, if the force depends on the position—that is, $F(x)$—then each incremental change in position, dx, results in an incremental amount of work $dw = F(x) \cdot dx$; the total work is the sum of all the increments of work done as the system moves. In other words, the total work is equal to the integral

$$w = \int_0^{x_0} F(x) \cdot dx \qquad (2.11)$$

The SI unit of work is the joule (J); because work is force times distance, it follows that 1 J = 1 N·m. If the force produces no movement, then the work is zero.

Energy is closely related to work and has the same units. A spring is a mechanical element that can store energy—the work done on it is converted into **potential energy**, denoted by U. Potential energy can also be stored in gravitational and electric fields. Another way of thinking about force is that it is the tendency for a system to move from high potential energy to low potential energy. In mathematical words, force is the negative of the gradient of the potential energy:

$$F_s = -\frac{dU}{dx} \qquad (2.12)$$

The steeper the gradient, the greater the force; there is a minus sign because the force is in the direction corresponding to a decrease in the potential energy (Figure 2.5). Now, as we saw before, the force exerted by a spring, F_s, is $-\kappa x$. Substituting this into Equation 2.12 and multiplying by dx gives $dU = -F_s \cdot dx = --\kappa x \cdot dx = \kappa x \cdot dx$. Upon integrating we obtain

$$U = \int_0^{x_0} \kappa x \cdot dx = \tfrac{1}{2} \kappa x_0^2 \qquad (2.13)$$

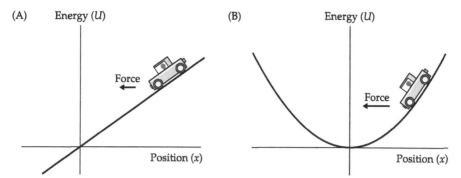

Figure 2.5 The force is the negative of the gradient of potential energy
(A) Constant force. (B) Force that depends on displacement.

When a constant force is applied to the mechanical system shown in Figure 2.4, the total work done on it is $Fx_0 = \kappa x_0^2$, where we have used the fact that the final displacement, x_0, is equal to F/κ. However, the potential energy gained by the spring, $\frac{1}{2} \kappa x_0^2$, is only half this amount. Where did the missing energy go? While the mass is actually moving, some of this energy is stored as **kinetic energy**,

$$K.E. = \tfrac{1}{2} m v^2 \tag{2.14}$$

But when the mass reaches its new steady-state position, the mean speed is zero and so the kinetic energy is also zero. The answer is that part of the work went to heating the dashpot. If we wish our system to remain at the same temperature, then this **heat**, denoted by Q, must be transferred to the surroundings, and we say that the heat has been dissipated. The rate at which the heat is dissipated is

$$\frac{dQ}{dt} = -F_{\text{drag}} v = -(-\gamma v) v = \gamma v^2 \tag{2.15}$$

and it can be shown that the total heat dissipated is $\frac{1}{2} \kappa x_0^2$ (Appendix 2.2). Thus we have

$$w = U + Q \tag{2.16}$$

Example 2.9 Energy of chemical bonds We can think of the energy of a chemical bond, the dissociation energy, as being approximately equal to the potential energy in the bond. This energy is $\sim \frac{1}{2} \kappa r^2$ where r is the extension required to break the bond, ~0.05 nm. For HCl, considered in Example 2.6, where the stiffness is ~517 N/m, the corresponding energy is $\sim 650 \times 10^{-21}$ J, in fairly good agreement with the bond energy of $\sim 720 \times 10^{-21}$ J (Moore, 1972). For a more accurate treatment, the non-Hookean stiffness of bonds must be taken into account.

This is a statement of the **Law of Conservation of Energy**, also known as the **First Law of Thermodynamics**.

> ***Example 2.10 Energy stored in protein conformational changes*** For a myosin molecule, the stiffness is thought to be about 4 pN/nm. For a conformational change of 5 nm, the total energy is 50 pN·nm = 50×10^{-21} J. This is approximately half the chemical energy derived from hydrolysis of the gamma phosphate bond of ATP (Chapter 14). We can generalize this argument to global conformational changes of other protein machines. The energies are on the order of 10 to 100×10^{-21} J, and the conformational changes are on the order of 1 to 10 nm. The corresponding stiffnesses are therefore on the order of 0.2 to 200 mN/m.

Summary: Generalizations to More Complex Mechanical Systems

By considering three mechanical elements—a mass, a spring, and a dashpot—we have introduced many of the mechanical concepts required to understand how forces influence proteins and cells. The mass and spring with damping illustrate that systems can respond to mechanical forces in two ways: They can oscillate or they can move monotonically.

The mechanical models considered in this chapter can be generalized in two ways. The first way is to increase the number of mechanical elements to include several masses, springs, dashpots, and even other elements such as latches and stops; the equations of motion are solved by balancing the forces across each element (Jaeger and Starfield, 1974). Molecular dynamics (McCammon and Harvey, 1987) is an example of such a generalization: Each atom in a protein and the surrounding fluid is represented by a point mass, each bond is represented by a spring (which need not have constant stiffness), and the damping is dropped from the equations (it is an "emergent" property of the system). The ensuing motion is complex and best solved numerically by computer.

The second way to generalize the model is to consider the mechanical behavior of "continuum" solids that have material properties such as elasticity, density, and viscosity. This "coarse" approach is taken in the next chapter.

Problems

2.1 Suppose that a force of 1 pN is applied to a globular 100 kDa protein. In the absence of damping, how fast will the protein be moving after 1 ns? During this time, how far will the protein have moved? Given the damping coefficient quoted in Table 2.2, what is the actual terminal velocity of the protein?

2.2 The motor protein kinesin can generate a force of 6 pN. Given that the viscosity of cytoplasm is ~1000 times that of water (for large objects like organelles), how fast could a single kinesin molecule move a bacterium through a cell (see Example 2.2)?

2.3 During mitosis, the chromosomes move several micrometers over the course of about 30 minutes. Calculate the average speed. If the viscosity of the cytoplasm is 1000 times that of water, what force is required (assuming a chromosome has the same drag coefficient as a bacterium)?

2.4 The probes used in atomic force microscopes (AFMs) typically have stiffnesses of ~1 N/m. Given that the mass is ~100 ng, what is the resonance frequency in vacuum (without damping)? The damping coefficient of a probe in water is ~1 µN·s/m. Is the motion in water overdamped or underdamped?

2.5 The chemical energy available from the hydrolysis of ATP is ~100 × 10^{-21} J. How far can a motor protein exert a force of 6 pN before 100 × 10^{-21} J of work is done?

2.6 In Example 2.2, it was stated that the bacterial motors must generate a force of 0.5 pN in order to propel the bacterium at a speed of 25 µm/s. What is the power output of the bacterium? How many equivalent ATPs must be hydrolyzed per second in order to power this movement?

2.7 Coulomb's law states that the force between two charges q_1 and q_2 separated by a distance r is

$$F = \frac{1}{4\pi\varepsilon_0\varepsilon}\frac{q_1 q_2}{r^2}$$

where ε_0 is the electric constant, also called the permittivity constant, equal to 8.854×10^{-12} C²/N·m² and ε is the dielectric constant (equal to 1 for a vacuum, ~3 for oils, and 80 for water). Calculate the force between two electronic charges separated by 1 nm in pure water. [Answer: 2.9 pN.] Note that the force would be smaller in a salt solution due to screening of the charge by the salt ions, but it will be larger in the interior of proteins where the dielectric constant is similar to that of oil.

2.8 If two springs are placed in parallel, show that their stiffnesses add. If they are placed in series, show that their compliances add (the **compliance** is the reciprocal of the stiffness). If two dashpots are placed in parallel, show that the total drag coefficient is the sum of the individual coefficients. If the dashpots are placed in series, show that reciprocals of the drag coefficients add.

2.9 A Voigt element consists of a spring and a dashpot in series. When one end is held fixed and a constant force is abruptly applied to the other at time zero, how does the system move?

2.10 Show that the motion of the Maxwell element (Figure A, below), in response to a force F, is

$$x(t) = x_0 - (x_0 - y_0)\exp\left(-\frac{t}{\tau}\right) \quad x_0 = \frac{F}{\kappa_2} \quad y_0 = \frac{F}{\kappa_1 + \kappa_2} \quad \tau = \gamma\left(\frac{1}{\kappa_1} + \frac{1}{\kappa_2}\right)$$

as plotted in Figure B.

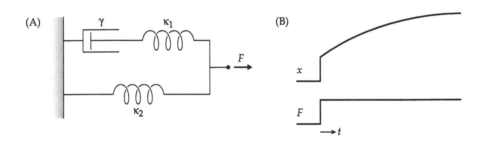

CHAPTER

3

Mass, Stiffness, and Damping of Proteins

The purpose of this chapter is to get a feeling for what proteins are like as mechanical devices. How rigid are they? How quickly do they move and change shape? And what is the quality of their motion: When a protein is struck by a force, does it ring like a tuning fork (underdamped motion), or does it creep monotonically into a new shape (overdamped motion)? To answer these questions, I begin this chapter with a discussion of the material properties of proteins—their density, their elasticity, and the frictional forces that damp their motion. Proteins have similar densities and rigidities to hard plastics and Plexiglas. However, owing to their small size, the viscous forces from the surrounding fluid are large compared to the inertial forces. Consequently, the global motions of proteins are overdamped, meaning that proteins relax monotonically into new conformations. Thus a protein, as a mechanical device, is like a little plastic toy. But if we were to scale it up by a factor of 10^7 so that a 5-nm-diameter protein becomes a 50-mm-diameter device that would fit into the palm of one's hand, then we would have to increase the viscosity by the same amount (i.e., bathe it in treacle) in order to damp out any tendency for oscillation.

Mass

Mass equals density, ρ, times volume, V:

$$m = \rho V \qquad (3.1)$$

The densities of various amino acids, proteins, organelles, and cells are given in Table 3.1. Proteins are composed of relatively light elements—carbon, oxygen, nitrogen, and hydrogen—and are about 40% denser than water, with different proteins having slightly different densities. We take the average density of proteins to be 1.38×10^3 kg/m^3.

Table 3.1 Densities of molecules, proteins, organelles, and cells relative to water

Substance	Density (relative to water[a])
Water	1.00
Glycerol	1.26
Glycine	1.16 (solid)
Alanine	1.40 (solid)
Glutamic acid	1.46 (solid)
Hemoglobin	1.33 (in solution)
Trypsin	1.38 (in solution)
Lysosyme	1.42 (in solution)
Chromosome	1.36
Virus	1.15
Mitochondrion	1.18
Synaptic vesicle	1.05
Erythrocyte	1.10
Fibroblast	1.05

Source: Rickwood, 1984; Kaye and Laby, 1986; Creighton, 1993.
[a] The density of water at 20°C is 998 kg/m^3.

The SI unit of mass is the kilogram (kg). However, in biochemistry the mass of proteins and other biomolecules is usually expressed as **molecular mass**, defined as the mass in grams of a mole of the molecules. The unit is the **dalton** (Da). According to this definition, a hydrogen atom has a molecular mass of 1 Da, corresponding to an actual mass of 1.66×10^{-24} g (1 g ÷ N, where N is the Avogadro constant given in the table on the endpapers) or 1.66×10^{-27} kg. A protein of molecular mass 100,000 Da, or 100 kDa, has a mass of 166×10^{-24} kg.

The density of proteins is such that each kDa of protein occupies a volume of about 1.2 nm^3. A 100 kDa protein therefore has a volume of 120 nm^3; if it were spherical its diameter would be 6 nm. Because the average molecular mass of an amino acid is 119.4 Da (weighted according to amino acid frequency in globular proteins; Creighton, 1993), there are ~7 amino acids per nm^3.

Elasticity

A solid is **homogenous** if its mechanical properties are identical throughout, and it is **isotropic** if these properties do not depend on direction. If a small tensile force F is applied to a homogenous, isotropic solid of uniform cross-sectional area A (Figure 3.1), it is found experimentally that the **strain**, the relative length change $\Delta L/L$, is proportional to the **pressure**, the force per unit area:

$$\frac{F}{A} = E \frac{\Delta L}{L} \qquad (3.2)$$

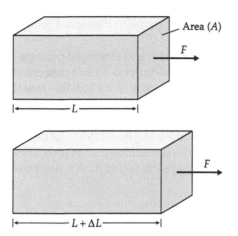

Figure 3.1 A solid strained by a tensile force

Because the extension (ΔL) is proportional to the force, this is an example of Hooke's law. The constant of proportionality in Equation 3.2, E, is known as the **Young's modulus**, or the **elastic modulus**. Because strain is the ratio of lengths and is therefore dimensionless, the Young's modulus has the same units as pressure, namely newtons per square meter (N/m²) or pascals (Pa). The Young's moduli of various materials and proteins are given in Table 3.2. Note

Table 3.2 Young's moduli and tensile strength of materials

Material	Young's modulus, E (GPa)	Tensile strength (GPa)[a]
Carbon nanotube	1300	14
Diamond	1200	—
Steel (stainless)	211	1.1 (wire)
Glass (quartz)	73	1 (fiber)
Wood (fir, along grain)	16	0.06
Plexiglas	3	0.05
Plastic (polypropylene)	2.4	0.035
Teflon (PTFE)	0.34	0.022
Rubber (polyisoprene)	0.02	0.017
Silk (*Bombyx mori*)	5–10	0.3–0.6
Keratin (hair)	2.4	0.2
Actin	2.3	0.03
Collagen	2	0.1
Tubulin	1.9	—
Elastin	0.002	0.002

Source: Data for nonproteins from Tennent, 1971; Kaye and Laby, 1986; Wong et al., 1997. Data on proteins from Table 8.5 (from Wainwright et al., 1976; Kaye and Laby, 1986) and from Fraser and Macrae, 1980; Tsuda et al., 1996.

[a] Note that drawing a material out into a wire or fiber increases its tensile strength (Gordon, 1984).

that the most rigid proteins have similar Young's moduli to Plexiglas (Perspex) and hard plastics!

For many materials—for example, metals, plastics, and structural proteins—Hooke's law applies only for forces that cause strains up to 0.1 to 1 percent. At higher forces the material yields; the yield pressure is called the **tensile strength**. By contrast, some resilient materials such as rubber and proteins like elastin and titin can be strained up to 100% or more.

The Young's modulus is a material property, meaning that it does not depend on the object's size or shape. On the other hand, the stiffness of an object does depend on its size and shape (as well as its Young's modulus). This is illustrated in the following examples.

Example 3.1 Stiffness of a rod under tension The longitudinal spring constant of the rod shown in the figure is:

$$\kappa = \frac{F}{\Delta L} = \frac{EA}{L}$$

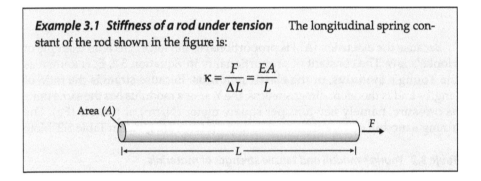

Example 3.2 The cantilever spring A rod subject to a bending force (as shown in the figure below) has a stiffness (see Equation 6.5 and Figure 6.2)

$$\kappa = \frac{4\pi}{3} \frac{Er^4}{L^3}$$

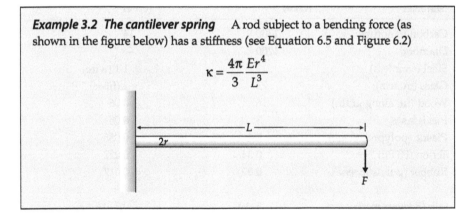

In general, proteins are neither homogenous nor isotropic due to their complex atomic structures. For this reason, care must be taken when considering their mechanical properties. For a nonhomogenous, nonisotropic solid, there are as many as 21 elastic parameters for *every* point in the material (Kittel, 1996): An exact description of the elasticity of a material could therefore be as complex as the full atomic description! By contrast, a homogenous and isotropic material has just two elastic parameters, the Young's modulus and Poisson's

> **Example 3.3 The coiled spring** The coil in the figure has a stiffness
>
> $$\kappa \cong \frac{1}{4N}\frac{Gr^4}{R^3}$$
>
> where the shear modulus, G, is defined in Problem 3.6.

ratio, the latter being a measure of how much the cross-sectional area changes as the material is stretched (Problem 3.6). Thus the description of the elastic properties of a material is greatly simplified if the material is homogenous and isotropic.

Is it valid to think of proteins as having material properties, or must we always think in terms of their atomic structures? This question is related to the domain concept of structural biology (Creighton, 1993) in which proteins are thought of as comprising fairly rigid domains joined by more flimsy connecting regions (Yguerabide et al., 1970; Mendelson et al., 1973; Gerstein et al., 1994). In this picture the hinging or twisting of domains is attributed to the less substantial thickness of the connections, in the same way that a rubber dumbbell bends about its linking rod not because the rod is composed of a weaker material, but because it has a reduced cross-section. Thus the domain concept encompasses the idea that proteins have material properties.

Are there experiments to support the notion that proteins can be thought of as mechanically isotropic, at least to a first approximation? In the case of the globular proteins actin and tubulin, which polymerize to form cytoskeletal filaments, the Young's moduli are found to be approximately independent of the direction of the applied force, indicating that there is no drastic departure from isotropy (Chapter 8). In addition, the Young's moduli of several filamentous proteins are similar, despite their quite different atomic structures (see Tables 3.2 and 8.5); this suggests the existence of a material property that is independent of the atomic details. On the other hand, wet hair has significant mechanical anisotropy: The Young's modulus measured using longitudinal forces is an order of magnitude greater than that measured using transverse forces, due to the orientation of the constituent coiled coils. Nevertheless, even in this case, the anisotropy may be described quite simply.

In summary, the concept that proteins have material properties derives support from both structural and mechanical studies. The simplicity of the material description over the atomic one makes it a useful conceptual tool for understanding protein mechanics. Furthermore, the material description can be readily tested and refined by mechanical experiments on proteins; by contrast, the tools necessary for relating mechanical measurements to atomic descriptions of proteins via molecular dynamics simulations are only now being developed (Krammer et al., 1999; Marszalek et al., 1999).

The Molecular Basis of Elasticity

The rigidity of most materials arises from the stiffness of the bonds that hold the constituent atoms together. In the case of proteins, there are strong, covalent chemical bonds, and weaker, noncovalent physical bonds that include electrostatic bonds (ion pairs and hydrogen bonds) and van der Waals bonds.

The energy of a bond holding together two atoms depends on the separation, r. At equilibrium, when there is no net force acting on the atom, the energy is a minimum. The separation for which the energy is a minimum is the bond length, r_0. The energy profile, $U(r)$, forms a well as shown in Figure 3.2, and at the bottom of the well the profile is approximately parabolic: $U(r) \cong U_0 + \frac{1}{2}\kappa(r - r_0)^2$, where U_0 is the bond energy and κ is the stiffness of the bond. To

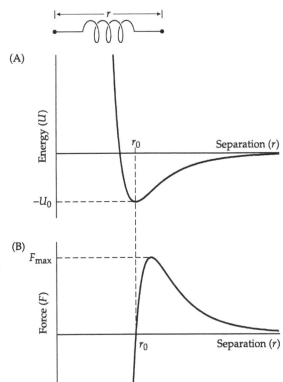

Figure 3.2 Energy and force of a van der Waals bond between two atoms
(A) The bond energy (U) plotted against the center-to-center spacing (r). (B) The force ($F = dU/dr$) required to stretch the bond. The energy function used is the **Lennard–Jones potential**,

$$U(r) = -U_0 \left[2\left(\frac{r_0}{r}\right)^6 - \left(\frac{r_0}{r}\right)^{12} \right]$$

The first term on the right-hand side corresponds to the attractive component, which falls off with the sixth power of the distance according to the van der Waals interaction. The second term is repulsive and corresponds to a "steric" force (Israelachvili, 1991). The potential is a minimum when the force is equal to zero, and this occurs when $r = r_0$. The force is a maximum when $r \cong 1.11 r_0$; when stretched beyond this distance, the bond breaks. The asymmetry of the energy profile means that the stiffness is not a constant, though for small strains (<1%) the stiffness changes by less than 10%. The asymmetry also means that the bond expands when heated, because, as the temperature increases, the bond vibrates more, and the atoms spend less time at the bottom of the well (at separation r_0), and more time at higher energy levels that are associated with a larger average bond length due to the asymmetry.

stretch the bond a small distance $(r - r_0)$ requires an applied force $F(r) = dU/dr \equiv \kappa(r - r_0)$. In other words, for small forces, the extension of the bond, $r - r_0$, is proportional to the force: Hooke's law holds. For larger tensile forces, Hooke's law breaks down and the bond becomes softer. Eventually a maximum force is reached; beyond this force the bond breaks (see Figure 3.2), and the material yields.

If we knew the spring constant of every bond in a material, then we could calculate its Young's modulus (and other elastic parameters). This approach is taken in molecular dynamics, in which the stiffness and length of each cova-

---

Example 3.4 The Young's modulus of a covalent solid The stiffness of the C–C single bond is ~550 N/m and the bond length is 0.14 nm (Tung et al., 1984). If carbon formed a cubic lattice, its Young's modulus would be ~4 × 10^{12} Pa, or 4000 GPa, along the [100] axis. This overestimates the Young's modulus of diamond by a factor of three (see Table 3.2). The overestimation occurs because carbon is tetravalent rather than hexavalent, and the tetrahedral coordination of each atom with its nearest neighbors (Moore, 1972) means that bond bending contributes additional compliance.

---

lent, ionic, and van der Waals bond is specified, as well as the bending and torsional stiffnesses of the covalent bonds (Levitt, 1974; Tung et al., 1984; McCammon and Harvey, 1987). To illustrate how the stiffness of the constituent bonds determines the rigidity of a material, consider a hypothetical material composed of a cubic lattice of identical atoms, shown in Figure 3.3, connected by bonds of stiffness κ and length r_0. Suppose that a force is applied perpendicularly to one face of the lattice such that each bond experiences a tensile force F. Each bond will be stretched a distance Δr according to Hooke's law: $F = \kappa \Delta r$. Dividing through by r_0^2, we have $F/r_0^2 = (\kappa/r_0)(\Delta r/r_0)$. Now F/r_0^2 is the force

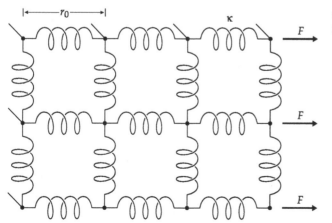

Figure 3.3 A cubic lattice of springs

per unit area and $\Delta r/r_0$ is the strain in each bond (and thus the strain in the whole material). Comparing this with the definition of Young's modulus (Equation 3.2), we see that for this hypothetical material, $E = \kappa/r_0$. Note that κ/r_0 has the same units as Young's modulus.

Table 3.2 shows that the Young's moduli of proteins are much less than that of solids that are held together by covalent bonds (e.g., diamond), metallic bonds, or electrostatic bonds (e.g., glass). This is because the compliance of proteins (the reciprocal of the stiffness) is dominated by the softer van der Waals bonds between the uncharged amino acids: The van der Waals bonds are the weak links in the structure. Evidence for this statement comes from the following "back of the envelope" calculation. The van der Waals potential energy between two solids whose planar surfaces are separated by a distance D is

$$U(D) = -U_0 \left[\frac{1}{3}\left(\frac{D_0}{D}\right)^2 - \frac{1}{3}\left(\frac{D_0}{D}\right)^8 \right] \text{ energy per unit area} \qquad (3.3)$$

(Israelachvili, 1991; and see Heinz and Hoh, 1999 for several other force laws used to interpret AFM measurements). D_0 is the resting (equilibrium) separation, which we take to be twice the van der Waals radius (Creighton, 1993): $D_0 \cong 0.3$ nm. U_0 is equal to twice the surface energy (also called the surface tension): $U_0 \cong 40$ mJ/m^2 = 10 kT/nm^2 = 40 pN/nm for molecules like oils and hydrocarbons that are composed of hydrogen, carbon, oxygen, and nitrogen (Israelachvili, 1991). The first term in the square brackets in Equation 3.3 corresponds to the attractive dipole–dipole interactions, whereas the second term arises from steric repulsion between adjacent atoms. In Appendix 3.1 it is shown that a solid composed of uncharged amino acids of diameter ~0.6 nm held together by van der Waals bonds is expected to have a Young's modulus of ~4 GPa. The experimental finding that the Young's moduli of filamentous proteins are in the range 1 to 5 GPa (see Table 3.2 and Chapter 8), close to this theoretical value, therefore supports the notion that the rigidity of proteins is primarily limited by the rigidity of the van der Waals bonds. A further corollary of the van der Waals model is that the maximum tensile strength of a protein ought to be 0.1 to 0.2 GPa (Appendix 3.1). These values are also similar to those of proteins (see Table 3.2), again supporting the van der Waals model for protein rigidity. A possible exception is silk: Its high Young's modulus and tensile strength are probably due to the strong hydrogen bonding along the β-sheet backbone of the silk molecules. In general, though, the van der Waals rigidity is expected to be an upper limit. The rigidity will be less if the protein is not well folded (i.e., the amino acids are not well packed together) or if the protein fluctuates between a number of different conformational states (Chapter 5).

Rubber and flexible proteins such as titin resist deformation for a completely different reason than do rigid materials. Deformations tend to align the constituent polymeric chains (Figure 3.4), and the loss of entropy associated with such alignment makes the deformation energetically unfavorable. We will con-

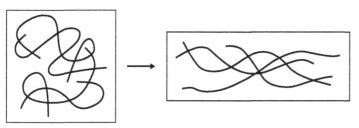

Figure 3.4 Deformation of a rubber-like material

sider the statistical physics of rubber-like elasticity in more detail in Chapter 6, and show that for small deformations, the stiffness is approximately constant, so Hooke's law again applies. At higher strains, the chains become taut, and the stiffness increases, in contrast to the behavior of rigid materials, which get softer before eventually yielding.

Viscous Damping

As a protein changes shape, it is subject to two types of damping forces that slow its motion. The first arises from the viscosity of the surrounding fluid. We call this solvent friction. The second arises from transitory interactions between amino acids that slide with respect to one another as the protein changes shape. We call this **protein friction**. The two types of damping have a similar molecular origin, as is shown below.

Viscosity is defined as follows. When two surfaces submerged in a fluid are moved slowly with respect to each other as shown in Figure 3.5, it is found experimentally that the force per unit area required to produce the shear, F/A, is proportional to the velocity gradient, dv/dx in the fluid between them:

$$\frac{F}{A} = \eta \frac{dv}{dx} \tag{3.4}$$

The constant of proportionality, η, is called the coefficient of viscosity, or simply the viscosity. Because the unit of pressure is the pascal and the unit for velocity gradient (also called shear rate) is s^{-1}, the unit of viscosity is Pa·s. A fluid for which the viscosity is independent of the velocity gradient is called

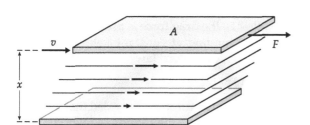

Figure 3.5 Velocity gradient in a viscous fluid between two surfaces

Table 3.3 Viscosities of various liquids

Liquid	Viscosity at 20°C (mPa·s)
Acetone	0.32
Water	1.00
N-hexadecane	3.34
Phenol	12.7
Motor oil (S.A.E. 30)	30
Olive oil	84
Ficoll 400 (50% w/v)	600
Glycerol	1408

Note: Data from Tennent, 1971; Kaye and Laby, 1986; and Resnick et al., 1992, except for Ficoll 400 (Pharmacia), a highly branched polysaccharide of molecular mass 400 kDa.

a **Newtonian fluid**. Most common liquids are Newtonian. The most common deviation from Newtonian behavior is **shear thinning**, the tendency for the viscosity to decrease at high velocity gradients. The viscosity of several fluids is given in Table 3.3.

An object moving through a viscous fluid will experience a drag force that opposes its motion. The magnitude of the drag force depends on the pattern of fluid flow around the object. This, in turn, depends on the **Reynolds number** defined by

$$Re = \frac{\rho L v}{\eta} \tag{3.5}$$

where ρ is the density of the liquid, L is the characteristic length of the object (in the direction of the flow), v is the speed of the object, and η is the viscosity. Note that the Reynolds number is dimensionless, and its physical meaning is that it is the ratio of the inertial and the viscous forces. The Reynolds number tells you what opposes the acceleration of an ocean liner or a bacterium (see Table 3.4). For the ocean liner ($Re \gg 1$), it is the mass; for the bacterium ($Re \ll 1$), it is the drag, and the mass does not matter. If the Reynolds number is less than 1, the flow is laminar and nonturbulent and is referred to as **creeping flow**. The Reynolds number scales with the size of the object, L: The smaller the object, the smaller the ratio of inertial to viscous forces. As illustrated in Table 3.4, microscopic objects like cells and proteins have a Reynolds number less than 1.

At low Reynolds number (i.e., $Re < 1$), the drag force is proportional to the speed, and for a sphere of radius r moving at constant velocity v in an unbounded fluid, the drag force is given by **Stokes' law**:

$$F_d = -\gamma v = -6\pi\eta r v \quad \text{or} \quad \gamma = 6\pi\eta r \tag{3.6}$$

Stokes' law also holds when the velocity is not constant, provided that the inertial forces are less than the viscous forces (i.e., the motion is overdamped, Appendix 3.2). A nearby surface significantly increases the drag coefficient

Table 3.4 Reynolds numbers

Object	Size	Speed	Density of the fluid (kg/m3)	Viscosity of the fluid (Pa·s)	Reynolds number
Ocean liner	100 m	30 m/s	1000	10^{-3}	3×10^9
Swimmer	2 m	1 m/s	1000	10^{-3}	2×10^6
Bee	10 mm	0.14 m/s	1.3	18×10^{-6}	100
Protein	6 nm	8 m/s	1000	10^{-3}	0.05
Bacterium	2 µm	25 µm/s	1000	10^{-3}	5×10^{-5}

(Happel and Brenner, 1983): For example, if the surface of a sphere is within a radius of a plane surface, then the drag is increased ~40%.

> **Example 3.5 Jar of honey** The viscosity of honey is ~100 Pa·s. According to Stokes' law, to pull a honey spoon, a sphere of diameter 20 mm, out of a jar of honey at a rate of 1 m/s requires a force of about 18 N. We can check that the Reynolds number is ~0.2, so Stokes' law does apply. This force corresponds to the weight of a 1.8 kg mass, so it is not surprising that we can actually pick up the jar with the viscous force.

The Molecular Basis of Viscosity

Although the molecular basis for the viscosity of gases is well understood, that of liquids is not. In an ideal gas, the force needed to shear two adjacent planes in the gas arises from the transfer of momentum due to the diffusion of the gas molecules from regions of high speed to regions of low speed. Because the rate of diffusion (and therefore the change in momentum) increases with temperature, the viscosity of gases actually increases with temperature! By contrast, the viscosity of liquids decreases with temperature; this suggests that the viscosity of liquids is due to intermolecular bonds that break more rapidly at higher temperatures.

I now derive a simple theory of viscosity based on molecular friction between two surfaces (Figure 3.6). Suppose that the molecules on one surface

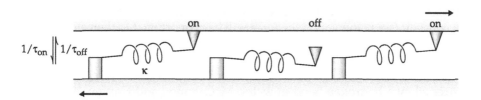

Figure 3.6 Protein friction due to transient crosslinks between two surfaces

make transitory crosslinks with the molecules on an adjacent surface, and that the surfaces slide past one another at a speed that is slow compared to the breaking of the crosslinks—that is, while the crosslink is attached, the surfaces move through a distance that is small compared to the molecular size. Let the rate of detachment be $1/\tau_{on}$ (the reciprocal of the time, τ_{on}, that they spend attached), and suppose the speed of movement is v. Then each molecule will be stretched by an average amount $\tau_{on}v$ during each attachment. If the stiffness of each molecular bond is κ, then the average force opposing the shear is $-\kappa\tau_{on}v$. If each molecule spends a fraction p of its time attached as a crosslink, then the average force per molecule is $-p\kappa\tau_{on}v$. Because this force is proportional to the speed, we can think of it as being a drag force. The associated drag coefficient (per molecule) is

$$\gamma = p\kappa\tau_{on} \tag{3.7}$$

This drag coefficient is independent of speed provided that $p\tau_{on}$ is independent of speed, which occurs if both the attachment and detachment rates are independent of the speed. For low speeds this is likely to be the case. A similar expression has been derived by Schoenberg (1985), Tawada and Sekimoto (1991), and Leibler and Huse (1993).

Equation 3.7 accounts for many features of viscous damping. As expected, the damping increases with the number of molecular bonds between the surfaces. The drag increases as the attached time increases because the crosslinks get more stretched and produce a greater force. The drag also increases with the stiffness of the crosslinks, again because a greater opposing force is generated. Because the attached time is expected to decrease with temperature, so too will the viscosity (p is expected to change little with temperature). Furthermore, for large speeds, corresponding to large shear rates, we expect the attached time to decrease as the shear disrupts the bonds. Thus the model predicts that the viscosity should eventually decrease as the velocity gradient is increased, giving rise to shear thinning.

We can develop this approach a little further to derive an approximate expression for the viscosity of a liquid in terms of molecular parameters. The number of molecules per unit area is $\sim 1/\delta^2$, where δ is the dimension of the molecule. The separation of the layers is also $\sim \delta$. We can therefore write

$$\frac{\text{Force}}{\text{Area}} = \frac{p\kappa\tau_{on}v}{\delta^2} = \left(p\kappa\delta^2\right)\left(\frac{1}{\delta^3}\right)\left(\tau_{on}\right)\left(\frac{v}{\delta}\right) \cong wc\tau_{on}\frac{dv}{dx} \tag{3.8}$$

where w is the intermolecular bond energy and c is the concentration (molecules/m^3). Provided that the lifetime of the intermolecular association, τ_{on}, is a constant (see below), then the force will be proportional to the velocity gradient, dv/dx, as expected for the viscous force in a Newtonian fluid (see Equation 3.4). The corresponding viscosity is $\eta \cong wc\tau_{on}$. For water, $c = 55{,}000$ moles/m^3 = 33×10^{27} molecules/m^3, $w = 8\,kT \cong 32 \times 10^{-21}$ J/molecule (the strength of a hydrogen bond in water; McCammon and Harvey, 1987), and $\tau_{on} \cong 10$ ps (Eisenberg and Kauzmann, 1969). The predicted viscosity is therefore ~10 mPa·s, which is close to (though a little higher than) the viscosity of water at room temperature, ~1 mPa·s. This analysis assumes that the distances through which the crosslinks are deformed are much less than the size of the bonds. This relationship is satisfied provided that the velocity gradient is much smaller than the detachment rate, $1/\tau_{on}$. In the case of water, the analysis should be valid for velocity gradients up to ~ 10^9 s^{-1}, which corresponds to a very high shear rate.

Because the sliding of protein domains past each other also entails the breaking and unbreaking of numerous weak bonds, we expect that protein movements will be slowed down by internal viscosity. Internal viscosity, or protein friction, has been measured in relaxed muscle fibers where the myosin heads make transitory crossbridges to the actin filaments (Brenner et al., 1982). Molecular dynamics modeling suggests that the interior of proteins is liquid-like and hence should possess viscosity (McCammon et al., 1977). Protein friction has been measured in myoglobin where it is found to contribute a damping force that is four times greater than that from the solvent (Ansari et al., 1992). The magnitude of the internal friction for other proteins is not known. However, because the viscosities of aromatics and light oils are in the range of 1 to 30 times that of water, we expect protein friction to be at least as important as solvent friction. If more long-lived crosslinks must be broken, then protein friction could be much larger. A challenge for experimentalists is to measure the internal viscosity, preferably at the single-protein level, to determine the extent to which protein friction limits the speed of conformational changes.

The Global Motions of Proteins Are Overdamped

In Chapter 2 I established criteria for whether the motion of an object in response to a mechanical force is oscillatory (underdamped) or monotonic (overdamped). It was shown that the behavior depends on the relative magnitudes of the inertial and viscous forces. These in turn depend on the material properties of the object—its mass, stiffness, and damping. In this chapter I have described the material properties of proteins. We are therefore now in a position to determine whether the global motions of proteins are underdamped or overdamped. The answer is arrived at through a scaling argument: As the dimension of an object gets smaller, the viscous forces increase relative to the inertial forces, and as a result, *the global motions of small, comparatively soft objects such as proteins in aqueous solution are expected to be overdamped*.

To develop the scaling argument, we consider first a crude mechanical model of a globular protein as a homogenous and isotropic cube with side L, density ρ, and Young's modulus E, damped by a fluid of viscosity η. The mass is $m = \rho V = \rho L^3$. The stiffness is $\kappa = EL$ (Example 3.1), assuming that the protein is globular rather than elongated. (See the end of this section for an analysis of elongated proteins such as cytoskeletal filaments.) The drag force associated with a global conformational change that alters the shape of a protein should be roughly given by Stokes' law, for which the drag coefficient is $\gamma \cong$

$3\pi\eta L$ (Appendix 3.3). In Chapter 2 I showed that the motion is overdamped if the ratio $4m\kappa/\gamma^2$ is less than 1. In the present case

$$\frac{4m\kappa}{\gamma^2} = \frac{4 \cdot \rho L^3 \cdot EL}{(3\pi\eta L)^2} = \left(\frac{2}{3\pi}\right)^2 \frac{\rho E}{\eta^2} L^2 \qquad (3.9)$$

The important feature of this equation is that it shows how the ratio scales with dimension, L: The smaller the object, the smaller the ratio, and the less is the tendency for oscillation. The reason for this scaling behavior is that while the damping and the stiffness decrease in proportion to the length, the mass decreases much faster (to the third power), so the inertial forces decrease more quickly than the viscous forces. While the numerical term on the right-hand side of Equation 3.9 depends on the particular model, the scaling behavior does not.

How small must a protein be to ensure that its motion is overdamped and that it does not oscillate when subject to an external force? The most rigid proteins have Young's moduli, E, on the order of 1 GPa (see Table 3.2). The density, ρ, is on the order of 10^3 kg/m^3, and the viscosity of water, η, is on the order of 1 mPa·s (see Table 3.3). Thus for a rigid protein in water, $\eta^2/\rho E \cong 1$ nm^2, and according to Equation 3.9, the motions of globular proteins or protein domains of diameter less than a characteristic length $L_c \cong (3\pi/2)\sqrt{\eta^2/\rho E} \cong 5$nm will be overdamped. This length corresponds to a medium-sized globular protein of ~1000 amino acids. Thus the model predicts that global motions of rigid globular proteins or protein domains of molecular weight less than 100 kDa should be overdamped. This conclusion is confirmed by the more accurate analysis presented in Appendix 3.2. The analysis in the Appendix also justifies the use of Stokes' law for the drag coefficient, which strictly applies only if the motion is overdamped. Although the quality of motion of proteins is difficult to measure experimentally, molecular dynamics modeling studies lend support to these arguments (McCammon et al., 1976).

There are several additional arguments for why the motions of even large proteins ($L>5$ nm) ought to be overdamped:

1. The rigidity of allosteric, energy-transducing proteins such as motor proteins and the ribosome (Example 3.6) is likely to be much less than that of rigid proteins like those of the cytoskeleton. Consider a protein that undergoes a fairly modest conformational change of 1 nm, corresponding approximately to the size of a nucleotide or an amino acid. Many proteins and protein complexes such as motors (Chapter 12), G-proteins, and ribosomes (Frank, 1998) undergo substantially larger conformational changes (for review, see Gerstein et al., 1994). Suppose that the conformational change is associated with a large amount of mechanical work, say 100×10^{-21} J (= 25 kT) equal to the free energy of hydrolysis of the gamma phosphate bond of one molecule of ATP (Chapter 14). Because the mechanical work done on the proteins is equal to $\frac{1}{2}\kappa x^2$ the stiffness is 0.2 N/m, only ~1% of the stiff-

> **Example 3.6 Ribosome** If a large protein were to oscillate, how fast and how large might these oscillations be? Consider the ribosome, a globular protein–RNA enzyme complex of diameter ~30 nm (Ban et al., 1999, Clemons et al., 1999). The ribosome is the molecular machine that synthesizes proteins. If the ribosome were very rigid ($E = 1$ GPa), and the only damping came from the surrounding fluid, then it would oscillate at a frequency of $\sim(\kappa/m)^{0.5}/2\pi$ Hz $= (E/\rho)^{0.5}/2\pi L \sim 5$ GHz, corresponding to a period of 200 ps. The oscillation would decay quickly, with a time constant of $2m/\gamma \sim (2/3\pi)\rho L^2/\eta \sim 200$ ps (Equation A2.1 in the Appendix). In other words, the oscillations would die out after only a few cycles. The magnitude of the oscillations would depend on the size of the force. Suppose that the force did work on the protein equal to 100×10^{-21} J ($= 25$ kT), the free energy associated with the hydrolysis of one molecule of ATP (Chapter 14). If we think of this chemical energy as being converted into mechanical potential energy within the protein during the protein synthesis reaction, then the amplitude, x, of the deformation would be only ~0.8 Å (energy $= \frac{1}{2}\kappa x^2$, and we assume that ribosome is as rigid as a cytoskeletal protein with $\kappa = EL = 30$ N/m). The oscillations, if they occurred, would be very small indeed. Considering that the lifetimes of different chemical states are in the order of microseconds to milliseconds, it is unlikely that such small oscillations, even it they were to occur, would play important roles in the chemistry of protein synthesis.

ness of a rigid protein of length 10 nm and Young's modulus 2 GPa. This low value of stiffness leads to a much greater characteristic length of 50 nm, implying that even the motion of a ribosome, one of the largest protein machines, would be overdamped. Because this calculation used a small value for the conformational change and a large value for the work, even this low stiffness is likely to be an overestimate; indeed, the stiffness of motor proteins is on the order of only ~1mN/m. This argues strongly that protein motions are overdamped.

2. We expect that protein friction due to the fluid-like nature of the interior of proteins (McCammon and Harvey, 1987) will further dampen out any tendency to oscillate. There is little data on the magnitude of this effect, though experimental data of Ansari et al. (1992) indicate that the internal damping of myoglobin is four times greater than the external damping from the fluid.

3. Elongated proteins are more highly damped than globular proteins of the same molecular weight (Appendix 3.4). This is because as the aspect ratio increases, the damping increases while the stiffness decreases. For example, if the aspect ratio is 10, then the characteristic length for stretching motions is ~100 nm, while that for bending motions is ~4000 nm. Thus the motions of proteins with large axial ratios will always be overdamped.

4. For a protein filament, the damping ratio actually *increases* as the length increases: The longer the filament the more highly damped (Appendix 3.4), a scaling behavior opposite to that of globular proteins. *This leads to the important conclusion that the motion of the cytoskeleton is overdamped.*

I have belabored this discussion of mechanical damping because the quality of motion of a protein is very important for understanding how it works. Overdamping rules out many wacky ideas about high-frequency resonances and long-distance information transfer and processing in proteins (e.g., Penrose, 1994). Instead, we have a relatively simple view of proteins as mechanical devices that move monotonically into new structural states in response to applied (and internally generated) forces.

To get a feeling for how proteins move, imagine that the size of a protein were increased by a factor of 10^7, so that a 5-nm-diameter protein became a mechanical device of diameter 50 mm, fitting nicely in the palm of one's hand. Let's keep the density and Young's modulus the same so our device could as well be built of plastic or Plexiglas. Now, if the viscosity of the fluid bathing the device were increased by the same factor by putting it in treacle, then the ratio of the inertial to the viscous forces will the same for both protein and device (Equation 3.9). The Reynolds number will be unchanged and the pattern of fluid flow will be preserved (just scaled in size). However, to deform the plastic device to the same relative extent will require a much larger force because the device has a much greater cross-sectional area: Whereas a force of only 1 pN might be needed to induce a protein conformational change of 1 nm (corresponding to a strain of 20%), a force of 100 N, corresponding to a weight of 10 kg, would be required to produce the same strain in the plastic device. In response to the respective forces, the protein and the mechanical device will move at the same initial speed, but because the protein conformational change is so much smaller, the relaxation of the protein will be complete in much less time: A relaxation that took an almost imperceptible 100 ns for the protein will take a leisurely 1 s for the device.

The Motions of the Cytoskeleton and Cells Are Also Overdamped

One might expect, based on the scaling argument for globular proteins, that cells, whose linear dimensions are some 1000 times larger than those of proteins, might undergo underdamped, oscillatory motions. However, experimental measurements show that this is not the case: The motions of cells are very highly damped. For example, the cytoplasm of macrophages that have ingested 1-µm-diameter magnetic particles can be perturbed using a weak external magnetic field. The particles reorient extremely slowly, with time constants of minutes. The apparent intracellular viscosity is very high, ~1000 Pa·s (a million times the viscosity of water) for velocity gradients of 0.01 s^{-1} (Valberg and Feldman, 1987). The motion is highly overdamped. Because actin gels crosslinked with the actin binding protein ABP have similar viscoelastic properties to cells (Zaner and Valberg, 1989; see Problem 3.6 for the definition of viscoelasticity), it is likely that the viscoelasticity of cells arises from the stiffness of and damping on the cytoskeletal filaments. And because the long cytoskeletal filaments are highly damped, as argued in the previous section, so too are cells.

The high apparent viscosity of cytoplasm measured in the Valberg and Feldman experiment is still consistent with the cytoplasm being an aqueous environment. Indeed, small fluorescent probes of diameter ~1 nm are highly mobile inside cells, having rotational and translational diffusion coefficients similar to those in aqueous solution (Fushimi and Verkman, 1991). Even particles of diameter 6 nm, corresponding to that of a 100 kDa protein, diffuse quickly inside cells, as though the viscosity were about three times that of water (Luby-Phelps et al., 1987; Seksek et al., 1997). But larger particles are not nearly so mobile: 50 to 500-nm-diameter particles diffuse very slowly inside cells, indicating an apparent viscosity 30 to 300 times that of water (Luby-Phelps et al., 1987; Alexander and Rieder, 1991). Thus it appears that the cytoskeletal filaments form a gel with a mesh size of ~50 nm. Small solutes and proteins can readily diffuse through the pores, but the motion of larger particles, such as ribosomes and organelles, is severely restricted.

Even in cases where the cytoskeletal filaments are highly aligned and tightly crosslinked for maximum rigidity, the viscous forces dominate the inertial ones; this is true for cilia and flagella, which are composed of microtubules, as well as for the stereocilia of hair cells, which are composed of tightly crosslinked actin filaments (Chapter 8). Thus, even though it is conceivable that the rigidity of the whole cell could be large enough to result in underdamped motion, the cytoskeletal filaments are too sparsely crosslinked to make the network sufficiently rigid. It is only at the organismal level, where large multicellular structures are in contact with air, that oscillations—of the belly of an elephant, for example—are possible.

Summary

The rigidity of cytoskeletal proteins such as actin, tubulin, and keratin, which serve structural roles in cells, is similar to that of hard plastics like polypropylene, but substantially less than that of materials such as metal, glass, and wood. This is because proteins are held together by relatively weak van der Waals bonds. The rigidity of protein machines, proteins that are designed to undergo large conformational changes as they transduce chemical energy into mechanical work, is expected to be substantially less than that of structural proteins. As proteins move and change shape, they experience damping forces from the surrounding fluid as well as from internal friction. These viscous forces arise from the rapid making and breaking of bonds. Owing to the small size of proteins, the viscous forces on proteins are generally much greater than the inertial forces. Consequently, the global motions of proteins, especially less rigid ones, are highly overdamped: They creep rather than oscillate when subject to applied forces. The motions of the long, thin cytoskeletal filaments are also overdamped, due to their large aspect ratios. This, in turn, causes the motion of cells to be overdamped.

Problems

3.1 The Young's modulus of an active muscle (in the longitudinal direction) is ~40 MPa (Bagshaw, 1993). What is the spring constant of a muscle of length 100 mm and cross-sectional area 1000 mm^2?

3.2 By how much would such a muscle be extended by the weight of a mass of 10 kg? What is the fractional extension?

3.3 Using the Young's modulus for actin in Table 3.2 and a cross-sectional area of 20 nm^2, calculate the stiffness of a 1-µm-long actin filament.

3.4 Given that there are ~10^{15} actin filaments per square meter of cross-sectional area of a muscle and that the cross-sectional area of an actin filament is 20 nm^2, calculate the fraction of the cross-section occupied by actin. If a muscle were composed solely of continuous actin filaments (stretching from one end of the muscle to the other) occupying this volume fraction, what would be its Young's modulus? How does this compare with the value stated in Problem 3.1? What conclusions can you draw about the molecular basis for the rigidity of muscle? (Note that the actin filaments are not continuous but instead alternate with the myosin-containing thick filaments [see Figure 1.1]. The compliance contributed by actin is similar to that contributed by the crosslinks [the myosin heads] between the thin and thick filaments [Chapter 8].)

3.5 Show that the relaxation time of a rigid globular protein whose motion is highly damped by the solvent is ~10 ps, independent of the size of the protein.

[Answer: $\tau = \dfrac{\gamma}{\kappa} \sim \dfrac{3\pi\eta L}{EA/L} = 3\pi\dfrac{L^2}{A}\dfrac{\eta}{E} \cong 3\pi\dfrac{\eta}{E} \sim 10$ ps using $\eta = 1$ mPa·s/m and $E = 1$ GPa.]

3.6 Viscoelasticity Another way to deform a material is to place it under shear (see the figure below). In this case:

$$\frac{F}{A} = G\theta$$

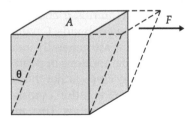

where G is the **shear modulus**. For a homogenous isotropic material, the shear modulus is related to the Young's modulus by $G = E/2(1 + \sigma)$. The parameter σ is Poisson's ratio, the relative amount of sideways contraction ($\Delta w/w$) of the material compared to the lengthwise strain ($\Delta L/L$): $\Delta w/w = \sigma(\Delta L/L)$. **Poisson's ratio** lies between -1 and $+0.5$; the lower value applies to a material that has constant shape, and the upper value applies to a material that has constant volume (i.e., incompressible). For an incompressible solid, $G = E/3$. For most materials, $0.2 \leq \sigma \leq 0.5$ (Kaye and Laby, 1986).

If the deformation of an elastic material requires the breaking and remaking of internal molecular bonds, then it will not respond instantaneously to an applied force, but instead it will relax more slowly to its new shape. We say that the material is **viscoelastic**. With the aid of Figure 3.5 and the figure opposite in Problem 3.6, we can write the equation of motion in response to a shearing force:

$$\frac{F}{A} = \eta \frac{d\theta}{dt} + G\theta$$

This is formally the same as the equation of motion of a damped spring—i.e., a spring in series with a dashpot (see Figure 2.3B). The material will deform with a time constant equal to η/G. Note that the relaxation time constant is also a material property because it does not depend on size.

If a protein has an internal viscosity (protein friction) 4 times that of water, a Young's modulus of 1 GPa, and a Poisson ratio of 0.25, calculate the relaxation time constant. Compare this to Problem 3.5.

CHAPTER

4

Thermal Forces and Diffusion

In addition to the mechanical forces discussed in Chapter 2, proteins and cells are subject to **thermal forces** that arise from collisions with water and other molecules in the surrounding fluid. During each short-lived collision, the change in momentum of the fluid molecule imparts an impulsive force on the object that it strikes. These collisional forces are called thermal forces because their amplitudes are proportional to the temperature of the fluid molecules. The resulting movement is called **thermal motion**, and the object is said to have **thermal energy**. Because the forces are randomly directed, the motion is characterized by frequent changes in direction and is called **diffusion**. The diffusion of a free particle or molecule is called **Brownian motion**.

Understanding thermal motion is crucial for molecular and cellular mechanics because the chemical reactions that drive biological processes have energies that are only a little higher than thermal energy; as a result, diffusive motions are quite large compared to directed ones, and thermal fluctuations are necessary for proteins to reach their transition states. It is the noisy, diffusive environment in which protein machines operate that distinguishes molecular machines from the macroscopic machines that we experience in our everyday world.

This chapter begins with Boltzmann's law, the fundamental physical law that describes how the probability of a molecule having a certain energy depends on the surrounding temperature. We then discuss some of the many corollaries of Boltzmann's law. These include the Principle of Equipartition of Energy, which states how much thermal energy a molecule has at a certain temperature, and the Einstein relation, which relates the diffusion coefficient of a molecule to its drag coefficient. Next we examine the diffusion of molecules under several different circumstances. One of these, diffusion up an energy gradient (i.e., against a force), is especially interesting and important because it makes predictions of how forces affect the rates of diffusion-limited reactions, a subject that is explored in more detail in Chapter 5. The present chapter ends with a discussion of the dynamics of a particle or molecule undergoing Brownian motion. We show that thermal motion can be exactly simulated by an applied mechanical force of a particular amplitude and time course that acts via the damping elements (the dashpots).

Boltzmann's Law

A particle or molecule always tends towards its lowest energy state. However, except at the absolute zero of temperature, the particles are agitated by molecular collisions. As a result, they do not spend all their time in the state with the lowest energy, but instead spend a fraction of their time in states with higher energy (Figure 4.1). **Boltzmann's law** says that if such a particle (or a group of particles or molecules in a larger system) is in thermal equilibrium, then the probability, p_i, of finding the particle (or group) in a state i that has energy U_i is

$$p_i = \frac{1}{Z}\exp\left[-\frac{U_i}{kT}\right] \qquad \text{where } Z = \text{constant} = \sum_i \exp\left[-\frac{U_i}{kT}\right] \qquad (4.1)$$

k is the Boltzmann constant and T is the absolute temperature. The Boltzmann constant is equal to 1.381×10^{-23} J/K. At the standard temperature of 25°C, corresponding to 298.15 K, kT is therefore equal to 4.116×10^{-21} J (Table 4.1). The value of kT is compared to other biologically relevant energies in Table 4.2. Z is a constant, sometimes called the **partition function**, whose value assures that the sum of all the probabilities adds up to 1 ($\Sigma p_i = 1$). Equation 4.1 is sometimes called Boltzmann's equation, Boltzmann's distribution, or Boltzmann's formula, and the exponential term in the equation is called the Boltzmann factor.

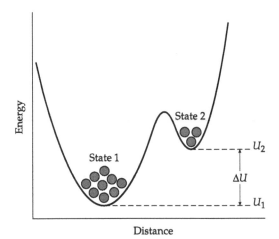

Figure 4.1 Molecules in an energy landscape
According to Boltzmann's law, the probability of finding a molecule in state 2 relative to state 1 is $\exp(-\Delta U/kT)$.

Table 4.1 Thermal energy

Quantity	25°C	37°C
kT	4.1164×10^{-21} J	4.2821×10^{-21} J
RT (= NkT)	2.4789 kJ/mol	2.5787 kJ/mol
RT	0.5921 kcal/mol	0.6159 kcal/mol
kT/e (= RT/F)	25.69 mV	26.73 mV

Note: Conversions: Temperature, 0 K = −273.15°C; Energy, 1 calorie = 4.1868 joules.

Boltzmann's law is very general. The energy could correspond to the particle's potential energy (gravitational, elastic, or electrical), its kinetic energy, or the energy associated with its phase, or electronic or chemical state. The state of a particle (or group of particles) is specified by the position and velocity of the constituent atoms as well as their electronic states. If there are just two states with energies U_1 and U_2 (and energy difference $\Delta U = U_2 - U_1$), then the ratio of the probabilities of being in the two states is

$$\frac{p_2}{p_1} = \exp\left(-\frac{\Delta U}{kT}\right) \qquad (4.2)$$

Boltzmann's law is a corollary of a postulate in statistical mechanics stating that each configuration of a closed system (a system of fixed total energy) is equally likely. The derivation of Boltzmann's law from this postulate can be found in Pauling (1970) and Berg (1993). Boltzmann's law is the most important physical law in biology and chemistry because it has so many consequences. Some of these consequences are illustrated in Example 4.1, some are illustrated in the next sections, and some in the next chapter. Boltzmann's law is also nonlinear, a point that should be heeded by those who seek linear approximations.

In our statement of Boltzmann's law we defined neither equilibrium nor temperature. These are deep concepts that might take an entire course in statistical physics to explore. Luckily, there is a simple way out. So fundamental is Boltzmann's law that we can use it to define temperature and equilibrium. We will say that a system is at **equilibrium** if Boltzmann's law holds, and we will define the temperature as the corresponding constant in the exponent of Equation 4.1. This definition of equilibrium is consistent with the more usual definition of equilibrium and temperature. First, if two systems are each in

Table 4.2 A comparison of energies

Energy	Formula	Value (10^{-21} J)
Thermal energy (25°C)	kT	4.1
Photon (green, λ = 500 nm)	$h\nu = hc/\lambda$	397
ATP hydrolysis in the cell	ΔG	100
Electron transport (180 mV)	eV	28.8

equilibrium with a third system, then the two systems are in equilibrium with each other. This follows from Boltzmann's law because the exponential function satisfies $\exp[-a] \times \exp[-b] = \exp[-(a+b)]$. And second, if a system is at equilibrium then it is at **steady state**, meaning that its (average) properties do not change with time. This follows from Boltzmann's law because if the energy U does not change with time (as we are implicitly assuming in our statement of the law), then the probability p also does not change with time. The difference between steady state and equilibrium is that in the former there can be a constant net flux of particles, but in the latter the net flux is zero.

Example 4.1 Applications of Boltzmann's law

EARTH'S ATMOSPHERE. The density of molecules in a gravitational field falls exponentially with the height (see figure below). The gravitational potential energy of a particle of mass m is $U = mgh$, where g is the acceleration due to gravity and h is the height above the Earth's surface. Let h_0 be the height at which this energy equals kT (i.e., $mgh_0 = kT$). According to Boltzmann's law, the probability of finding the particle at this height is 37% ($1/e$) that of finding it at zero height. For an oxygen molecule of molecular mass 32, $h_0 \cong 7.5$ km, the approximate height of Earth's atmosphere.

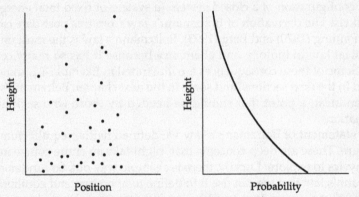

SETTLING OF BEADS. Instead of an oxygen molecule, consider a glass sphere of density (ρ) twice that of water (ρ_w), and radius r. The corresponding height of the "atmosphere" is

$$h_0 = kT/mg = kT/(\rho-\rho_w)Vg = 3kT/4\pi(\rho-\rho_w)gr^3$$

For a 200-nm-diameter bead, this corresponds to ~100 μm. For a 2-μm-diameter bead, the height decreases 1000-fold to 100 nm.

ANALYTIC CENTRIFUGATION. A molecule in a centrifuge experiencing a centrifugal acceleration, a_c, has potential energy $U = (m - m_w)a_c h$, where $m - m_w$ is the additional mass over that of the displaced solvent and h is the height above the bottom of the centrifuge tube. For a 100 kDa protein in water ($m - m_w \cong 66 \times 10^{-24}$ kg) spinning at $a_c = 10{,}000\,g$, $h_0 \cong 6$ mm. In an analytic ultracentrifuge, the protein "atmosphere" is measured and h_0 estimated from the exponential decrease in the protein concentration; this allows measurement of the mass of the protein.

THE NERNST EQUATION. If molecules with charge q are free to equilibrate between two compartments at electrical potential 0 and V volts, then the concentrations in the two compartments, C_0 and C_V are related by

$$\frac{C_V}{C_0} = \frac{p_V}{p_0} = \exp\left[-\frac{U}{kT}\right] = \exp\left[-\frac{qV}{kT}\right]$$

This is known as the **Nernst equation** (Hille, 1992). At 37°C, kT/e equals 26.7 mV (see Table 4.1), where e is the charge of the electron. The Nernst equation says that for each 26.7 mV increase in voltage, the concentration of monovalent cations decreases e-fold.

Equipartition of Energy

Boltzmann's law allows one to calculate the average thermal energy of a molecule (or system of molecules). Suppose that the molecule is at equilibrium in an **energy landscape**, $U(x)$, that varies with position, x, but not with time. For example, the molecule could be connected to a spring with potential energy $U(x) = \tfrac{1}{2}\kappa x^2$, where x is the extension (Figure 4.2). Due to thermal agitation, the molecule is constantly changing position. Now there are two ways that we can calculate the statistical properties of the molecule's position, such as its mean or its variance. First, we could follow the molecule over a long period of time, T, and measure its time-averaged **mean** position or **mean-squared** position

$$\langle x \rangle_T \equiv \frac{1}{T}\int_0^T x(t)\cdot dx \qquad \langle x^2 \rangle_T \equiv \frac{1}{T}\int_0^T x^2(t)\cdot dx$$

Alternatively, we could use Boltzmann's law to calculate the probability, $p(x)$, of finding the molecule at position x, and then calculate the **expected value** of the position or position squared according to

$$E(x) \equiv \int_{-\infty}^{\infty} x p(x)\cdot dx \qquad E(x^2) \equiv \int_{-\infty}^{\infty} x^2 p(x)\cdot dx$$

If we measure for a long enough time, then these two estimates of the average position should agree with each other

$$\langle x \rangle \equiv \langle x \rangle_\infty = E(x) \qquad \langle x^2 \rangle \equiv \langle x^2 \rangle_\infty = E(x^2) \qquad (4.3)$$

In this way, we can relate measurements (time averages) to the expectations based on Boltzmann's law. Equation 4.3 is the link between experiment and theory! Equation 4.3 holds generally for any function of x: $E[f(x)] = \langle f(x) \rangle$. In particular it holds for the **variance** of x,

$$\sigma_x^2 \equiv \langle (x - \langle x \rangle)^2 \rangle = \langle x^2 \rangle - \langle x \rangle^2$$

Often the mean of a variable is zero (e.g., the mean extension of a spring, the mean velocity), in which case the variance is equal to the mean square.

This approach can be used to calculate the average energy of a molecule. For example, consider the spring in Figure 4.2. The average energy is

$$\langle U \rangle = \tfrac{1}{2} \kappa \langle x^2 \rangle = \tfrac{1}{2} \kappa \int_{-\infty}^{\infty} x^2 p(x) \cdot dx = \tfrac{1}{2} kT \qquad (4.4)$$

where we have used Boltzmann's law for $p(x)$ (Equation 4.1) and evaluated the integral in Appendix 4.1. This result is remarkable because the average energy does not depend on the stiffness of the spring! It only depends on the temperature. This is a special case of a general theorem known as the **Principle**

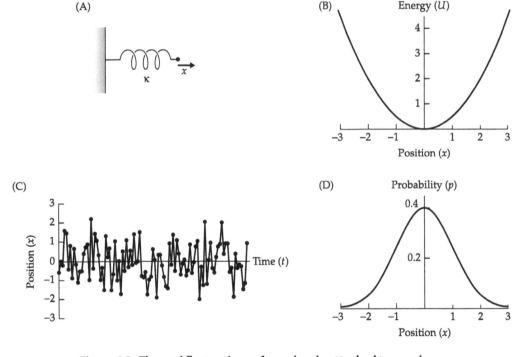

Figure 4.2 Thermal fluctuations of a molecule attached to a spring
A molecule attached to a spring (A) sits in a parabolic potential well (B). While it fluctuates (C), it spends more time near the center of the well (the deepest region) than at the periphery; its probability distribution (D) is a Gaussian (or Normal) distribution, which is peaked at the center.

of **Equipartition of Energy,** which states that if the energy of a molecule depends on the square of a parameter such as position or speed, then the mean energy associated with the degree of freedom measured by that parameter is $\langle U \rangle = \frac{1}{2}kT$. Another example of the principle is that the average kinetic energy of a molecule (in one direction) with mass m is $\langle K.E. \rangle = \frac{1}{2}m\langle v^2 \rangle = \frac{1}{2}kT$. If the molecule (or system of molecules) has two (or more) degrees of freedom that are independent, as defined in Appendix 4.1, then each degree of freedom contains $\frac{1}{2}kT$ of energy. An example of independent degrees of freedom are the components of the velocity of a molecule in the x-, y-, and z-directions: There is $\frac{1}{2}kT$ of kinetic energy in each, so the total kinetic energy is $\frac{3}{2}kT$. The **root-mean-square** speed, v_{rms}, of a molecule in three dimensions is therefore

$$v_{rms} = \sqrt{\langle v^2 \rangle} = \sqrt{\frac{3kT}{m}} \tag{4.5}$$

For example, the root-mean-square speed of a water molecule is 640 m/s (in air or water), that of a 100 kDa protein is 8.6 m/s at 25°C, and that of a bacterium of volume 1 μm³ is 3.5 mm/s.

The Equipartition principle breaks down in a number of circumstances. First, it is generally true only if the energy dependence is quadratic (see, e.g., Problem 4.7). Second, it breaks down if the thermal energy kT is small compared to the energy levels between different quantum states. In this case, the degree of freedom is said to be "frozen out." But for proteins at room temperature, thermal energy is large compared to the mechanical vibrational energy levels because proteins are relatively soft materials, and so the Equipartition principle applies to elastic deformations of proteins (Appendix 4.1).

Diffusion as a Random Walk

The forces that agitate molecules cause diffusion. Diffusion is a form of random motion that is characterized by frequent, abrupt changes in direction. The randomness is the result of the collisions with surrounding molecules, which themselves are moving in random directions. Some examples of diffusive motion are shown in Figure 4.3. The aim of this section is to derive the diffusion equation, which describes how the average concentration of a collection of molecules changes over time due to the diffusive motion of the individual molecules.

Diffusion plays a crucial role in many physical and chemical processes. Einstein was cited in his Nobel prize award for elucidating the molecular mechanism of Brownian motion. His original papers (see Einstein, 1956) are still among the clearest treatments of the subject. Careful measurements of the diffusion of micron-sized particles by Perrin and Svedberg, who won Nobel prizes in 1926 in physics and chemistry, respectively, confirmed Einstein's theory and permitted the measurement of Boltzmann's constant, k. Because the ideal gas constant $R = Nk$ was already known, the measurements allowed the determination of the Avogadro number N to within a few percent. Thus the study of

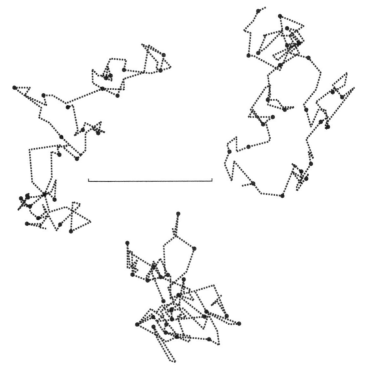

Figure 4.3 Examples of Brownian motion in two dimensions
Each simulated walk consists of 10,000 steps starting from the origin; each step has size +1 or –1 in both the horizontal and vertical directions. The positions at 100-step intervals are connected by dotted lines and every 400th position is marked by a filled circle. Note that the trajectories between the points are themselves as highly convoluted, but this is not shown. The scale bar corresponds to 100 step sizes. If the particle is a 100 kDa protein diffusing in water (diffusion coefficient 50 μm²/s; see Table 2.2) and the step interval were 1 second (10,000 s total time), then the scale bar would equal 1 mm. If the intervals were 100 μs (1 s total time), then the scale bar would equal 10 μm.

Brownian motion confirmed the atomic theory of gases and liquids and bridged the gap between visible objects and invisible molecules. The rich history means that virtually every equation has a name!

The first result needed to derive the diffusion equation is the following interesting consequence of random motion. If molecules are moving in random directions, then, on average, they will tend to move from areas of high concentration to areas of low concentration. The proof is in Appendix 4.2. The prediction, confirmed experimentally, is that the **concentration flux**, $J(x)$, the rate of movement of molecules per unit area, is proportional to the concentration gradient, dc/dx:

$$J(x) = -D \frac{dc}{dx}(x) \tag{4.6}$$

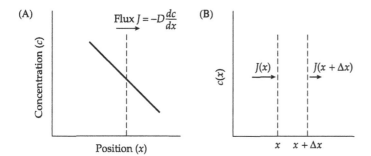

Figure 4.4 Flux and concentration change
(A) According to Fick's law, molecules diffuse from regions of high concentration to regions of low concentration. (B) Change in concentration due to change in flux. If the flux is not uniform—for example, if there is less flux out of a region than into it—then the concentration will change.

This equation is known as **Fick's law** and is shown in Figure 4.4A. The constant of proportionality, D, is called the **diffusion coefficient**. It is related to the size and frequency of the steps underlying the random motion: The larger and more frequent the steps, the greater the diffusion coefficient (Appendix 4.2). The negative sign reflects the tendency of molecules to move from regions of high concentration to regions of low concentration. The **concentration**, c, will usually be expressed in units of molecules per cubic meter, though sometimes moles per cubic meter or moles per liter will be used. The flux has units of molecules per unit area per second, so the diffusion coefficient has units of m^2/s.

To derive the diffusion equation we need to relate the flux back to the concentration. Consider Figure 4.4B. If fewer molecules leave a region to the right than enter it from the left, then there will be a net increase in the concentration in that region. In other words, provided that there are no sinks or sources of molecules, the change in concentration over time at any point equals the negative of the flux gradient at that point

$$\frac{\partial c}{\partial t}(x,t) = -\frac{\partial J}{\partial x}(x,t) \tag{4.7}$$

This is also proved in the Appendix. One application of this equation is that if the system is in the steady state—that is, if there is no change in concentration over time ($dc/dt = 0$)—then the flux is the same everywhere in the solution ($dJ/dx = 0$). Conversely, if the flux does not change from one position to another, then the concentration does not change with time.

Substituting Equation 4.6 into Fick's law gives:

$$\frac{\partial c}{\partial t}(x,t) = D\frac{\partial^2 c}{\partial x^2}(x,t) \tag{4.8}$$

This is known as the **diffusion equation**. Usually we are thinking about single molecules and want to know the probability, $p(x,t)$, of finding a molecule at position x at time t, rather than the concentration, $c(x,t)$, of a large number of molecules. Because the probability is proportional to the concentration (it is the concentration divided by the total number of molecules) and because differentiation is a linear operation ($d[a \cdot f(x)]/dx = a \cdot df/dx$), it follows that the probability, $p(x, t)$, also satisfies the diffusion equation.

Einstein Relation

Boltzmann's law allows us to derive an expression that relates the diffusion coefficient to the drag coefficient, introduced in Chapter 2. Suppose that an external force, $F(x)$, acts on a diffusing molecule. The force could be due to gravity or to an attached spring. The force will cause the molecule to move with velocity $v(x) = F(x)/\gamma$, where γ is the drag coefficient (Chapter 2). This "drift" velocity is an average speed superimposed on the diffusive motion. It is straightforward to show that the external force increases the flux by $v(x) \cdot c(x,t)$, or by $v(x) \cdot p(x,t)$ if we are thinking of the **probability flux**, $j(x)$:

$$j(x) = -D\frac{dp}{dx}(x) + \frac{F(x)}{\gamma}p(x) \tag{4.9}$$

Thus, in the presence of a force, the probability satisfies

$$\frac{\partial p}{\partial t}(x,t) = D\frac{\partial^2 p}{\partial x^2}(x,t) - \frac{\partial}{\partial x}\left[\frac{F(x)}{\gamma}p(x,t)\right] \tag{4.10}$$

This equation, which describes diffusion with drift, is known as the forward diffusion equation, or the **Fokker–Planck equation** (Papoulis, 1991).

If the system is in equilibrium, the probability does not change with time. The Fokker–Planck equation can then be solved to obtain $p(x)$. When this solution is compared to Boltzmann's law (Equation 4.1), it is found that the flux must be equal to zero everywhere, and that the diffusion coefficient is related to the drag coefficient by

$$D = \frac{kT}{\gamma} \tag{4.11}$$

This is known as the **Einstein relation**. It is proved in Appendix 4.2. By relating a molecular parameter, the drag coefficient, to a macroscopic parameter, the diffusion coefficient, the Einstein relation provides the link between the microscopic and macroscopic theories of diffusion. With the help of Stokes' law (Equation 3.6), this equation allows one to estimate the diffusion coefficient from the size of the particle and the viscosity of the solution

$$D = \frac{kT}{6\pi\eta r} \tag{4.12}$$

where r is the radius of a spherical particle, and η is the viscosity. Conversely, knowledge of the viscosity and the diffusion coefficient permits an estimate of the size of the particle, as shown in the following example.

> **Example 4.2 Diffusion of ions** A sodium ion has a diffusion coefficient of 1.33×10^{-9} m^2/s at 25°C (Hille, 1992). From Einstein's relation and Stokes' law, this corresponds to an apparent radius of 1.8 Å, where we used $\eta = 0.89$ mPa·s and $kT = 4.12 \times 10^{-21}$ J at this temperature. This is about two times the ionic radius of 0.95 Å measured in crystals (Hille, 1992). A useful rule of thumb is that a diffusion coefficient of 10^{-9} m^2/s corresponds to 1 μm^2/ms, so a small ion diffuses ~1 μm in 1 ms.

In this chapter we consider only the case where there is no chance of the molecule being destroyed. In other words, the total probability is unity at all times, or

$$\int_{-\infty}^{\infty} p(x,t) \cdot dx = 1 \tag{4.13}$$

In later chapters, when we consider the polymerization of cytoskeletal filaments and the movement of motor proteins, we will relax this condition by allowing chemical reactions to convert one type of molecule into another, or to destroy or create molecules. When these reactions also depend on position, the motion becomes very rich and is described by the **reaction–diffusion equation**

$$\frac{\partial p_i}{\partial t}(x,t) = D \frac{\partial^2 p_i}{\partial x^2}(x,t) - \frac{\partial}{\partial x}\left[\frac{F(x)}{\gamma} p_i(x,t)\right] \tag{4.14}$$
$$+ \sum_j \left[k_{ji}(x) \cdot p_j(x,t) - k_{ij}(x) \cdot p_i(x,t)\right]$$

where p_i, p_j, and so forth are the probabilities of the molecule being in various chemical states i, j, and so on, and k_{ij} is the rate constant for the transition between the i and j states.

Some Solutions to the Diffusion Equation

The utility of the diffusion equation is that it allows one to calculate how quickly, on average, it takes for a molecule to diffuse through a certain distance. This information can be used to evaluate the efficiency of diffusion as a transport process within cells. Furthermore, with the aid of the Fokker–Planck equation, we can calculate the time that it takes for a molecule to diffuse against an applied force. By turning this argument around, one can then gain insight into how forces affect chemical rates, a subject dealt with in detail in the next chapter. In this section we solve the diffusion equation in a few special cases that are relevant to cellular and molecular mechanics. Solutions for a wide variety of other cases can be found in Carslaw and Jaeger (1986).

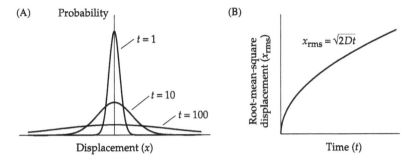

Figure 4.5 Diffusion of a molecule released at time 0 at the origin
(A) The curves show the probability of finding a molecule at increasing times.
(B) The growth in the root-mean-square displacement (x_{rms}) as a function of time.

Free Diffusion from a Point Source

If a molecule is released at the origin and allowed to diffuse in one dimension, then the probability of finding it at position x at time t later is

$$p(x,t) = \frac{1}{\sqrt{4\pi Dt}} \exp\left[-\frac{x^2}{4Dt}\right] \quad t>0 \qquad (4.15)$$

The solution is illustrated in Figure 4.5A. Note that the total probability is unity for all times greater than 0. The probability distribution corresponds to a normal, or Gaussian, distribution whose variance, σ^2, is $2Dt$. The **root-mean-square displacement**, x_{rms} (which equals the standard deviation, σ), therefore increases in proportion to the square root of time (Figure 4.5B). This is in contrast to motion with constant velocity, v, where the displacement increases in proportion to time ($x = vt$) and is always in one direction.

Example 4.3 The efficiency of diffusion as a cellular transport mechanism
Consider a 3-nm-radius protein (corresponding to a molecular mass of 100 kDa) diffusing through water. Its diffusion coefficient is ~100 µm²/s (at 37°C). The table on page 61 shows how long it takes for the protein to diffuse various distances. Because the average distance a particle diffuses increases only with the square root of time, diffusion of proteins becomes slow, greater than ~1 minute, over distances greater than ~100 µm. This might explain why the diameters of eukaryotic cells are usually smaller than 100 µm. On the other hand, a 1-µm-diameter organelle (such as a mitochondrion) in water has a diffusion coefficient of only ~0.5 µm²/s (at 25°C). Diffusion through 100 µm in this case takes ~3 hours; in a real cell it would take much, much longer because the cytoplasm is more like a gel with a mesh size of only ~50 nm (Luby-Phelps et al., 1987), so that

organelles larger than 50 nm are almost immobile. Thus even small eukaryotic cells will require motor proteins to move organelles from one place to another. The low mobility of organelles does have a benefit, though: Large organelles will stay where they are put and the internal structure of the cell will therefore be reasonably stable. For highly elongated cells—an extreme example are the neurons of the sciatic nerve, whose processes bridge from the spinal cord to the foot—active transport is essential, even for small metabolites and proteins.

Times for one-dimensional diffusion in aqueous solution

Object	Distance diffused			
	1 μm	100 μm	10 mm	1 m
K^+	0.25 ms	2.5 s	2.5×10^4 s (7 hrs)	2.5×10^8 s (8 years)
Protein	5 ms	50 s (~1 min.)	5×10^5 s (6 days)	5×10^9 s (150 years)
Organelle	1 s	10^4 s (~3 hr.)	10^8 s (3 years)	10^{12} s (30 millennia)

Note: K^+: Radius ~0.1 nm, $T = 25°C$, $D \cong 2000$ μm²/s.
Protein: Radius = 3 nm, viscosity = 0.6915 mPa·s⁻¹, $T = 37°C$, $D \cong 100$ μm²/s.
Organelle: Radius = 500 nm, viscosity = 0.8904 mPa·s⁻¹, $T = 25°C$, $D \cong 0.5$ μm²/s.

First-Passage Times

In the previous section we answered the question: How far, on average, does a molecule diffuse in a given time? However, a more relevant question is: How long, on average, does it take a molecule to diffuse through a given distance? We call this time the **first-passage time**. The rephrased question is more relevant because it allows us to calculate the rate of a process that is limited by diffusion. The **diffusion-limited** rate is the reciprocal of the first-passage time.

The first-passage time can be calculated by solving the diffusion equation for the particular geometry of the problem. In the absence of an external force, the first-passage time for one-dimensional diffusion through a distance x_0 is

$$t = \frac{x_0^2}{2D} \tag{4.16}$$

(Appendix 4.2). Not coincidentally, this is the same answer that we got when we approached the problem the other way round by considering the average distance diffused!

Example 4.4 Diffusion over molecular dimensions The first-passage time is very small when the distances are small. For example, consider a 1-μm-diameter spherical organelle diffusing in aqueous solution near a microtubule. If the sphere were kept from diffusing away from the microtubule but free to diffuse along it, then diffusion through 8 nm, the distance between adjacent tubulin dimers along the microtubule lattice, would only take 64 μs! Thus it is an appealing idea that motor proteins exploit this rapid motion by somehow allowing diffusion to occur in only one direction. But as we will see in later chapters, such a model for the motor mechanism is ruled out by several types of experiments.

The first-passage time becomes more interesting from a biological point of view when we consider diffusion in the presence of an external force. For example: How long does it take for a molecule to diffuse over an energy barrier at $x = x_0$? When the force is constant—that is, when the potential energy is $U(x) = -Fx$—the first-passage time is

$$t = 2\left(\frac{x_0^2}{2D}\right)\left(\frac{kT}{Fx_0}\right)^2\left\{\exp\left(-\frac{Fx_0}{kT}\right) - 1 + \frac{Fx_0}{kT}\right\} \qquad (4.17)$$

This equation is plotted in Figure 4.6. When the diffusion is steeply downhill—that is, the force is large and positive—the first-passage time approaches x_0/v, where v is the drift velocity ($v = F/\gamma$), as expected for a molecule drifting at constant speed. When the diffusion is steeply uphill—that is, the force is large and negative—the first-passage time increases approximately exponentially as the opposing force is increased. In this case, the diffusion rate, the inverse of the first-passage time, decreases approximately exponentially as the force

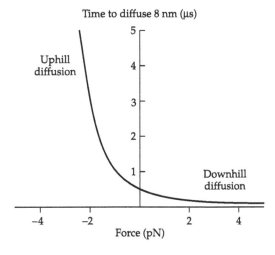

Figure 4.6 Time for a 100 kDa protein to diffuse 8 nm
When diffusion is uphill (against an opposing force) the first-passage time is long (left-hand side). When diffusion is downhill (in the direction of the force), the first-passage time is short (right-hand side).

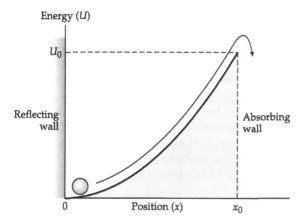

Figure 4.7 Diffusion of a molecule out of a hemiparabolic energy well

times the distance increases. In other words, the diffusive process behaves as though there is a barrier with energy $U_0 = |F| \cdot x_0$ (Chapter 5).

If the force opposes the motion and has amplitude proportional to position (Figure 4.7)—that is, the molecule is attached to an elastic element so that $F = -\kappa x$—the first-passage time is

$$t_K = \tau \sqrt{\frac{\pi}{4}} \sqrt{\frac{kT}{U_0}} \exp\left(\frac{U_0}{kT}\right) \qquad (4.18)$$

where $\tau = \gamma/\kappa$ is the drag coefficient divided by the spring constant. This equation is derived in Appendix 4.2 and assumes that the height of the energy barrier is high—that is, $U_0 \equiv U(x_0) \equiv \frac{1}{2}\kappa x_0^2 \gg kT$. t_K is called the Kramers time, after Kramers, who first derived it (Kramers, 1940). This equation forms the basis of the Kramers rate theory, which postulates that the rate of reactions is limited by diffusion over a high-energy transition state (Chapter 5).

Correlation Times*

So far I have been vague about the nature of the thermal forces that drive diffusion and Brownian motion. Indeed, we didn't need any information about the thermal forces to derive the Principle of Equipartition of Energy, only that Boltzmann's law was satisfied, and we only needed to assume that the thermal forces were randomly directed to derive the diffusion equation and to relate the diffusion coefficient to the drag coefficient. However, there are several "microscopic" details of diffusive motion that are important. For example, how long, on average, will a free molecule keep moving in one direction before

---

*An asterisk next to a heading denotes a more advanced section.

the thermal forces randomize its direction of motion? In other words, what is the **persistence time** (or **correlation time**) of the velocity? Another important question is: How long, on average, will it take for a molecule in a potential well to explore the different energy levels? In particular, how long will the molecule spend at each energy level? In other words, what is the persistence time, or correlation time, of the position? Finally, what are the amplitudes and statistical properties of the thermal forces? In this chapter, we will answer these questions. Because the answers require quite advanced mathematics, namely concepts from Fourier analysis, this section is more advanced than the other sections in this book. For those without this mathematical background, the next three sections can be skipped, except for Examples 4.5 and 4.6, which illustrate the main results.

Let $F(t)$ be the thermal force acting on a molecule due to collisions with surrounding solvent molecules: It comprises very brief impulses with random direction, occurring at random times. The equation of motion of the molecule in response to this force is

$$m\frac{d^2x}{dt^2}(t) + \gamma\frac{dx}{dt}(t) + \kappa x(t) = F(t) \quad (4.19)$$

and is known as the **Langevin equation** (Langevin, 1908). Because the thermal force is a random one, the most we can hope for is a description of the statistical properties of the resulting motion. These properties are described by the **autocorrelation function**, $R_x(\tau)$, of the position $x(t)$ of a molecule, which is defined by

$$R_x(\tau) = \langle x(t) \cdot x(t-\tau) \rangle \equiv \lim_{T \to \infty} \left\{ \frac{1}{T} \int_{-T/2}^{T/2} x(t)x(t-\tau) \cdot dt \right\} \quad (4.20)$$

The autocorrelation at delay τ is calculated by multiplying the position at a given time by the position at time τ earlier, and averaging over all times.

The autocorrelation function $R_x(\tau)$ has the properties that coincide with our intuitive notion of temporal correlation or persistence time. The autocorrelation function has its maximum value (equal to the variance of the signal $\langle x^2 \rangle = \sigma^2$) when $\tau = 0$ and it falls to zero when $\tau = \infty$ (we are assuming that the signal has mean $\langle x \rangle = 0$). This accords with the positions at two closely spaced times being highly correlated, but the positions at two widely separated times being uncorrelated. In addition, the autocorrelation is symmetric, $R_x(\tau) = R_x(-\tau)$. This relationship means that the correlation drops equally quickly whether we compare the signal with itself at earlier or later times.

The crucial additional property is that the autocorrelation function satisfies the equation of motion!

$$m\frac{d^2R_x}{d\tau^2}(\tau) + \gamma\frac{dR_x}{d\tau}(\tau) + \kappa R_x(\tau) = 0 \quad \tau > 0 \quad (4.21)$$

(Appendix 4.3). This means that the autocorrelation function has the same form as the response of the molecule to an impulsive external force. For example, in the case of overdamped motion appropriate for protein dynamics (Chapters 2 and 3), Equation 4.21 yields an autocorrelation function that is the sum of two exponentials, a fast, small-amplitude one with time constant m/γ (on the order of ps) and a slow, large-amplitude one with a time constant equal to γ/κ. This result is elaborated in Examples 4.5 and 4.6.

Example 4.5 Diffusion of a free protein   Consider a protein of molecular mass 100 kDa whose velocity (in one dimension) is simulated in the figure below. It has a mass of 166×10^{-24} kg (see Table 2.2) and a root-mean-square speed, $v_{rms} = (3kT/m)^{0.5} \cong 8.6$ m/s. The time constant is $\tau = m/\gamma$, where γ is the drag. If the protein is globular with a radius of 3 nm, then the damping is $6\pi\eta r \cong 60$ pN·s/m (see Table 2.2) and the time constant is only ~3 ps! This is the persistence time, or correlation time, of the velocity. Over this time the protein will move 0.24 Å, only ~1/300th its diameter! Thus even though the speeds of molecules are large, the high damping that they experience in water means that their inertia carries them for extremely small distances, so that after a fraction of an angstrom, they are likely to be moving in a different direction. Thus the model of diffusion as a random walk is a good one, provided that the time between the postulated steps is much longer than the 3 ps inertial time.

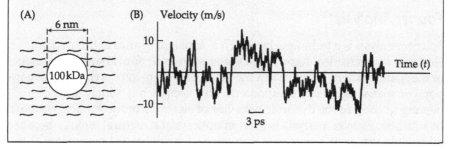

An important consequence of Equation 4.21 is that it suggests a strategy for estimating a molecule's molecular properties, such as stiffness and damping. Measure the thermal motion, calculate the autocorrelation function via Equation 4.20, and compare it to the theoretical autocorrelation function predicted from a particular model (via Equation 4.21) to obtain the molecular parameters that give the best fit. However, rather than calculating the autocorrelation of a fluctuating signal, it is more common to perform a Fourier analysis on the signal, and compare that to the predicted behavior, as described in the next section.

Example 4.6 Diffusion of a tethered protein Suppose that the protein of the previous example is attached to a spring. Its position is simulated in the figure below. If the stiffness is 1 pN/nm, it will have a root-mean-square displacement, $x_{rms} = \sqrt{\langle x^2 \rangle} = \sqrt{kT/\kappa}$ ≅ 2 nm. This is independent of its molecular mass! How long will it take such a protein to relax to a new position? The time constant is $\tau = \gamma/\kappa \approx 60$ ns, which is ~20,000 times greater than the inertial time constant (see the figure in Example 4.5). This is the persistence time, or correlation time, of the protein's position. The protein's position will be correlated on the nanosecond time scale: After times less than 60 ns, the protein will still be "quite near" to where it was. But over much longer times (>> 60 ns) the protein's position will be uncorrelated, and the probability of finding the protein in a certain position will depend only on its potential energy and not on time.

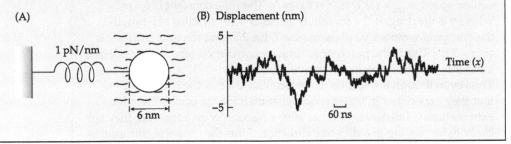

Fourier Analysis*

Fourier analysis is a technique by which a signal that varies in time, such as an electrical or mechanical signal, or in space, such as an optical signal (an image) or the spatial pattern of atoms in a molecule, is split up into its constituent temporal or spatial frequency components. This approach is widely used in engineering to analyze the dynamic behavior of electrical or mechanical systems. In addition, Fourier analysis is used in optics and structural biology because when light waves or X rays pass through and are diffracted by a material, the light or X rays that exit the specimen in a particular direction correspond to the scattering due to a particular spatial frequency of structures in the material. In other words, diffraction is a physical way of separating the different spatial frequency components. Likewise, the graphic equalizer on a stereo system is an analogue circuit that separates electrical signals into different temporal frequency components. The ear is another example of an analogue Fourier analyzer: Different frequencies of sound are separated into mechanical vibrations at different spatial locations along the cochlea. We are going to apply Fourier analysis to the dynamics of a molecule undergoing thermal motion. But the underlying principles are generally applicable and can be used to interpret shape fluctuations of polymers, to understand various contrast techniques used in light microscopy, and to see how diffraction patterns are used to deduce the structures of proteins.

Fourier analysis is based on the following mathematical property: A signal $x(t)$ whose total duration is T can be expressed as a **Fourier series**

$$x(t) = \sum_{n=1}^{\infty} a_n \cos \frac{2\pi n t}{T} + b_n \sin \frac{2\pi n t}{T} \qquad (4.22)$$

where the amplitudes of the cosine and sine components, a_n and b_n, are calculated from the time averages

$$a_n = 2\left\langle x(t) \cos \frac{2\pi n t}{T} \right\rangle = \frac{2}{T} \int_{-T/2}^{T/2} x(t) \cos \frac{2\pi n t}{T} \cdot dt \qquad n \geq 1$$

$$b_n = 2\left\langle x(t) \sin \frac{2\pi n t}{T} \right\rangle = \frac{2}{T} \int_{-T/2}^{T/2} x(t) \sin \frac{2\pi n t}{T} \cdot dt \qquad n \geq 1$$

As in the last section, we are assuming that the mean of the signal is zero (if not, the mean can simply be subtracted). There are a number of fast algorithms for calculating the Fourier series from digitized time traces (e.g., the Cooley–Tukey algorithm; Bendat and Piersol, 1986; Press, 1997).

For our purposes, the crucial function obtained from the Fourier analysis of a signal is the **power spectrum**. The power spectrum, $G_x(f)$, of a signal, $x(t)$, is defined so that $G_x(f) \cdot \Delta f$ is the mean-square displacement, or variance, of the signal in the frequency range $(f, f + \Delta f)$. The power spectrum has the following physical meaning. If a signal $x(t)$ is passed through a filter with center frequency f Hz, and bandwidth Δf Hz (this corresponds to a "notch" filter) to obtain a signal $x_{f,\Delta f}(t)$, then $G_x(f) \cdot \Delta f$ is the mean-squared value of this filtered signal (Figure 4.8). The power spectrum can be calculated directly from the Fourier series. If we pass $x(t)$ through a filter of bandwidth $\Delta f = 1/T$, and take the variance we obtain

$$G_x(f) \cdot \Delta f \equiv \langle x_{f,\Delta f}^2(t) \rangle = \tfrac{1}{2}\left(a_n^2 + b_n^2\right) \qquad f = n/T > 0, \; \Delta f = 1/T \qquad (4.23)$$

This is derived in Appendix 4.3. There are a number of slightly different ways of defining the power spectrum (Bendat and Piersol, 1986; Bracewell, 1986; Papoulis, 1991): The definition of Equation 4.23 corresponds to the "one-sided

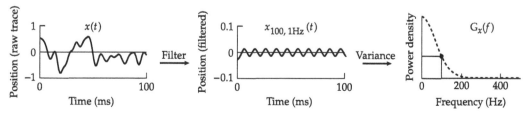

Figure 4.8 Definition of the power spectrum of a signal x(t)

power spectrum," for which only positive frequencies are considered. If the signal corresponds to a displacement, then the units of $G_x(f) \cdot \Delta f$ are m². The unit of $G_x(f)$ is therefore m²/Hz. If the signal were a force or a voltage, the units would be N²/Hz or V²/Hz. Thus the power spectrum is not a true distribution of power, which would have units of W/Hz. The misnomer arose because the power spectrum was first used for electrical signals, in which case the square of the voltage is proportional to the power ($V^2 = V \cdot iR = P \cdot R$ by Ohms law). The power spectrum is analogous to the intensity distribution of a diffraction pattern.

The reason why the power spectrum is so useful for analyzing thermal motion is that the power spectrum is the Fourier transform of the autocorrelation function

$$G_x(f) = 2 \int_{-\infty}^{\infty} R_x(\tau) e^{-2\pi i f \tau} \cdot d\tau = 4 \int_0^{\infty} R_x(\tau) \cos(2\pi f \tau) \cdot d\tau \qquad f > 0 \qquad (4.24)$$

where we have a factor of two in the middle expression because we are using the one-sided power spectrum, and the right-hand expression follows because the autocorrelation function is symmetrical (Appendix 4.4). This relationship is significant because it allows one to compare the power spectrum, computed directly from the experimental recordings via the Fourier series, with a theoretical expression for the power spectrum, which is calculated with the help of Equation 4.24 from the predicted autocorrelation function (Equation 4.21). Figure 4.9 summarizes this approach. First, we measure the thermal motion of a molecule and calculate the power spectrum. Then we compare this power spectrum with the power spectrum predicted by a model equation of motion, which has variables of mass, damping, and stiffness. The model is refined by adjusting the variables until the best fit is found. In this way, the molecular parameters—the mass, damping and stiffness—can be deduced from the thermal motion. This process is shown in the following example. An analogous approach is used in X-ray crystallography; the power spectrum corresponds to the intensity of the diffracted pattern, which is compared to theoretical diffraction patterns predicted by structural models.

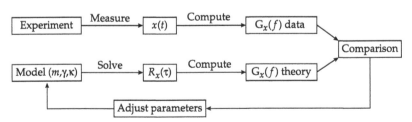

Figure 4.9 Using Fourier analysis to measure molecular parameters

Example 4.7 Power spectrum of a damped spring Figure A shows the position of the tip of a flexible glass fiber undergoing thermal motion. The noisy curve in Figure B shows the power spectrum of this motion calculated via the Fourier series (Equation 4.23). The smooth curve is the theoretical curve obtained as follows. First, the autocorrelation function is found as the solution to Equation 4.21 with zero mass

$$R_x(\tau) = \frac{kT}{\kappa}\exp(-|\tau|/\tau_0) \qquad \tau_0 = \gamma/\kappa$$

where we have used $R_x(0) = \langle x^2 \rangle = kT/\kappa$ from the Equipartition principle. The associated power spectrum is calculated via Equation 4.24 to be

$$G_x(f) = \frac{4kT\gamma}{\kappa^2}\frac{1}{1+(2\pi f\tau_0)^2}$$

This curve is called a **Lorenzian** and corresponds to the smooth curve in Figure B with parameters $\kappa = 0.043$ mN/m and $\gamma = 0.16$ μN·s/m. By this means, the stiffness and damping are deduced from the thermal motion. There are other approaches to estimating the stiffness and the drag coefficient. For example, measuring the mean-square displacement, $\langle x^2 \rangle = kT/\kappa$, and the correlation time, $\tau = \gamma/\kappa$ (or cutoff frequency $f_0 = 1/2\pi\tau$), allows one to solve for $\kappa = kT/\langle x^2 \rangle$ and $\gamma = \tau\kappa$. In this example, $\tau = 3.7$ ms.

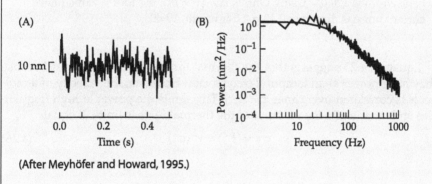

(After Meyhöfer and Howard, 1995.)

The Magnitude of the Thermal Force*

Now that we know the power spectrum of the molecule's position, it should be possible to calculate the power spectrum of the equivalent force necessary to produce the motion. This is the thermal force, and it is shown in Appendix 4.3 to have a power spectrum equal to

$$G_F(f) = 4kT\gamma \qquad (4.25)$$

This equation has a number of important implications. First, it says that to account for the thermal motion, we simply redraw the mechanical circuit to include a force generator that has a variance $4kT\gamma$ N^2/Hz in each frequency interval (Figure 4.10). Second, the variance of the thermal force is independent of the frequency: This is equivalent to the force being due to impulsive collisions, as we have claimed (see also Appendix 4.3). The third implication is that the amplitude of the force depends only on the drag coefficient, and not on the stiffness (or on the mass if it had been included). This finding is quite general and is known as the **Fluctuation–Dissipation theorem** (Landau et al., 1980). Another way of stating this result is that the fluctuations come from the dissipative elements, namely the dashpots. In general, to simulate the thermal motion of any linear mechanical circuit, we place a random force generator with power spectrum $4kT\gamma_i$ *in parallel* with each damping element γ_i.

> **Example 4.8. Thermal noise also occurs in electrical circuits** The analogue of the thermal forces that originate in the dashpots of mechanical circuits is the **Johnson noise** that originates in the resistors of electrical circuits. Johnson noise is due to thermally driven voltage fluctuations. Its power spectrum is $G_V(f) = 4kTR$, where R is the resistance. In an electrical circuit with a feedback resistor of 1 GΩ, typical of the feedback resistors in patch-clamp headstages, the root-mean-square voltage noise over a 1000 Hz bandwidth is 126 µV. Using Ohm's law ($V = iR$), the root-mean-square current noise is therefore 0.13 pA (Sigworth, 1995).

Equation 4.25 suggests that the thermal forces are instantaneous because they have power at all temporal frequencies. However, the velocity of a molecule is correlated over times $\tau = m/\gamma$. This limits the power at high frequencies and, as a result, the variance of the thermal force is finite. It equals

$$\sigma_F^2 = \langle F^2 \rangle = \gamma^2 \langle v^2 \rangle \tag{4.26}$$

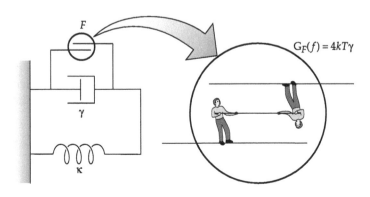

Figure 4.10 Representation of the thermal force
The thermal force is represented as a random force in parallel with the dashpot.

(Appendix 4.3). This equation shows that the root-mean-square thermal force is proportional to the root-mean-square velocity times the damping, in accordance with the notion that the thermal force has the same physical origin as the damping force.

Summary

Because molecules are agitated by collisions with other molecules, they are not always found in their lowest energy state. The probability of a molecule being in a higher energy state is given by Boltzmann's law. Boltzmann's law has several corollaries. One corollary is the Principle of Equipartition of Energy, which says that each degree of freedom of the molecule (which meets certain criteria) has $\frac{1}{2}kT$ of energy associated with it. k is the Boltzmann constant and T is the absolute temperature, so $kT \cong 4 \times 10^{-21}$ J at room temperature. An example of this principle is that the average potential energy of a spring is $\frac{1}{2}kT$. Another example is that the average kinetic energy of a molecule is $\frac{3}{2}kT$ (the 3 appears because there are three components of velocity and therefore 3 degrees of freedom). The randomly directed collisions with the surrounding molecules cause particles to diffuse, and the statistical properties of diffusing particles can be determined by solving the diffusion equation. A second corollary of Boltzmann's law is that the diffusion coefficient (D) is related to the drag coefficient (γ) via the equation $D = kT/\gamma$, known as the Einstein relation. A detailed analysis of diffusion of a particle shows that the velocity persists (or is correlated) over times shorter than m/γ, where m is the mass. For a protein, this time is on the order of picoseconds (10^{-12} s), after which time the protein is likely to be moving in a different direction and with a different speed. If the protein is tethered by a spring of stiffness κ, then the position is correlated over times shorter than γ/κ, which ranges from 1 nanosecond to 1 microsecond for spring constants between 0.016 and 16 pN/nm. Using Fourier analysis, it is possible to deduce the molecular properties of a mechanical system, such as the stiffness, the damping, and the mass, from an analysis of the thermal motion.

Problems

4.1 Consider a stack of ten shelves and suppose that the spacing of the shelves is such that the potential energy on each shelf is 1 kT higher energy than that on the shelf below. Suppose that 1000 books are placed on the shelves according to Boltzmann's law. How many books are there on each shelf? [Answer: From the bottom there are 632, 233, 85, 31, 12, 4, 2, 1, 0 and 0 books on each shelf.]

4.2 A solution of gold spheres is stored on a shelf. After a week or so, it is noticed that the spheres have settled stably in the container and that the density (as estimated from the color) decreases e-fold every 20 mm from bottom to top. Given that the density of gold is 19.3 times that of water, what is the diameter of particles?

4.3 A protein solution was placed in an analytic centrifuge and spun until equilibrium was reached. The measurement was then repeated. But it was realized, too late, that prior to the second run, the protein had been denatured by the rotating graduate student. Remarkably, it was found that the distribution was the same for both runs. Why is this?

4.4 The Stokes' radius of a sodium ion, the apparent radius obtained from the diffusion coefficient, is about twice the ionic radius, found in crystals (see Example 4.2). Why might this be?

4.5 Calculate the root-mean-square velocity of a bacterium (assume it is spherical of radius 1 μm and that it is 10% heavier than water). What is the correlation time of the velocity? How far will the bacterium move during this correlation time?

4.6 Consider our canonical 100 kDa protein (see Table 2.2). How long will it take to diffuse 40 nm? Suppose there is a force of 1 pN. How long will it take to diffuse 40 nm in the direction of the force? How long will it take to diffuse 40 nm against the force? [Answers: 24 μs, 4 μs, 11 ms.]

12   2.2   4.2

4.7* If $U(x) = F|x|$, that is, a particle is trapped between two linear inclines, show that $\langle U \rangle = kT$ (and not $\tfrac{1}{2}kT$).

4.8* If a particle is released at the origin and there are absorbing walls at $x = -a$ and $x = b$, show that the probability of being absorbed at $-a$ divided by the probability of being absorbed at b is b/a (Goel and Richter-Dyn, 1974).

4.9* The flow of heat also satisfies Fick's law and the diffusion equation. The flux of heat is proportional to the thermal gradient, where the constant of proportionality is the thermal conductivity, K. The flow of heat then changes the temperature of a unit volume of solid by $c\rho$, where c is the heat capacity and ρ is the density. For the case of heat flow in solids, the equivalent to the diffusion coefficient is

$$D_{thermal} = K/c\rho$$

For water at 25°C, $K = 0.606$ W/m·K, $c = 4.18$ kJ/kg·K, and $\rho = 997$ kg/m³.

(a) What is $D_{thermal}$ for water? (Check that the units are m²/s.) How does this compare with the diffusion coefficient for an ion? [Answer: 0.15 mm²/s.]

---

* The asterisk denotes more advanced problems.

(b) Suppose, hypothetically, that heat were carried by little particles (called calorics). Use Stokes' law to calculate the radius of the heat particle.

(c) How long does it take heat to diffuse 3 nm, roughly the distance from the center of a globular protein to the surrounding fluid? [Answer: ~10 ps for diffusion in 3 dimensions.]

CHAPTER

5

Chemical Forces

In addition to mechanical and thermal forces, proteins are also subject to chemical forces. By **chemical forces**, we mean the forces that arise from the formation of intermolecular bonds. For example, consider what happens when a protein first comes in contact with another molecule such as a small ligand or another protein. As energetically favorable contacts are made, the protein may become stretched or distorted from its equilibrium conformation. If the protein can adopt two different structures, the binding could preferentially stabilize one of these structures. Chemical forces also arise from changes in bound ligands. One example is the *cis*-to-*trans* isomerization of retinal bound to the opsin protein: Following the adsorption of light, the all-*trans*-retinal is initially in a highly strained conformation, and its relaxation drives the slower structural changes of the opsin (Peteanu et al., 1993). Another example is the hydrolysis of the gamma phosphate bond of ATP bound to myosin (Chapter 14). In all the above cases, the chemical change produces a local distortion that in turn pushes the protein into a new low-energy conformation.

To understand how protein machines work, it is essential to understand how proteins move in response to these chemical forces. Just as a chemical force might cause a protein to move in one direction, an external mechanical force might cause the protein to move in the opposite direction. For example, the binding of a ligand might stabilize the closure of a cleft, whereas an external tensile force might stabilize the opening of the cleft; as a result, the mechanical force is expected to oppose the binding of the ligand. Thus mechanical forces can oppose chemical reactions, and, conversely, chemical reactions can oppose mechanical ones. If the chemical force is strong enough, the chemical reaction will proceed even in the presence of a mechanical load. In this case we say that the reaction generates force.

The purpose of this chapter is to elucidate the general principles by which applied forces affect both the rates and equilibria of chemical reactions, and, by extension, how chemical reactions generate force.

Chemical Equilibria

A central question is how force affects the equilibrium between two structural states of a protein

$$E_1 \rightleftharpoons E_2$$

However, we immediately encounter a difficulty: What is a structural state of a protein? The difficulty arises because thermal fluctuations cause a complex molecule like a protein to occupy, sequentially, an enormous number of different **conformational states**, where a conformational state is a set of positions or coordinates of all the atoms. (We ignore the velocities of the atoms, which will average out over the picosecond timescale, as argued in the last chapter.) If the fluctuations in the positions of the atoms are not too large (or if the atoms do not spend too much time at large distortions), then we can think of the different conformations as small deviations about a stable, minimum-energy state. We call the time-average of these conformational states a **structural state**; it will be similar to the minimum-energy state (and the two will be the same if the fluctuations are symmetrical). Structural states can be solved by X-ray crystallography or NMR. According to our definition, a structural state is an ensemble of a large number of the individual conformational states that do not vary too much from the mean. An unfolded protein, by contrast, has such large fluctuations that the distances between atoms in different amino acids vary by more than the size of the amino acids themselves; in this case, we say that the protein is unstructured.

It is possible that the conformational states segregate into two ensembles with different means. In this case we say that the protein adopts two structural states. We denote the states as E_1 and E_2. The different structural states might have different functional properties. An enzyme in one state might be catalytically active whereas in the other state it might be inactive. In this case, the structural states are "on" and "off." An ion channel might be open or closed. Different states might have different affinities for another protein, for DNA, or for a ligand. In the case of a motor protein, the lever might be up or down, corresponding to the pre- and post-working strokes.

To determine the probability of finding the protein in one of the two structural states E_1 and E_2, we would like to apply Boltzmann's law. However, as stated in the last chapter, Boltzmann's law applies only to individual conformational states and not to ensembles of states. There is, however, a way around this. For a structural state, E, which is really an ensemble of states with energies $\{U_i\}$ (Figure 5.1), we define the **free energy** as

$$G = U - TS \tag{5.1}$$

where U is the average potential energy, T is the temperature, and S is the entropy (Appendix 5.1). U comprises the internal energy—the energy associ-

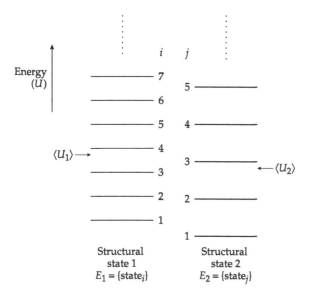

Figure 5.1 Ensembles of states
Suppose a molecule can be in one of two structural states E_1 and E_2, each comprising an ensemble of individual conformational states with different potential energies. Each conformational state has energy U_i or U_j and, at equilibrium, the probability of finding the molecule in conformational state i, p_i, is proportional to $\exp[-U_i/kT]$. The probability of finding the molecule in structural state E_1 is the sum of all the probabilities p_i. Likewise, the probability of finding the channel in structural state E_2 is the sum of all the probabilities p_j. The result of this calculation, given in Appendix 5.1, is a generalized form of Boltzmann's law that relates the probabilities to the free energies (see Equation 5.2).

ated with all the bonds (covalent, electrostatic, van der Waals)—plus other terms corresponding to potential energies arising from external variables such as pressure, force, electrical fields, or gravity. The entropy is a measure of disorder: The larger the number of conformations in an ensemble, the greater the entropy. It is shown in the Appendix that Boltzmann's law holds for ensembles of conformational states, E_1 and E_2, if the energies are replaced by the free energies, G_1 and G_2. In other words,

$$\frac{[E_2]}{[E_1]} = \frac{p_2}{p_1} = \exp\left[-\frac{\Delta G}{kT}\right] = \text{constant} \equiv K_{eq} \qquad \Delta G = G_2 - G_1 \qquad (5.2)$$

This equation is important because it shows that there is such a thing as an **equilibrium constant**, K_{eq}. It is a constant in the sense that it does not depend on the concentrations of E_1 and E_2, though in general it will depend on the temperature, ionic strength, and other variables. Equation 5.2 is known as the **Law of Mass Action** because if E_1 and E_2 are in equilibrium and more protein in the E_1 form is added, then the amount of E_2 will increase as the system returns to the equilibrium ratio. Likewise, adding E_2 pushes the reaction back toward E_1. The law applies equally well to the equilibrium between small molecules such as substrate and product as between protein conformations E_1 and E_2.

If we ignore the external potential energy terms, G is the **Helmholtz free energy**. If the only potential energy corresponds to pressure–volume work, PV, then G is the **Gibbs free energy** and U is called the **enthalpy**. If there are other potential energy terms arising from mechanical forces or electrical potentials, then we simply call G the free energy. Usually, the difference between the

Helmholtz and Gibbs free energies can be ignored because the volume changes associated with structural changes in proteins are quite small, no more than a few percent of the total volume. Because the total volume of proteins is small anyway, on the order of 120 nm^3 for a 100 kDa protein, the volume changes are only on the order of 1 nm^3. At the standard pressure of 1 atmosphere (~100 kPa), the $P\Delta V$ work is only ~0.1×10^{-21} J, which, because it is much less than kT (4×10^{-21} J), will have very little effect on the equilibrium. Volume changes are expected to be important only in the depths of the ocean where pressures may exceed 100 atmospheres. On the other hand, as we shall see, the potential energy due to mechanical external forces can be very significant.

The Effect of Force on Chemical Equilibria

Boltzmann's law allows us to calculate how a force influences the equilibrium between two (or more) structural states. If the difference between two structural states is purely translational—that is, if state E_2 corresponds to a movement through a distance Δx with respect to state E_1, as occurs when a motor moves along a filament against a constant force—then the difference in free energy is $\Delta G = - F \cdot \Delta x$, where F is the magnitude of the force *in the direction* of the translation (Figure 5.2A). If the length of a molecule changes by a distance Δx as a result of a conformational change (Figure 5.2B), then the difference in free energy is

$$\Delta G \cong \Delta G^0 - F \Delta x$$

Figure 5.2 Displacements associated with structural changes
(A) Translation. (B) Lengthening. (C) Lengthening with constant stiffness.

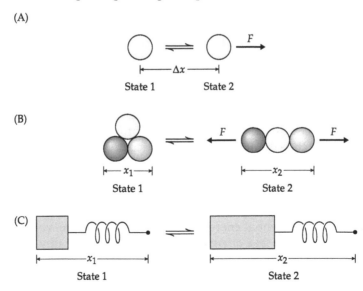

where F is the tension across the molecule and ΔG^0 is the free energy difference in the absence of tension. The equality is exact if the molecule is composed of rigid domains that undergo relative translation as shown in Figure 5.2B, or if the two structural states have equal stiffness (Figure 5.2C) (see below). At equilibrium

$$\frac{[E_2]}{[E_1]} = \exp\left[-\frac{\Delta G}{kT}\right] \cong \exp\left[-\frac{\Delta G^0 - F\Delta x}{kT}\right] = K_{eq}^0 \exp\left[\frac{F\Delta x}{kT}\right] \quad (5.3)$$

where K_{eq}^0 is the equilibrium constant in the absence of the force. The crucial point is that an external force will couple to a structural change if it is associated with a length change in the direction of the force. If the change in length of the molecule is 4 nm, then a force of 1 pN will change the free energy by 4 pN·nm $\cong kT$. According to Equation 5.3, this will lead to an e-fold change in the ratio of the concentrations. Because protein conformational changes are measured in nanometers, and energies range from 1 kT (thermal energy) to 25 kT (ATP hydrolysis) (see Table 4.2), it is expected that relevant biological forces will be on the scale of **piconewtons**.

An example of how forces modulate the state of a protein—in this case a mechanically sensitive ion channel—is explored in Example 5.1.

An expression analogous to Equation 5.3 holds for voltage-gated ion channels (Hille, 1992). In this case, the structural change associated with the opening of such a channel is coupled to movement of charge, Δq, across the electric field caused by the transmembrane potential, V. Mutagenesis studies indicate that the moving charges include positively charged arginine residues in the S4 transmembrane helix (Hille, 1992). From a physical viewpoint, charge (Δq) is analogous to displacement (Δx), and potential (V) is analogous to force (F). The energy difference between the open and closed states therefore includes a term $V\Delta q$, and this makes the opening sensitive to the voltage. The openings of the voltage-dependent Na and K channels that underlie the action potential are strongly voltage dependent: Classic experiments by Hodgkin and Huxley showed that the ratio of the open probability to the closed probability increased approximately e-fold per 4 mV (Hodgkin, 1964). This indicates that the opening of each channel is associated with the movement of about six electronic charges across the membrane ($\Delta q = kT/V \cong 6e$, where e is the charge on the electron). The predicted movement of these electronic charges has been directly measured as a nonlinear capacitance of the membrane (Armstrong and Bezanilla, 1974) that is analogous to the nonlinear stiffness of the hair bundle in Example 5.1.

Protein conformational changes are sensitive to many other "generalized" forces including membrane tension, osmotic pressure, hydrostatic pressure, and temperature. Sensitivity to these forces requires that conjugate structural changes occur in the protein (Howard et al., 1988). In the case of membrane tension, σ, the conjugate variable is area, Δa, and the energy difference equals $\sigma\Delta a$: The sensitivity of stretch-activated ion channels to membrane tension (Guharay and Sachs, 1984; Chang et al., 1998; Batiza et al., 1999) suggests that the conformational change associated with channel opening leads to an increase

Example 5.1 Mechanically sensitive ion channels in hair cells

The sensory hair cells of the inner ear underlie the perception of sound, linear acceleration, angular acceleration, and gravity. When the hair bundle is deflected by an external force, there is shear between adjacent stereocilia within the bundle (see Figure 8.2). This shear in turn tenses elastic tip links that pull on and open ion channels at the tips of the stereocilia (Figures A and B). Because the opening of a channel shortens the tip link, the open state is stabilized by deflections that increase the tension in the tip link. As a result, the open probability increases as the hair bundle is displaced to the right, defined as the positive direction (Figure B). Flow of potassium ions through the channels alters the electrical potential across the membrane, producing the receptor potential that can be measured by inserting a glass microelectrode into the cell (Figure C). The receptor potential triggers synaptic release and eventually leads to perception by the central nervous system. Due to the extra degree of freedom corresponding to the opening and closing of the channels, the stiffness of the hair bundle is not constant, but depends on the displacement. The stiffness can be measured by displacing the bundle with a flexible glass fiber (see Figure 15.3) and measuring the flexion in the fiber using a photodiode detector (see Figure 15.5). As seen in Figure D, the stiffness has a minimum when the channels are open 50% of the time. The displacement dependence of the stiffness indicates that the "swing" of the gate (Δx) is 2 to 4 nm and that there are 1 or 2 channels per stereocilium (Howard and Hudspeth, 1988; Hudspeth et al., 1990; Denk et al., 1995).

The open probability is

$$p_{open} = \frac{1}{1+\exp\left[-\dfrac{F\Delta x}{kT}\right]} \qquad F = a\kappa(X-X_0)$$

where a ($\cong 0.14$) is a geometric factor that relates the displacement of the bundle to the shear between stereocilia, κ is the stiffness of the tip link, X is the average displacement of the hair bundle, and X_0 is the displacement at which the channels are open 50% of the time. This equation follows from Equation 5.3, and provides a good fit to the receptor potential (Figure C). The stiffness of the hair bundle is

$$K = \frac{d^2 G}{dX^2} = K_S + N\kappa a^2 - N\frac{(a\kappa \Delta x)^2}{kT}p(1-p)$$

where the free energy of the bundle, G, includes an entropy term corresponding to the opening and closing of the channels (see Appendix 5.1) (Hudspeth et al., 1990; Markin and Hudspeth, 1995). K_S is an additional constant term corresponding to stiffness arising from the bending of the bases of the stereocilia, and N is the number of channels. This equation provides a good fit to the stiffness data, with $\Delta x = 4$ nm and $\kappa = 0.5$ pN/nm.

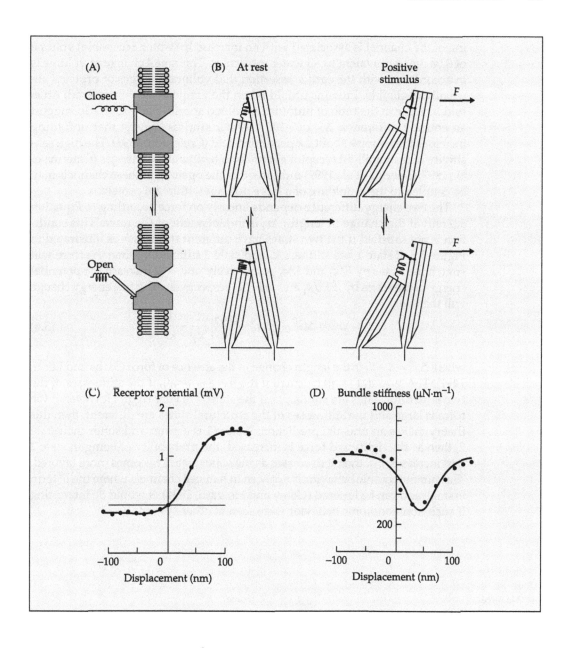

in membrane area of 4 to 8 nm². Sensitivity of channels to changes in osmotic pressure, π, induced by addition of sucrose or sorbitol, suggests that their opening is associated with a change in "solute accessible volume": ΔV is ~1 nm³ for the potassium channel from nerve (Zimmerberg et al., 1990) and ~30 nm³ for the large anion channel from mitochondria (Zimmerberg and Parsegian, 1986). These volumes correspond to 30 to 1000 water molecules. The sensitivity of the acetylcholine receptor channel to pressure (Heinemann et al., 1987), suggests that the binding of acetylcholine and subsequent opening of the channel is associated with an increase in (water-accessible) volume of 0.08 nm^3, equivalent to ~3 water molecules. This small change in volume is in accordance with the earlier assertion that volume changes of proteins are usually negligible. Proteins unfold when the temperature is increased: An e-fold increase in the ratio of unfolded to folded species per 3°C would suggest an entropy difference, ΔS, of ~100 k. Other studies suggest that unfolding increases the entropy about 2 k per amino acid (Creighton, 1993). The high sensitivity of the vanilloid receptor channels to temperature changes (Caterina et al., 1997; Caterina et al., 1999) indicates that the opening of these channels may be coupled to the unfolding of a large domain within the protein.

The free energy difference depends linearly on force according to Equation 5.3 only if the change in length, Δx, is independent of the force. This condition is not satisfied if the two states have different stiffnesses as illustrated in Figure 5.3. If state 1 has stiffness κ_1 and state 2 stiffness κ_2, then the force will stretch the states by F/κ_1 and F/κ_2, respectively, and will increase the potential energy of the states by $\frac{1}{2}F^2/\kappa_1$ and $\frac{1}{2}F^2/\kappa_2$, respectively. The free energy change will then be

$$\Delta G = \Delta G^0 - F\Delta x^0 - \tfrac{1}{2}F^2\left[\frac{1}{\kappa_2} - \frac{1}{\kappa_1}\right] \tag{5.4}$$

where $\Delta x^0 = x_2 - x_1$ is the length change in the absence of force (Sachs and Lecar, 1991). The term in F^2 will be small if the force is small, if the stiffnesses of the two structural states are similar, or if the states are very rigid. However, if the force is large and the stiffnesses of the structural states are different, then this theory makes an unusual prediction: If state 1 is shorter and softer than state 2, then as the rightward force is increased, the probability of being in state 2 first increases, but then it decreases as the softer state 1 becomes more favored. This nonmonotonic behavior is not seen in hair cell channels where the F^2 term is small and can be ignored (Corey and Howard, 1994); it would be interesting if such nonmonotonic behavior were seen in other systems.

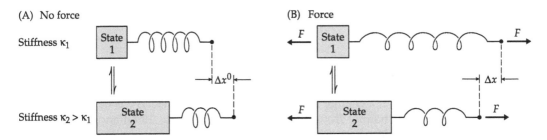

Figure 5.3 Differences in compliance between structural states
(A) No force. (B) Force. The effect of force on the free energy is calculated in detail in Appendix 5.1 and summarized in Equation 5.4.

Rate Theories of Chemical Reactions

Forces also affect the rates of chemical reactions. In order to understand how this occurs, we must first discuss theories of chemical reactions.

The simplest chemical reaction is the interconversion between two species that satisfies

$$E_1 \underset{k_{-1}}{\overset{k_1}{\rightleftharpoons}} E_2 \qquad \frac{d[E_1]}{dt} = -k_1[E_1] + k_{-1}[E_2] \qquad (5.5)$$

This reaction is said to obey first-order kinetics because the rate of change depends linearly on the concentration of the species. The constants of proportionality, k_1 and k_{-1} are called **rate constants** and they have units of s^{-1}. The solution is

$$\frac{[E_1](t)}{[E_1]+[E_2]} = [E_1](0) + \left\{ \frac{k_{-1}}{k_1+k_{-1}} - [E_1](0) \right\} \{1 - \exp[-(k_1+k_{-1})t]\} \qquad (5.6)$$

This equation says that the reactant (E_1) approaches its equilibrium concentration with an exponential time course. The time constant is $(k_1 + k_{-1})^{-1}$. The product (E_2) also approaches its equilibrium concentration exponentially (with the same time constant). Such exponential time courses, which are characteristic of first-order reactions, were first measured in 1850 as changes in optical rotation associated with the acid-induced hydrolysis of sucrose (Eyring and Eyring, 1963).

When the reaction described by Equation 5.5 reaches equilibrium, it is at steady state with the forward and reverse reactions exactly balanced ($d[E_1]/dt = 0$). At equilibrium, we therefore have

$$\frac{k_1}{k_{-1}} = \frac{[E_2]}{[E_1]} = K_{eq} = \exp\left[-\frac{\Delta G}{kT}\right] \qquad (5.7)$$

This shows that the equilibrium constant (K_{eq}) defined in Equation 5.2 is equal to the ratio of the forward and reverse rate constants. This makes sense, because if the forward rate constant is increased, then the equilibrium concentration of product will also increase as expected for a larger equilibrium constant. On the other hand, if Equation 5.7 is satisfied, then the reaction is at equilibrium because Boltzmann's law is satisfied. If the free energy difference between product and reactant, ΔG, depends on the force, then either the forward or the reverse rate (or both) must depend on force.

The physical picture behind a first-order reaction is that it corresponds to a very rapid (almost instantaneous) transition between two structural states. In other words, the duration of the transition is very much shorter than the average lifetimes of the states ($1/k_1$ for E_1 and $1/k_{-1}$ for E_2). How fast might the transition be? Covalent chemical changes occur very rapidly, on the 0.1 picosecond timescale of molecular vibrations (corresponding to optical wavenumbers of ~1000 cm^{-1}). But global protein conformational changes occur much more slowly. The speed is ultimately limited by the speed of sound (Appendix 5.2), so the fastest relaxations of very rigid proteins have time constants of ~10 ps (see Problem 3.5). For more typical, softer proteins such as motor proteins, the relaxations are even slower, with time constants on the order of 10 ns (see Example 2.5 and Figure 2.4D). Nevertheless, even a transition lasting 10 ns is a very short time compared to the lifetimes of structural states, which are typically 1 ms or longer.

Several properties of first-order reactions can be understood using the idea that the reaction proceeds via a high-energy **activated state**, or **transition state** (Figure 5.4). Because the activated state has a free energy, G_a, much greater than that of the initial or final states, the probability that the protein will be in the activated state is very low. The short-lived activated state accords with the transition itself being very rapid. Because the activated state occurs at an energy maximum, it differs from a structural state in that it is not stable: We do not expect to be able to crystallize a protein in one of its transition states.

The activated-state concept leads naturally to the idea that chemical reactions can indeed be described using rate constants. To make this connection requires two additional assumptions. First, it is assumed that the reactant (E_1) is in equilibrium with the activated state (E_a). And second, it is assumed that the activated state is equally likely to break down to reactant or product with some rate A. By Boltzmann's law, the probability of being in the activated state is $[E_a]/[E_1] = \exp[-(G_a - G_1)/kT]$ and so the rate of formation of product is $A[E_a] = A[E_1]\exp[-(G_a - G_1)/kT]$. Thus the rate is linearly proportional to reagent concentration, $[E_1]$, as expected for a first-order reaction. The rate constant is

$$k_1 = A \exp\left[-\frac{\Delta G_{a1}}{kT}\right] \qquad \Delta G_{a1} = G_a - G_1 \qquad (5.8)$$

A similar expression holds for the reverse reaction; the ratio, $k_1/k_{-1} = K_{eq}$, accords with the Law of Mass Action. Equation 5.8 is called the **Arrhenius equation** and the constant A is called the **frequency factor**, or **pre-exponential factor**.

The activated-state concept also accounts for the strong temperature dependence of chemical reactions. Because the frequency factor, A, is not expected to depend strongly on temperature, the Arrhenius equation predicts that the rate constant depends on temperature according to $\sim\exp(-\Delta G_{a1}/kT)$. This agrees with the experimental results that biochemical reactions have strong temper-

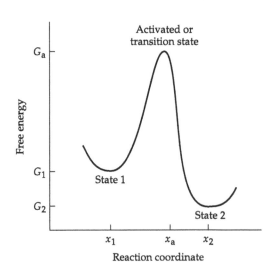

Figure 5.4 The activated state
The activated state corresponds to a position (x_a) along the reaction coordinate, intermediate between the initial (x_1) and final (x_2) positions.

ature dependencies. Typically, the rates of biochemical reactions double to quadruple for every 10°C increase in temperature: We say that the Q_{10} is between 2 and 4. Such Q_{10}s imply that the enthalpy, ΔU_a, of the transition state is some 20 to 40 kT above the initial state (Appendix 5.2). Thus, as argued by Arrhenius, the strong temperature dependence of chemical reactions supports the concept of a high-energy activated state.

To predict the absolute rate of a biochemical reaction, a more detailed theory is needed because the Arrhenius theory provides no information about the frequency factor, A. Two such detailed theories are the Eyring rate theory and the Kramers rate theory. Both require that the **reaction coordinate**, the parameter that measures the progression of the reaction, be specified. If a protein changes overall length as a result of the $E_1 \to E_2$ transition, then we could make length the reaction coordinate, though many other reaction coordinates are possible; indeed, the distance between any two atoms that move relative to one another during the reaction could be used as a reaction coordinate. *If the protein is subject to a force, then a natural reaction coordinate is the length of the protein in the direction of the force.*

In the **Eyring rate theory**, the reaction is assumed to correspond to the breakdown of a single quantum-mechanical vibration of the protein. In this case the frequency factor is $\sim kT/h \cong 6 \times 10^{12}$ s^{-1}, where h is the Planck constant (Atkins, 1986). The absolute rate is then the frequency factor reduced by the exponential term. For example, a reaction with a rate constant of 10^3 s^{-1} would have an activation free energy (ΔG_{a1}) of 22 kT. The Eyring theory is expected to apply to covalent changes of proteins and their ligands. However, it is not expected to apply to global conformational changes of proteins in which a large number of bonds are made and broken, because in this case the reaction does not correspond to a single mode of vibration of the protein.

In the case of global protein conformational changes, a more physically realistic model is the **Kramers rate theory**. According to this model, the protein diffuses into the transition state with a rate that is the reciprocal of the diffusion time

$$k_1 = \frac{\varepsilon_1}{\pi} \frac{1}{\tau_1} \sqrt{\frac{\Delta G_{a1}}{kT}} \exp\left[-\frac{\Delta G_{a1}}{kT}\right] \qquad \Delta G_{a1} = G_a - G_1 \qquad (5.9)$$

ε_1 is an "efficiency factor" equal to the probability of making the transition when at the transition state (Appendix 5.2). According to Kramers rate theory, the frequency factor is approximately equal to the inverse of the relaxation time, $\tau^{-1} = \kappa/\gamma$ (the other pre-exponential terms are close to unity). This makes intuitive sense: We can think of the protein sampling a different energy level every τ seconds, because τ is the time over which the protein's shape becomes statistically uncorrelated. The protein can react only when it attains the energy of the transition state, and the probability of this occurring is proportional to $\exp(-\Delta G_{a1}/kT)$.

The Eyring and Kramers rate theories represent two extreme views of the mechanism of global conformational changes of proteins. In the Eyring model, the transition state is like the initial state (Figure 5.5). A sudden, local chemical change (such as the binding of a ligand or the chemical change in a bound ligand) creates a highly strained protein that then relaxes into a new stable conformation. The relaxation is along the quadratic energy curve and has time

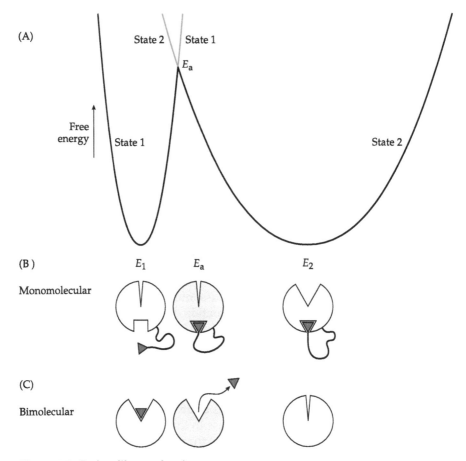

Figure 5.5 Eyring-like mechanism
(A) Energy diagram. (B) Monomolecular model. (C) Bimolecular model. The reaction coordinate is the extent of opening or closing of the cleft. The shaded structures are the strained transition states.

constant $\tau = \gamma/\kappa$. This formulation has been used to model the working stroke of myosin (Eisenberg and Hill, 1978; see Figure 16.3): After myosin has bound to actin in its pre-powerstroke state, the phosphate rapidly dissociates, leaving the protein in a highly strained post-powerstroke state. The relaxation of this highly strained state drives the sliding of the filaments and the shortening of the muscle. If the filaments are prevented from sliding, the strained state will maintain the tension in the muscle.

The energy profile associated with an Eyring-like protein conformational change is analogous to that associated with a spectroscopic change. When a small molecule absorbs light and undergoes an electronic change, the transition is drawn as a vertical line in accordance with the Franck–Condon principle, which states that the more massive nuclei take much longer to move than the lighter electrons. Consequently, the newly formed state is strained and relaxes slowly (though still on the subpicosecond timescale) as the nuclei move into their new stable positions. The analogous principle for proteins is that global structural or "physical" changes of proteins are much slower than local chemical changes, because structural changes are slowed by protein and solvent viscosity. The fast local changes leave the protein in an unstable global conformation which then relaxes more slowly into a new stable state. (These ideas are expanded in Figure 16.2.) The transition is not really vertical: it is actually a very steep parabola, with curvature appropriate for the high rigidity of the local bonds.

In the Kramers view, the protein undergoes a global diffusion into the activated state. When a sufficiently large conformational change has been achieved, the protein converts to the final state (Figure 5.6). This is reminiscent of the mechanism postulated by Marcus to explain electrochemical reactions in solution (Marcus, 1996). In the extreme, the protein diffuses all the way to the final state, which is then locked in by a subsequent chemical change (Figure 5.7). This extreme case has been called a **thermal ratchet** mechanism on account of the prominent role played by diffusion in reaching the transition state (Hunt et al., 1994; Peskin and Oster, 1995). The application of these ideas to motor proteins is discussed in Example 5.2. Of course, if the forward process is purely diffusive, then the reverse is Eyring-like. However, it should be pointed out that even in the Eyring-like mechanism, the activated state is also reached by a ther-

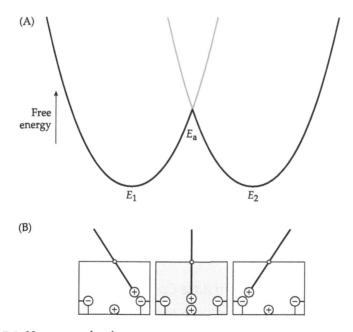

Figure 5.6 Marcus mechanism
(A) Energy diagram. (B) Physical model that gives the reaction profile. The reaction coordinate is the position of the end of the lever. The shaded structure is the high-energy transition state.

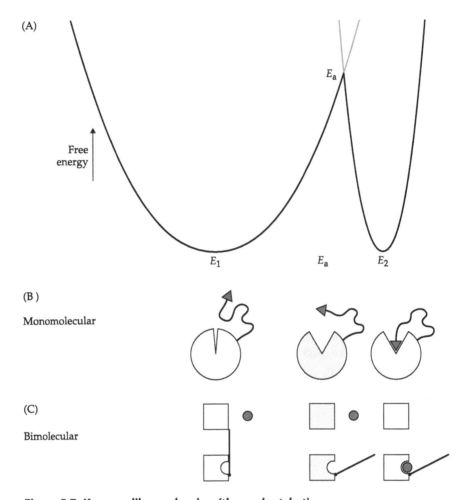

Figure 5.7 Kramers-like mechanism (thermal ratchet)
(A) Energy diagram. (B) Monomolecular model (C) Bimolecular model. The physical picture is of a "foot-in-the-door" mechanism such that the global conformational change must be completed before the structural change can occur. The reaction coordinate is the extent of opening of the cleft or the gate. The shaded structures are the strained transition states.

mal fluctuation, just one that is more localized. The role of these different mechanisms in force generation by motor proteins is discussed in Chapter 16.

Effect of Force on Chemical Rate Constants

The activated-state concept makes specific predictions of how rate constants depend on external force. If the protein structures are very rigid and the transitions $E_1 \to E_a \to E_2$ are associated with displacements x_1, x_a, and x_2 *in the direc-*

tion of the force, F, then the energies of the states will be decreased by Fx_1, Fx_a, and Fx_2, respectively. This implies that

$$k_1 = A \exp\left[-\frac{\Delta G_{a1} - F\Delta x_{a1}}{kT}\right] = k_1^0 \exp\left[\frac{F\Delta x_{a1}}{kT}\right] \quad (5.10)$$

where $\Delta G_{a1} = G_a - G_1$ and $\Delta x_{a1} = x_a - x_1$. An analogous expression holds for k_{-1}. Another way of writing this is to let $\Delta x_{a1} = \theta(x_2 - x_1)$, where θ is the fraction of the distance of the transition state toward the final state (Hille, 1992). Note that the ratio of the forward and reverse rate constants must give the correct force dependence for the equilibrium (Equations 5.3 and 5.7).

A useful way of thinking about the effect of force on the reaction rates is that it tilts the free energy diagram of the reaction (Figure 5.8). If the displacement of the activated state is intermediate between the initial and final states ($x_1 < x_a < x_2$), then a negative external force (a load) will slow the reaction, whereas a positive external force (a push) will accelerate the reaction. However, if $x_a = x_1$—that is, if the transition state is reactant-like—then force will have little effect on the forward rate constant. On the other hand, if $x_a = x_2$—that is, if the transition state is product-like—then the force will have little effect on the reverse rate constant. If the displacement of the activated state is not intermediate, it is even possible that a load could actually increase the forward rate constant (if $x_a < x_1$), though in this case the backward rate would be increased even more.

Example 5.2 Thermal ratchet models for motor proteins
Consider a hypothetical motor protein with $\kappa = 4$ pN/nm and radius 3 nm (Example 2.4) and where the total free energy available from ATP hydrolysis at physiological ATP, ADP, and P_i concentrations is 25 kT. The drag coefficient, γ, is 60 pN·s·m^{-1} (see Table 2.2), and the relaxation time is ~15 ns. According to the Kramers theory (Equation 5.9), it would take about 10 s to pick up 20 kT of energy by a purely diffusive process. But for myosin, the complete ATP hydrolysis reaction only takes about 0.05 s (Chapter 14). Therefore, if the ATP hydrolysis reaction has an efficiency of 80% (20 kT/25 kT), such a diffusive step could not be on myosin's reaction pathway. This was the argument used by Eisenberg and Hill (1978) to rule out a Kramers-like mechanism postulated by A. F. Huxley (1957). However, a more reasonable efficiency for myosin is 50% (Chapter 16), corresponding to an energy of 12.5 kT. To pick up this amount of energy would take only 7 ms, less than the time that myosin spends detached from the actin filament. Thus the kinetics of myosin is not inconsistent with a Kramers-like mechanism after all.

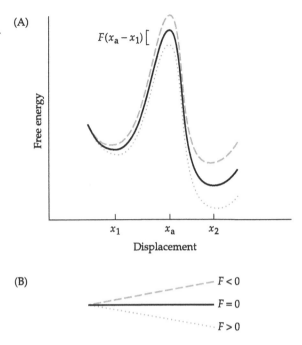

Figure 5.8 Force tilts the energy profile (A) The energy profile in the presence of negative (dashed) and positive (dotted) external forces. The external force is equal to the slope of the tilt shown in (B). A negative external force is a load that slows down the reaction. A positive external force is a push that accelerates the reaction.

The activated-state concept makes several predictions for how forces affect the rates of chemical reactions. It is a challenge for experimentalists to test these predictions. For example, the rate constant for the opening of the transduction channels in hair cells is strongly force dependent, but that for closing is not (Corey and Hudspeth, 1983). This suggests that the transition state is more like the open state: The gate has to move almost all the way towards its open position before the conductance becomes significant (Howard et al., 1988). The rate constant for the opening of voltage-dependent ion channels is also more voltage dependent than the closing (Hille, 1992). Another example of force-dependent rate constants is the detachment of myosin from actin. Negative force (load) slows down the detachment rate (Finer et al., 1994), suggesting that the transition state is associated with the movement towards the post-powerstroke position. Another example is the unfolding and folding of titin (Erickson, 1994; Kellermayer et al., 1997; Rief et al., 1997; Tskhovrebova et al., 1997), illustrated in Example 5.3.

In general, the transition state could correspond to a distortion of the initial state in any of three directions, not just in the direction corresponding to the final state (Lecar and Morris, 1993). In this general case, the rate constant will be affected by any force that has a component in the direction of the transition state. This can be appreciated by making an analogy to a door handle or a latch, where the transition state (handle down) corresponds to movement in a direction perpendicular to the direction of opening of the door (Figure 5.9). An example of this occurs with motor proteins: Under certain circumstances, the speed of movement of kinesin along a microtubule can be accelerated by

Example 5.3 Unfolding titin using an atomic force microscope An atomic force microscope can be used to reversibly unfold immunoglobulin modules, ~100 amino acid domains found in a variety of proteins including the muscle protein titin. This is shown in Figure A, where the one end of the protein is attached to a solid support, and the other end is attached to the tip of an AFM and pulled. The unfolding rate depends only weakly on force (e-fold increase per 16 pN), indicating that the transition state for unfolding is of similar length to the folded state ($\Delta x = kT/F = 0.25$ nm). On the other hand, the folding is strongly dependent on force (e-fold slowing per 1.6 pN). These results support the energy profile shown in Figure B: A small strain of 0.25 nm, about 5% of the length of the folded protein (5.1 nm), is enough to completely destabilize the structure and lead to unfolding. But the folding of the protein requires the formation of a nearly fully folded transition state. The free energies in Figure B are calculated from the rate constants using Eyring rate theory (though this may not be valid, as argued above). The abbreviations in the figure are: N = native state, A = activated state, CD = compact disordered state, and ED = extended state (Carrion-Vazquez et al., 1999).

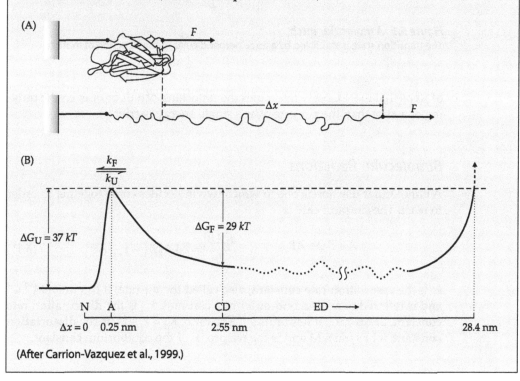

(After Carrion-Vazquez et al., 1999.)

forces perpendicular to the direction of motion (Gittes et al., 1996). Thus it is important to realize that force is a vector quantity and that force will couple to a reaction whenever it has a component that is in the direction of the reaction pathway. In the most general case, the reaction pathway and its force dependence could be very complicated. It could involve a highly convoluted sequence

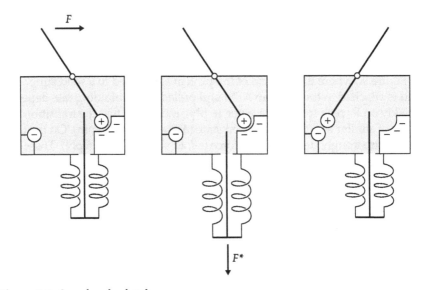

Figure 5.9 A molecular latch
The transition state is stabilized by a force perpendicular to the direction of motion.

of steps that more closely resembles the unlocking of a door or even the untying of a knot than the passage over a single energy barrier.

Bimolecular Reactions

A bimolecular reaction is one in which two molecules come together in order to react. The simplest case is

$$A + B \underset{k_{-1}}{\overset{k_1}{\rightleftharpoons}} AB \qquad \frac{d[AB]}{dt} = k_1[A][B] - k_{-1}[AB] \qquad (5.11)$$

k_1 is the **association rate constant**, also called the **on-rate**; it has units $M^{-1} \cdot s^{-1}$ and is referred to as a second-order rate constant. k_{-1} is the **dissociation rate constant**, or **off-rate**; it has units s^{-1}. The ratio, $K_d = k_{-1}/k_1$, is the **dissociation constant**; it has units M and is the reciprocal of the equilibrium constant.

Association Rates

Diffusion sets an upper limit on the association rate constant. By solving the diffusion equation in three dimensions, it can be shown that the rate constant for **diffusion-limited** collisions between spheres of equal diameter is ~8 × $10^9 \, M^{-1} \cdot s^{-1}$ (Appendix 5.3). If one of the spheres is larger than the other, the rate constant will be even larger. If one of the spheres, the target, is thought of as a fixed point, then the collision rate constant is ~2 × $10^9 \, M^{-1} \cdot s^{-1}$.

Only the very fastest atomic and diatomic reactions have on-rates that approach these diffusion-limited collision rate constants. Typical ligand–protein on-rates are in the range of 10^6 to 10^8 $M^{-1} \cdot s^{-1}$ (e.g., pyruvate, H_2O_2, CO_2, ACh, GTP, tRNA; Fersht, 1985; ATP, Chapter 14). The fastest protein–protein on-rates are in the range 10^6 to 10^7 $M^{-1} \cdot s^{-1}$: Examples include the polymerization of actin and tubulin (Chapter 10) and the binding of motors to their filaments (Chapter 14).

It is not surprising that the measured on-rates for protein reactions are less than the diffusion-limited rate constants that describe the collision of spheres. The reason is that the binding of ligands to proteins and proteins to proteins is **stereospecific**, meaning that it depends on the precise position and orientation of the reacting species. These steric considerations lead to a reduction in the diffusion-limited collision rate. The reduction is by a factor of $\sim(s/2R)^4$ (Doi, 1975; Bell, 1978; Berg and von Hippel, 1985; Appendix 5.3), where s is the positional precision and $2R$ is the diameter of the binding species (~1 nm for a ligand, ~6 nm for a protein). If the positional precision is 1Å (as judged from the crystal structures of proteins in which the atoms typically have root-mean-square fluctuations of 0.3 to 0.5 Å; Creighton, 1993), these steric considerations reduce the diffusion-limited on-rate by $\sim10^4$-fold for a ligand and up to $\sim10^7$-fold for a protein. But now the measured on-rates are too high! This suggests that a more appropriate value for the positional precision is 5–10 Å, similar to the Debye length describing the screening of electrostatic forces by ions. Thus it is likely that electrostatic (Schurr, 1970a,b; Berg and von Hippel, 1985; Gilson et al., 1994) and other (Northrup and Erickson, 1992) long-range forces accelerate on-rates.

Michaelis–Menten Equation

The bimolecular equation is very important in biochemistry. In enzyme kinetics we can think of A being the enzyme (E), B being the **substrate** (S), and AB being the intermediate that breaks down into enzyme plus **product** (P):

$$E + S \underset{k_{-1}}{\overset{k_1}{\rightleftharpoons}} ES \overset{k_2}{\longrightarrow} E + P \qquad [E] + [ES] = [E_t] \qquad (5.12)$$

Often, the substrate, also called the **reagent**, is well in excess of the enzyme, in which case the steady-state rate of product formation is

$$\text{Rate} = \frac{d[P]/dt}{[E_t]} = k_{cat} \frac{S}{K_M + S} \qquad k_{cat} = k_2 \qquad K_M = \frac{k_{-1} + k_2}{k_1} \qquad (5.13)$$

(Appendix 5.4). This is the **Michaelis–Menten equation**, and K_M is the Michaelis–Menten constant. k_{cat} is the maximum rate per enzyme molecule and K_M is the concentration of substrate for which the rate is half maximal. If ES is in equilibrium with E and S ($k_{-1} \gg k_2$), then the reaction is said to follow a Michaelis–Menten kinetic mechanism. In this case, the Michaelis–Menten constant is equal to the dissociation constant $K_d = k_{-1}/k_1$, and the half-maximal rate occurs when the enzyme is 50% occupied by substrate. If $k_{-1} \ll k_2$, the reaction is said to follow a Briggs–Haldane mechanism. In this case, the K_M is the substrate concentration at which the enzyme spends half its time waiting for the substrate to bind.

Protein Complexes

Another important application of the bimolecular equation (Equation 5.12) is the case where the complex AB is an active species that catalyzes another reaction

$$A + B \underset{k_{-1}}{\overset{k_1}{\rightleftharpoons}} AB \begin{array}{c} C \\ \downarrow r \\ D \end{array} \tag{5.14}$$

For example, AB might be a motor (A) walking down a filament (B) catalyzing the hydrolysis of ATP (C) to products (D) with some rate constant r. Or AB might be an active signaling complex that in turn activates another molecule or catalyzes the formation of a messenger molecule. Or it might be an active transcription factor. The difference between this scheme and the Michaelis–Menten scheme is that the complex can catalyze multiple additional reactions that do not necessarily lead to the dissociation of the complex AB. The solution is given in the Appendix.

Cyclic Reactions and Free Energy Transduction

If we permit the reverse reaction to occur, then the Michaelis–Menten scheme becomes a cyclic reaction

$$E + S \underset{k_{-1}}{\overset{k_1}{\rightleftharpoons}} ES \underset{k_{-2}}{\overset{k_2}{\rightleftharpoons}} E + P \tag{5.15}$$

A general cyclic reaction (an *n*-cycle) is

$$E_1 \underset{k_{-1}}{\overset{k_1^*}{\rightleftharpoons}} E_2 \underset{k_{-2}}{\overset{k_2}{\rightleftharpoons}} \ldots E_n \underset{k_{-n}^*}{\overset{k_n}{\rightleftharpoons}} E_1 \tag{5.16}$$

where we have absorbed the concentration of [S] into the rate constant so k_1^* is now a first-order rate constant. At steady state, the average rate of flow of substrate through the reaction, the **flux**, is given by

$$\text{Flux} = \frac{k_i[E_i] - k_{-i}[E_{i+1}]}{[E_t]} \tag{5.17}$$

Note that the flux is the same at each step; otherwise, there would be build-up of one of the species, contradicting the steady-state assumption. If we define the transition $E_n \to E_1$ as the completion of a cycle, then the flux is equal to the net number of complete cycles per second.

If the reaction is at equilibrium, then the average flux equals zero (though there will always be fluctuations in the flux just as there are thermal fluctuations in the position of a particle). This is a consequence of the **Principle of Detailed Balancing**: If a system in which several chemical reactions take place is at equilibrium, then each of the individual reactions is separately at equilibrium. The principle follows from Boltzmann's law because, at equilibrium, the ratio of the concentrations of the consecutive states, $[E_{i+1}]/[E_i]$, depends only on the relative energies of the two states and not on the absolute reaction rates. Therefore, the ratio remains unchanged if we "freeze out" all the other transitions except those between the two states in question. In this case the steady-state condition implies that $k_i[E_i] = k_{-i}[E_{i+1}]$, and so the flux is zero. This corollary of Boltzmann's law is also called the **Principle of Microscopic Reversibility** because at equilibrium a reaction is equally likely to be going in the forward or reverse direction. Zero flux at equilibrium means that there can be no perpetual motion machines. Conversely, if the average flux is zero, then each step is at equilibrium (because it satisfies Boltzmann's law, Equation 5.7) and thus the entire system is in equilibrium.

Another consequence of equilibrium is that the ratio of the products of the forward and reverse rate constants is unity

$$1 \equiv \frac{[E_1]}{[E_2]}\frac{[E_2]}{[E_3]}\cdots\frac{[E_n]}{[E_1]} = \frac{k_1[S]}{k_{-1}}\frac{k_2}{k_{-2}}\cdots\frac{k_n}{k_{-n}[P]} = \frac{k_1[S]k_2 k_3 \ldots k_n}{k_{-1}k_{-2}k_{-3}\ldots k_{-n}[P]}$$

Conversely, if this ratio is unity, then the reaction is in equilibrium (this follows by writing $k_i/k_{-i} = [E_{i+1}]/[E_i] + \text{flux} \cdot ([E_t]/[E_i])$). Thus there are several necessary and sufficient conditions for a cyclic reaction to be in equilibrium.

The steady-state flux through a cycle can be solved in terms of the rate constants using analytical methods (Hill, 1989) or matrix-inversion numerical methods (Press, 1997). If there are many steps, then the general solution is a complicated function of the rate constants. However, if each step is irreversible, then there is a particularly simple formula for the steady-state flux: The average duration of one cycle (1/flux) is the sum of the average durations of each of the steps: $(\text{rate})^{-1} = (k_1[S])^{-1} + (k_2)^{-1} + \cdots + (k_n)^{-1}$ or

$$\text{Flux} = k_{cat}\frac{[S]}{K_M +[S]} \qquad \frac{1}{k_{cat}} = \frac{1}{k_2}+\cdots+\frac{1}{k_n} \qquad K_M = \frac{k_{cat}}{k_1}$$

A cyclic reaction that has a nonzero flux is an example of a process that is at steady state but not at equilibrium. A specific example is a 2-cycle with $k_1 = 2 \times 10^6$ $M^{-1}\cdot s^{-1}$, $k_{-1} = k_2 = 1$ s^{-1}, $k_{-2} = 1 \times 10^6$ $M^{-1}\cdot s^{-1}$, $[S] = [P] = 1$ μM. The flux is equal to 0.2 s^{-1}. Evidently, the steady state is maintained by the replenishment of substrate molecules, which provides an energy source to maintain the system away from equilibrium.

The energy to drive a cyclic chemical reaction comes from the change in free energy associated with converting one molecule of S to a molecule of P

$$\Delta G^0_{S \to P} = \Delta G^0 + kT \ln \frac{[P]}{[S]} = kT \ln \frac{K}{K_{eq}} = kT \ln \frac{[P][S_{eq}]}{[P_{eq}][S]} \quad (5.18)$$

where ΔG^0 is defined as the **standard free energy**, the difference in free energy when the substrate (or substrates) and product (or products) are both in their standard states (Appendix 5.5). Because the standard state has both the substrates and the products at 1 M concentration, another way to view the standard free energy is that it is the free energy associated with the reaction when all the reactants are at 1 M concentration. Note that the free energy of a reaction is not really dependent on the standard free energy, but rather, it depends on the extent to which the substrates and products are out of equilibrium. For example, the standard free energy for the hydrolysis of ATP is -54×10^{-21} J, but at typical cellular concentrations of ATP, ADP, and P_i (1 mM, 10 μM, and 1 mM, respectively) the free energy is even more negative, -101×10^{-21} J (Chapter 14). However, if the ATP concentration were 1 aM (10^{-18} M) and the ADP and P_i concentrations were each 100 mM, then the free energy would have a similar magnitude but would be of opposite sign ($\sim +100 \times 10^{-21}$ J). In this case we could view ADP, rather than ATP, as the energy source.

If a reaction is coupled to movement, such as a motor moving along a filament, then the free energy of the reaction will be affected by an external force. If one cycle results in a net displacement through distance Δx, then

$$\Delta G_{S \to P} = \Delta G^0 - kT \ln \frac{[S]}{[P]} - F \cdot \Delta x \quad (5.19)$$

Because the substrate and product molecules are free in solution and so are not affected by a force, ΔG^0 is independent of force. The force at which the free energy change is zero is the **equilibrium force**, $F_{eq} = (\Delta G^0 - kT\ln[S]/[P])/\Delta x$. When $F = F_{eq}$, the free energy change is zero and the reaction is at equilibrium. The work done by the reaction is $w = F\Delta x$, which has a maximum value at equilibrium equal to $F_{eq}\Delta x = \Delta G^0 - kT\ln[S]/[P]$. At equilibrium the flux is zero and so the velocity of movement, $v = $ flux$\cdot\Delta x$, is also zero. The equilibrium force is also the **reversal force**. It is the force about which the velocity changes sign; when $F < F_{eq}$, the velocity is positive, whereas for $F > F_{eq}$, the velocity is negative.

For small changes in force about the equilibrium force, $|(F - F_{eq})\Delta x| << kT$, the velocity will depend linearly on $(F - F_{eq})$. But for larger forces, the more common case, the effect of force on the velocity is more complicated. To determine how the velocity depends on force, the force dependencies of all the rate constants need to be specified, and the steady-state solution for the flux solved

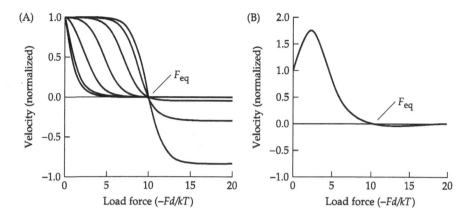

Figure 5.10 Force–velocity curves
(A) Families of velocity curves, $v(F)$, obtained by varying a single force-sensitive rate constant. (B) Nonmonotonic velocity curve obtained if the cycle has a small force-sensitive backward step followed by a large force-sensitive forward step.

in terms of these rate constants. Even for a 2-cycle with only one force-dependent rate constant, the velocity can depend on the force in a number of qualitatively different ways (Figure 5.10A). If there are two force-dependent steps, then the force–velocity curve can be even more interesting. Indeed, the speed need not even be monotonic: If there is a small backward step followed by a large forward step, then the speed can be nonmonotonic with a small load increasing the speed and a large load eventually decreasing it (Figure 5.10B).

Summary

Forces can influence the rates and equilibria of chemical reactions. If a protein has two different structural states, then at equilibrium the probability of finding it in one of these states is related to the difference in free energy according to Boltzmann's law. If work is done on the molecule as a result of the interconversion between the states, then the difference in free energy is altered and the equilibrium is shifted toward the state most stabilized by the force. The change in equilibrium could be due to a change in the forward rate constant, to a change in the backward rate constant, or both.

Different kinetic models make different predictions of how force affects rates of reactions. In Eyring-like models, highly localized conformational changes of the initial state occur prior to and drive slower global conformational changes into the final state. Because there is little distortion of the protein in the transition state, little work will be done on it by an external force, so the rate constant will not depend strongly on force. On the other hand, in Kramers-like models, global conformational changes of the initial state occur prior to more localized changes that lock the protein into the final state. In this case, the transition state is highly distorted, and the rate constant will depend strongly on force.

At equilibrium the forward and reverse reaction rates are the same. However, if the interconversion between different states of a protein is coupled to a source of chemical energy (the breakdown of a chemical substrate that is not at equilibrium with its product), then it is possible that the interconversion rate in one direction is greater than that in the other. In this case there is a flux, which, if coupled to a displacement, will lead to movement of the enzyme. If the chemical free energy associated with the conversion of a substrate molecule to product is greater than the mechanical work done by an external force, then the flux will continue even against the mechanical load. In this way, chemical reactions can generate force.

Problems

5.1 Suppose that one structural state, the T state ("tense"), is 1% denser than the other state, the R state ("relaxed"). How could the different densities be measured? If the molecular mass is 100 kDa, what is the volume difference? At what pressure would you expect the equilibrium between the states to be affected?

5.2 Suppose that one could pull directly on the gate of an ion channel, and that the gate swings through 2 nm as it goes from the closed to the open position. If, in the absence of force, the channel spends half its time open and half its time closed, how much force is needed to increase the open probability to 0.9?

5.3 Suppose that a protein has a stiffness of 2 pN/nm in state 1 and a stiffness of 1 pN/nm in state 2, but that the two states have the same resting length (the length in the absence of a force). If there is initially a very low probability of being in state 2, how much force is needed to increase the open probability e-fold?

5.4* Suppose that the ratio of substrate to product in a mixture is ten times greater than the ratio at equilibrium. How much mechanical work could be obtained by converting one molecule of substrate to one molecule of product? Suppose that you have a total of N substrate plus product molecules and that the equilibrium ratio is 1. What is the total amount of mechanical work that could be done with the mixture before it becomes completely spent?

---

*More difficult

CHAPTER

6

Polymer Mechanics

This chapter is about the mechanical properties of slender rods—that is, rods whose lengths are much greater than their diameters. Slender rods are good models for biological polymers such as DNA and the protein filaments that make up the cytoskeleton. I ask: How do slender rods bend in response to mechanical and hydrodynamic forces? How much force is needed to buckle a slender rod? How quickly will it bend or buckle? And how much does the shape of a slender rod fluctuate in response to thermal forces?

Because the cytoskeletal filaments are the most important structural elements within cells, knowing the forces required to bend and buckle slender rods should allow one to model the mechanical properties of cells. Such cellular models are the starting point for modeling the mechanical properties of tissues and organs. In this way, the theory of slender rods developed here allows us to relate the mechanical properties of individual molecules to the mechanical properties of cells and tissues.

After discussing the statics and dynamics of more rigid polymers such as actin filaments and microtubules, I turn to the mechanics of flexible polymers such as DNA and unfolded protein chains. We arrive at the interesting notion of an entropic spring: Even a completely unstructured polymer chain resists being straightened because of its tendency to return to its more disordered configurations. Thus a very flexible polymer behaves like a spring and we say it has entropic, or rubber-like, stiffness. The entropic stiffness of segmented proteins or even random polypeptides is remarkably large. We arrive at the counterintuitive result that the entropic stiffness due to stretching a disordered domain may be as high as the enthalpic stiffness due to bending a well-ordered domain; This finding must be incorporated into our thinking about the molecular mechanics of motors and other proteins.

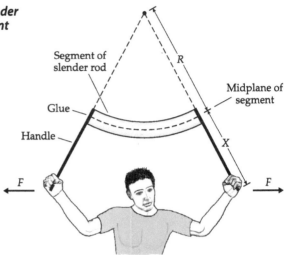

Figure 6.1 Bending of a slender rod due to a bending moment

Flexural Rigidity and the Beam Equation

Suppose that we grab a small length of rod and bend it into a circular arc, as shown in Figure 6.1. Provided that we don't apply too much force, the **curvature** of the bend, $1/R$ (R is the **radius of curvature**), will be proportional to the torque, or **bending moment**, $M = F \cdot X$ (force times the distance). In other words, $M \propto 1/R$ or

$$M = EI\frac{1}{R} \tag{6.1}$$

where the constant of proportionality, EI, is called the **flexural rigidity**. This equation is called the **beam equation** and it is analogous to Hooke's law for a spring: The bending moment is analogous to force, the curvature is analogous to elongation, and the flexural rigidity is analogous to the spring constant.

The concept of flexural rigidity applies to any slender rod and does not presuppose that the rod is isotropic or homogenous, or that the cross-section is circular. It merely relates the bending to the applied moment, and, provided that the moment is small, the flexural rigidity is expected to be a constant that does not depend on the size of the moment. For a protein polymer such as a microtubule or an actin filament, this constant will be completely determined by the properties of the bonds between the atoms within each protein subunit and the properties of the bonds that hold the subunits together in the polymer (ben Avraham and Tirion, 1995). In general, if the rod is nonisotropic or has a noncircular cross-section, then the flexural rigidity will depend on the direction of the bend—for example, a ruler can be bent easily about one axis but not about the other. Because cytoskeletal filaments have approximate circular or helical symmetry, their flexural rigidities ought to be approximately independent of the direction of the bend.

If a rod is made of an isotropic homogenous material, the flexural rigidity can be separated into two terms—the Young's modulus, E, which is a property of the material (Equation 3.2), and the **second moment of inertia of the cross-section**, I, determined by the shape of the rod. This separation was anticipated by writing flexural rigidity as EI in the initial definition. To see how the material and geometric contributions to rigidity arise, we have to consider the forces within the bent rod. On one side of the midplane (the outside of the bend), the material is stretched; on the other side (the inside of the bend), the material is compressed (see Figure 6.1). The resistance of the material to these tensile and compressive forces counterbalances the external bending moment, and therefore determines the flexural rigidity. By summing all the restoring moments due to these internal stresses, it can be shown (Appendix 6.1) that the second moment is

$$I = \int_{\text{area}} y^2 \cdot dA \qquad (6.2)$$

where y is the distance from the middle of the section. The second moments for rods of various cross-sections are shown in Table 6.1. Because I depends

Table 6.1 Second moments of inertia of slender rods

Cross-section	Second moment (I)		Notes
circle, r	$\dfrac{\pi}{4} r^4$		
ellipse, a, b	$\dfrac{\pi}{4} ab^3$	$\dfrac{\pi}{4} a^3 b$	About the minor and major axes, respectively
rectangle, a, b	$\dfrac{ab^3}{12}$	$\dfrac{a^3 b}{12}$	About the narrow and wide axes, respectively
annulus, r_1, r_2	$\dfrac{\pi}{4}\left(r_2^4 - r_1^4\right)$		
bundle of n circles, $2r$	$\left(\dfrac{2}{\pi^2} n^3 + n\right)\dfrac{\pi}{4} r^4$		
ellipse at angle θ	$I_1 \cos^2\theta + I_2 \sin^2\theta$		I_1 and I_2 are the second moments for bending about the two principal axes
Helical filament	$\sqrt{I_1 I_2}$		E.g., actin filament
Helical filament with elliptical section	$\dfrac{A^2}{4\pi}$		A is the cross-sectional area (I is independent of eccentricity)
Helical filament with rectangular section	$\dfrac{A^2}{12}$		A is the cross-sectional area (I is independent of eccentricity)

only on the geometry, *the bending of a rod made of an isotropic and homogenous material depends only on the Young's modulus and the shape.*

The beam equation holds provided that the internal strains remain below the maximum for the particular material. For plastic-like materials such as proteins, the maximum strain is a few percent (see Table 3.2). Therefore, if the radius of the filament divided by the radius of curvature is less than 0.01, the strain will nowhere exceed the maximum strain and the beam equation will hold. For a microtubule of radius ~12 nm, the theory should hold provided the radius of curvature is greater than ~1 μm; with this curvature the maximum distortion in the 4-nm-long tubulin monomer is less than 0.04 nm = 0.4 Å. Actin filaments have a smaller diameter, and so the beam equation should hold for even tighter bends with radii of curvature as small as ~250 nm, the resolution of the light microscope. For both microtubules and actin filaments, the forces required to produce such pronounced bends are larger than the forces likely to be encountered in vivo or in vitro—thus the beam equation should generally be applicable. By the same argument, very large forces are necessary to bend and snap an actin filament or a microtubule.

Applications of the Beam Equation: Bending and Buckling

To apply the beam equation (Equation 6.1) we need to put it into a more useful form. One way to do this is to describe a rod of length L constrained to lie in a plane as a set of points $\{x(s), y(s)\}$ where s is the distance along the contour of the rod (Figure 6.2A). s is called the **arc length**. All the information about the shape of the rod is contained in the **tangent angle**, $\theta(s)$, at each position s. The tangent angle is related to the spatial coordinates by $dx/ds = \cos\theta$ and $dy/ds = \sin\theta$ (Figure 6.2A). Furthermore, with the help of Figure 6.2B, it follows that $d\theta/ds = 1/R$; in other words the curvature is the derivative of the tangent angle with respect to the arc length. Using these relations, we can rewrite the beam equation in terms of tangent angles and arc lengths:

$$\frac{d\theta}{ds}(s) = \frac{1}{EI}M(s) \tag{6.3}$$

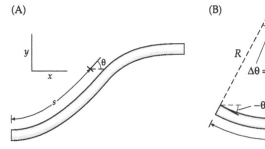

Figure 6.2 Geometry of bent rods
(A) Definition of arc length and tangent angle. (B) The radius of curvature is equal to the change in tangent angle with respect to arc length. Over an arc length Δs, the tangent angle increases through $\Delta\theta = \theta_2 - \theta_1$. For small angles and arc length, $1/R \approx \Delta\theta/\Delta s$.

If we know the bending moment at each point along the length of the filament, as well as appropriate boundary conditions, we can use the beam equation to calculate the resulting shape of the beam, $\theta(s)$.

The beam equation is often difficult to solve. However, if the tangent angle, θ, is always small, then $x \cong s$, $dy/dx \cong dy/ds = \sin\theta \cong \theta$, and $d^2y/dx^2 \cong d\theta/ds$. In this case, the beam equation reduces to

$$\frac{d^2y}{dx^2} = \frac{M(x)}{EI} \qquad (6.4)$$

The Cantilevered Beam

Let us solve the small-angle beam equation for a rod that is clamped at one end, and has a transverse force F applied to its free end (Figure 6.3). In this case $M(x) = F(L - x)$ and Equation 6.4 is quite readily solved (see Figure 6.3). The interesting point is the free end, which is deflected through a distance $y(L) = FL^3/3EI$. Because this deflection is proportional to the force, we can think of the clamped beam as a cantilever spring with spring constant

$$\kappa = \frac{F}{y(L)} = \frac{3EI}{L^3} \qquad (6.5)$$

The stiffness increases as the flexural rigidity increases and as the length decreases.

Example 6.1 A glass fiber Consider a glass rod of radius 0.25 μm and length 100 μm. $E \cong 70$ GPa, $I = (\pi/4)r^4 \cong 3 \times 10^{-27}$, and so the spring constant is 0.64 pN/nm, similar to that of a motor protein. Glass rods of these dimensions are therefore useful force transducers for studying motor proteins. Note that the spring constant can be decreased by increasing the length or decreasing the radius: Doubling the length will decrease the stiffness 8-fold to 0.08 pN/nm, whereas halving the radius will decrease the stiffness 16-fold to 0.04 pN/nm.

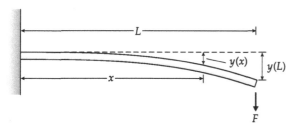

Figure 6.3 Cantilevered beam
In the small-angle approximation, the deflection (y) of the beam is given by $y(x) = (F/EI)(Lx^2/2 - x^3/6)$. This satisfies Equation 6.4, together with the boundary conditions $y = 0$ when $x = 0$ and $dy/dx = 0$ when $x = 0$ (the angle is zero at the clamped end).

> **Example 6.2 A microtubule** The flexural rigidity of a microtubule is ~30 × 10^{-24} N·m² (see Table 8.2). The spring constant of a cantilevered microtubule of length 10 μm is ~10^{-7} N/m. This is very soft and a force of only 0.1 pN would be needed to deflect it through 1 μm (Gittes et al., 1993).

> **Example 6.3 A coiled coil** A coiled coil has a flexural rigidity on the order of 400 × 10^{-30} N·m² (Chapter 8). An 8 nm length of coiled coil clamped at one end will therefore have a stiffness of ~2 pN/nm. The regulatory domain of myosin II is composed of an 8 nm α-helix that is stabilized by the light chains. If this regulatory domain has a similar flexural rigidity to a coiled coil, then much of the compliance of myosin may reside in this domain (Howard and Spudich, 1996).

Buckling

Another important application of the beam equation is to a beam under compression (Figure 6.4A). The bending moment is $M(x) = -Fy$, and, if the ends are free to pivot, application of a compressive force exceeding the **critical force** or the **Euler force**,

$$F_c = \pi^2 \frac{EI}{L^2} \tag{6.6}$$

causes the beam to buckle into the sinusoidal shape shown in Figure 6.4A. The remarkable thing is that the amplitude of the buckle does not depend on the magnitude of the force (Appendix 6.1). This means that the beam collapses. This phenomenon, which is important in engineering, can be readily demonstrated with a flexible plastic ruler—once the ruler has buckled, little additional force is required to increase the bend. The buckling forces for rods under various constraints are shown in Figure 6.4B–D.

> **Example 6.4 Motor force required to buckle a microtubule** Suppose that only one end of a microtubule is clamped and that a motor protein is at the other end. The force required to buckle a 10-μm-long microtubule with $EI = 30 \times 10^{-24}$ is 6.1 pN. Gittes et al. (1996) used this technique to estimate that a single molecule of kinesin could generate a force of ~6 pN.

The above analysis of a buckled beam only holds for small angles. For large buckles, the restoring force increases, and so an ideal bent beam will not completely collapse when the Euler force is just exceeded. In practice, however, a

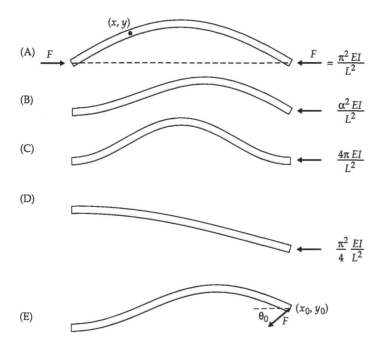

Figure 6.4 Buckling of beams under compressive forces
(A) Ends free to rotate but not move laterally. (B) One end clamped. In this case, α is the smallest positive number that satisfies tan α = α; α² ≈ 20.19. (C) Both ends clamped. (D) One end clamped, but the other free to rotate and translate laterally. (E) One end clamped and the other subject to a large force of direction and magnitude indicated.

real beam may become so bent during the collapse that the material's maximum strain is exceeded: In this case, the beam will fail. At higher curvature the exact form of the beam equation is

$$\frac{d^2\theta}{ds^2} = -\frac{F}{EI}\sin[\theta(s) - \theta_0] \quad (6.7)$$

where θ_0 is the angle of the force whose magnitude is F (Figure 6.4E). The solutions are expressed in terms of elliptic integrals, and can be found, for example, in Byrd and Friedman (1971). The beam equation also allows one to determine how a slender rod will bend when forces act on it at various positions along the rod's length. Conversely, if the shape of a slender rod is known, then the magnitudes and directions of all the forces acting on it can be deduced—the "inverse" problem can be solved and this has been used to calculate the force generated by the motor protein kinesin (Gittes et al., 1996).

Drag Forces on Slender Rods

When a force is applied to a rod, it might translate, rotate, bend, or buckle. A crucial question is: How fast are these motions? The answer depends on the

Example 6.5 Drag on a sperm A human sperm has a head of length 5.8 µm, a width of 3.1 µm, and a tail of length $L = 36$ µm, which tapers from a diameter of ~1 µm down to ~0.16 µm (see figure below). The ratio of the volumes of the tail and the head is ~0.3. Yet for movement parallel to the tail's axis, the ratio of the drag coefficients is $\cong 1.5$. This illustrates the rule of thumb that the drag on an object is determined primarily by the largest dimension, even when motion is parallel to the long axis. For a sperm moving at $v = 50$ µm/s, the drag force given by $\sim(\gamma_{head} + \gamma_{tail}) \cdot v \cong 5$ pN. More exact formulas for the drag coefficients of complex shapes can be found in Garcia de la Torre and Bloomfield (1981).

Example 6.6 Drag forces in gliding assays In an in vitro motility gliding assay (Chapter 13), a single kinesin molecule can move a microtubule of radius ~15 nm at a speed, v, of ~1 µm/s, independent on the microtubule's length (for lengths up to $L = 20$ µm). Thus the motor must be able to exert a force well in excess of the drag force $F_{drag} = c_\parallel \cdot L \cdot v = 2\pi\eta L v / \cosh^{-1}(h/r) \cong 0.1$ pN, assuming that the axis of the microtubule is height $h = 25$ nm above the surface. Indeed, to appreciably slow the movement of a 10-µm-long microtubule (to 0.25 µm/s), the viscosity must be increased 100-fold (Hunt et al., 1994). Using this approach, Hunt et al. (1994) deduced that a kinesin molecule could exert forces up to ~5 pN against a viscous load.

hydrodynamic properties of the rod. As it moves or bends, the rod experiences a drag force from the surrounding fluid and this drag force opposes the motion, slowing it down. Because the flow of fluid around a rod is influenced by nearby surfaces, the drag coefficients are larger near surfaces than free in solution. The drag coefficients for cylindrical rods, prolate ellipsoids (Perrin, 1934, 1936), and spheres are given in Table 6.2. Drag coefficients per unit length for cylinders near plane surfaces (Jeffrey and Onishi, 1981; Hunt et al., 1994) are given in Table 6.3.

Dynamics of Bending and Buckling

Now that we have an expression for the drag per unit length (see Table 6.3), we can calculate the speed of bending of a long filament. When a filament bends there is motion perpendicular to the axis of the filament (Figure 6.5).

Table 6.2 Drag coefficients in an unbounded solution

Parameter	Direction	Cylinder ($L \gg r$)	Ellipsoid ($b \gg a$)	Sphere
$\gamma_\parallel$	→	$\dfrac{2\pi\eta L}{\ln(L/2r) - 0.20}$	$\dfrac{4\pi\eta b}{\ln(2b/a) - 0.5}$	$6\pi\eta r$
$\gamma_\perp$	↑	$\dfrac{4\pi\eta L}{\ln(L/2r) + 0.84}$	$\dfrac{8\pi\eta b}{\ln(2b/a) + 0.5}$	$6\pi\eta r$
γ_r	↻	$\dfrac{\tfrac{1}{3}\pi\eta L^3}{\ln(L/2r) - 0.66}$	$\dfrac{\tfrac{8}{3}\pi\eta b^3}{\ln(2b/a) - 0.5}$	$8\pi\eta r^3$
γ_a	↻ (axial)	$4\pi\eta r^2 L$	$\dfrac{16}{3}\pi\eta a^2 b$	$8\pi\eta r^3$

Note: The parallel, $\gamma_\parallel$, and perpendicular, $\gamma_\perp$, drag coefficients are defined by $F = \gamma_\parallel v$ and $F = \gamma_\perp v$, where F is the force and v is the velocity. The rotational drag coefficients are defined by $T = \gamma_r \omega$ and $T = \gamma_a \omega$, where T is the torque and ω is the angular velocity. The values for cylinders are from Tiraldo and Garciá de la Torre, 1981, and those for prolate ellipsoids are from Perrin, 1934.

Table 6.3 Drag coefficients (per unit length) for a cylinder near a plane surface

Definitions	Drag coefficient	Force or torque
	$c_\parallel = \dfrac{2\pi\eta}{\cosh^{-1}(h/r)} \cong \dfrac{2\pi\eta}{\ln(2h/r)}$	$F = c_\parallel L v$
	$c_\perp = 2 c_\parallel$	$F = c_\perp L v$
	$c_v = 1/\left(c_\perp^{-1} - c_a^{-1}\right)$	$F = c_v L v$
	$c_a = \dfrac{4\pi\eta}{\left[1 - (r/h)^2\right]^{1/2}}$	$T = c_a L r^2 \omega$
	$c_r = \dfrac{1}{3} c_\perp$	$T = c_r \omega (L_1^3 + L_2^3)$

Source: From Hunt et al., 1994.

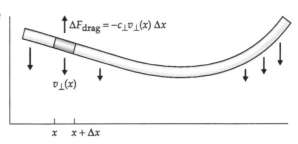

Figure 6.5 Drag force on a bent rod as it straightens out

Thus at each position along the length of the rod, there will be a drag force per unit length

$$f_\perp(x) = -c_\perp v_\perp(x) = -c_\perp \frac{\partial y}{\partial t}(x)$$

We are assuming that the bending is not too large, so that the small-angle approximation applies. These drag forces create bending moments from which the **hydrodynamic beam equation** can be derived (Appendix 6.2)

$$\frac{\partial^4 y}{\partial x^4} = -\frac{c_\perp}{EI}\frac{\partial y}{\partial t} \tag{6.8}$$

We can now answer the question: How long does it take for a bent rod to relax back to its straight conformation? This is analogous to asking how long it takes for a stretched spring to relax back to its resting length. The answer is found by solving the hydrodynamic beam equation with appropriate boundary and initial conditions. For an unconstrained rod, the solutions are in the form of hydrodynamic modes (Figure 6.6). The physical interpretation of the modes is that if the rod is initially bent into the shape of one of these modes and released, then the shape of the rod will be maintained as it straightens. While the rod straightens, the amplitude of the bend decreases exponentially with a time constant that depends on the mode number according to the formula:

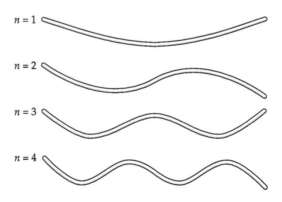

Figure 6.6 The first four hydrodynamic modes of an unconstrained slender rod

$$\tau_n \cong \frac{c_\perp}{EI}\left(\frac{L}{\pi(n + \frac{1}{2})}\right)^4 \qquad n = 1, 2, 3, \qquad (6.9)$$

If the initial shape of a bent rod does not correspond to one of the hydrodynamic modes, then the shape will change as the rod straightens. The precise time course of the shape change can be deduced using the hydrodynamic modes because any shape can be written as the sum or superposition of the hydrodynamic modes, in analogy to Fourier analysis in which any time-varying signal can be written as the sum of cosine and sine modes. Because the higher modes relax more quickly, according to Equation 6.9, the sharp bends will quickly straighten out and the shape of a bent rod will resemble the first mode during the slowest phase of the relaxation.

Example 6.7 Relaxation of microtubules and actin filaments For a microtubule of length 50 µm and radius 15 nm, of flexural rigidity 30×10^{-24} N·m², and height 1000 nm above a surface immersed in an aqueous solution (viscosity 1 mPa·s), the relaxation time for the first mode is 1.1 s, and that for the second mode is 0.14 s. A microtubule of length 5 µm has a relaxation time of ~100 µs, ~10,000 times smaller. Actin filaments relax considerably more slowly due to their greater flexibility: With a flexural rigidity of ~60×10^{-27} N·m² (see Table 6.4 and Equation 6.11), a 50-µm-long actin filament will have a first-mode relaxation time of ~500 s (ignoring the small difference in drag coefficients due to the different radii of the filaments) (Gittes et al., 1993).

Example 6.8 Time constant of a force fiber The dynamics of a slender rod that is clamped at one end and is either free at the other end or attached to another spring is solved in Appendix 6.2. This is useful because it applies to a cantilever force probe that may be attached to a molecule or cell of interest. For a glass rod with radius 0.25 µm, length 200 µm, and $E = 70$ GPa, the stiffness at the tip is 0.08 pN/nm (Equation 6.5 and Example 6.1). If the viscosity is 0.89 mPa·s and the fiber is 2 µm above the surface, then the relaxation time of the slowest mode is $\tau_1 = 3c_\perp L / \beta_1^4(0)\kappa_f = 2.45$ ms. This agrees well with the measured time constants (see the figure in Example 4.7). If the fiber is attached to an elastic load such as a cell or a motor protein, then the system will relax more quickly because of the increased stiffness.

Example 6.9 Dynamics of buckling For a 10-µm-long microtubule clamped at one end, the critical force is ~6 pN using $EI = 30 \times 10^{-24}$ N·m². For $c_\perp = 12$ mPa·s (Hunt et al., 1994) and a compressive force of 7 pN (i.e., just 1 pN above the critical force), the time constant is 60 ms (Appendix 6.2).

Figure 6.7 Fluctuations in shape
Two snapshots of slender rods of length L bent by thermal forces. The more flexible the rod (the smaller the persistence length), the greater the curvature, and the smaller the end-to-end length.

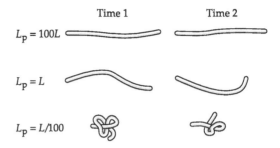

Thermal Bending of Filaments

Just as thermal forces cause a spring to undergo fluctuations in length, thermal forces cause a flexible filament to undergo fluctuations in shape (Figure 6.7). Shape fluctuations are important because if something is to have a structure, be it a molecule or a cell or an organism, its shape must be maintained over time. If an object is very flexible, such as a long piece of DNA or an unfolded chain of amino acids, thermal fluctuations lead to such large shape fluctuations that we say that the object is unstructured. By contrast, a structured object—such as a straight actin filament or microtubule, or a highly ordered set of amino acids in a protein—must be rigid enough that thermal fluctuations do not perturb the shape too much. The question is: How rigid must a filament be to have a structure? And conversely: If the shape of a filament fluctuates, what is its flexural rigidity?

Persistence Length

The most important parameter describing a filament's resistance to thermal forces is the persistence length, L_p. Its intuitive meaning is the length of filament over which thermal bending becomes appreciable. If the length of a filament is much greater than the persistence length, the tangent angles at the two ends will be uncorrelated. For a filament in two dimensions (Figure 6.8), the persistence length can be defined precisely in the following way. It can be shown that the time average of the cosine of $\theta(s) - \theta(0)$ decreases exponentially as the arc length, s, increases (Appendix 6.3). This accords with our intuition because as $\theta(s)$ differs from $\theta(0)$, $\cos[\theta(s) - \theta(0)]$ decreases, and when the two angles become uncorrelated, the difference is equally likely to be any angle

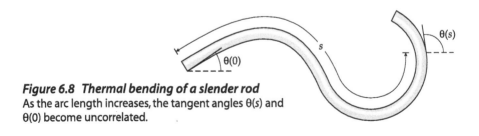

Figure 6.8 Thermal bending of a slender rod
As the arc length increases, the tangent angles $\theta(s)$ and $\theta(0)$ become uncorrelated.

between 0 and 2π, and so $\cos[\theta(s) - \theta(0)]$ will have an average value equal to zero (it is equally likely to be positive or negative). The persistence length is defined in terms of this characteristic distance by

$$\langle \cos[\theta(s) - \theta(0)] \rangle = \exp\left(-\frac{s}{2L_p}\right) \quad (6.10)$$

Using the Principle of Equipartition of Energy, it can also be shown that the persistence length is proportional to the flexural rigidity

$$L_p = \frac{EI}{kT} \quad (6.11)$$

(Appendix 6.3). Again this makes intuitive sense—the more rigid a rod, then the longer the distance before the thermally induced bends become large. The inverse dependence on temperature also makes sense because at a higher temperature, the thermal forces are also larger (see Equation 4.25), and so the thermal bends will become more pronounced. The proportionality between persistence length and flexural rigidity is a consequence of the principle that thermal forces behave just like other mechanical forces, and so the bending in response to thermal forces should depend solely on the flexural rigidity. The persistence lengths of DNA, actin filaments, and microtubules are given in Table 6.4.

The factor of 2 in Equation 6.10 was chosen not simply to give a "nice" relationship between persistence length and flexural rigidity (see Equation 6.11). There is another reason. Suppose that we had considered a filament in three dimensions rather than two. Such a filament can bend in two different directions, with angles $\Delta\theta_1$ and $\Delta\theta_2$ with respect to the filament axis. Because the angles fluctuate independently, the cosine of the total angular change, $\Delta\theta_{3D}$, decays twice as quickly; in other words

$$\langle \cos[\Delta\theta_{3D}(s)] \rangle = \exp\left(-\frac{s}{L_p}\right) \quad (6.12)$$

This is the more usual definition of the persistence length.

Thermal Bending of Semiflexible Polymers

The persistence length of a polymer can be measured in several different ways. One way is to measure the exponential decay of the cosine according to the

Table 6.4 Persistence lengths of biological polymers

Polymer	Persistence length	Reference
DNA	50 nm	Hagerman, 1988
Actin	15 µm	Yanagida et al., 1984
Microtubule	6 mm	Mickey and Howard, 1995

definition of Equation 6.10. Another way is to measure the end-to-end length. Any bending in the filament will cause the end-to-end length to drop below the contour length. The persistence length can be calculated from the formula for the mean-squared end-to-end length

$$\langle R^2 \rangle = 2L_p^2 \left[\exp(-\frac{L}{L_p}) - 1 + \frac{L}{L_p} \right] \quad (6.13)$$

(Appendix 6.3). This approach is practical for **semiflexible** polymers, whose length is approximately equal to the persistence length. In this case, the end-to-end distance falls appreciably below the contour length. Measurement of the end-to-end length has been used to estimate the persistence length of actin filaments (see Table 6.4; Nagashima and Asakura, 1980; Chapter 8).

For more rigid filaments (such as microtubules) that are considerably shorter than their persistence length (~10 μm vs. ~10 mm; see Table 6.4), the persistence length can be estimated from the fluctuations in shape. This can be done by considering the tangent angle, θ(s), which specifies the shape, to be the superposition of bending modes, as explained in the previous section: As the thermal forces bend the rod, the amplitudes of the modes fluctuate and the persistence length can be calculated from the variance of these fluctuations (Appendix 6.3). This is analogous to measuring the stiffness and damping of a spring from its thermal fluctuations (see the figure in Example 4.7). In the case of a bending polymer, each of the modes fluctuates independently and so several independent measurements of the persistence length (and therefore the flexural rigidity) can be obtained from observations on a single polymer (Gittes et al., 1993).

Entropic Elasticity of a Freely Jointed Chain

We complete the discussion of the bending of filaments by considering the so-called **freely jointed chain**, a filament composed of n segments (each of length b), which are completely free to swivel in three dimensions (Figure 6.9). Such a chain behaves as an **entropic spring** and is a good model for rubber-like elasticity. The freely jointed chain model has several important biological

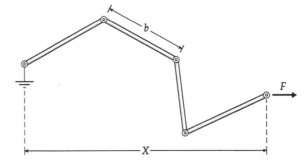

Figure 6.9 A freely jointed chain
Extension of a freely jointed chain under the influence of a force F. The joints between the rods, and between the leftmost rod and the fixed point, are assumed to be completely free to swivel.

applications. It can be used to understand the mechanics of proteins that have segmental flexibility, which means that they contain globular or rigid coiled-coil domains linked by flexible regions. Examples of such proteins include antibodies (Yguerabide et al., 1970) as well as motor proteins (Mendelson et al., 1973). Segmental flexibility provided important early evidence for the domain hypothesis, namely that proteins are composed of discrete functional modules joined by looser, often protease-sensitive, connections. Another application of the freely jointed chain is to the structure (or lack of structure) of unfolded proteins, and to the structure and mechanical properties of DNA (Smith et al., 1992; Bustamante et al., 1994; see Figure 8.6) and flexible proteins like titin (Kellermayer et al., 1997; Rief et al., 1997; Tskhovrebova et al., 1997).

The freely jointed chain has several interesting mechanical properties. Its force-extension curve is measured by determining the force, F, required to extend the chain an average distance $\langle X \rangle$. This can be compared to the theoretical curve

$$\langle X \rangle = nb \cdot L\left(\frac{Fb}{kT}\right) \qquad (6.14)$$

(Appendix 6.3) where $L(Fb/kT)$ is the **Langevin function** plotted in Figure 6.10. The Langevin function increases from zero to one as F increases; consequently, the average extension increases from zero to nb, the length of the taut chain.

The remarkable thing about freely jointed chains is that they have nonzero stiffness, even for very small forces when they are "slack"! This counterintuitive property follows because, for small F, $L(Fb/kT) \cong Fb/3kT$, so for small forces we can write

$$F = \frac{3kT}{nb^2}\langle X \rangle \qquad (6.15)$$

Thus the freely jointed chain is an entropic spring with spring constant equal to $3kT/nb^2$. The resistance to extension arises because extension decreases the disorder and the corresponding decrease in entropy costs free energy. The stiff-

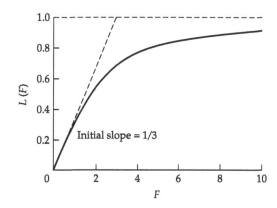

Figure 6.10 The Langevin function
The initial slope is 1/3 and the function asymptotes to 1 as F gets large and the chain becomes taut.

ness increases as the temperature increases, but decreases as the number and length of the segments increase. The freely jointed chain has a number of other properties in common with rubber-like materials: As the temperature increases, the spring constant increases and the chain tends to shorten; and the chain gets stiffer as it is stretched.

> **Example 6.10 The spring constant of a freely jointed chain** The spring constant of a freely jointed chain is large if the segments are short.
>
> 1. The spring constant of a chain made up of just one segment of length 4 nm is ~1 pN/nm. This is interesting because this is similar to the stiffnesses of motor proteins such as myosin and kinesin (Chapters 15, 16). Thus the dimerization domain of kinesin, or the S2 domain of myosin, might act as a freely jointed chain that contributes to the compliance of the motor.
> 2. A random peptide chain comprising 12 residues has a stiffness of ~10 pN/nm when it is "slack." The stiffness is considerably higher as the chain approaches its taut length of 4 nm. Thus random peptide linkers, though disordered, have considerable entropic stiffness.

Worm-Like Chain

A freely jointed chain with many segments ought to behave qualitatively like a very flexible slender rod, a slender rod whose length is much greater than its persistence length. We call such a rod a **worm-like chain**. In both cases, the chains should trace out highly convoluted paths in the absence of external forces (Figure 6.11). For the case of a freely jointed chain with n segments each of length b, the mean-square end-to end distance, $\langle R^2 \rangle$, is equal to nb^2 (Appendix 6.3), whereas for a slender rod in three dimensions, the mean-square end-to-end displacement approaches $2LL_p$ when the length, L, is much greater than the persistence length, L_p (see Equation 6.13). If the total length of the freely

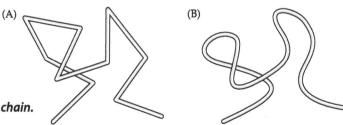

Figure 6.11 (A) A freely jointed chain. (B) A worm-like chain.

jointed chain, nb, is the same as the length of the worm-like chain, L, then it must be that

$$2L_p = b \tag{6.16}$$

This means that in the absence of a force, the worm-like chain is similar to a freely jointed chain whose segments are of length twice the persistence length. The equivalent segment length of a worm-like chain, $b = 2L_p$, is called the **Kuhn length**. There are small differences in the way that worm-like chains and freely jointed chains elongate under the influence of external forces (Bustamante et al., 1994).

Summary

The resistance of a slender rod to bending forces is determined by its flexural rigidity. The higher the flexural rigidity, the greater the resistance to bending. For an isotropic and homogenous rod, the flexural rigidity is proportional to the Young's modulus, E, and the constant of proportionality is called the second moment of inertia, I. I is determined by the shape of the cross-section. If the bending moments on a rod are known, then the beam equation can be used to calculate the resulting shape. If the rod is in a fluid, then the bending will be opposed by drag forces, and the shape will change slowly in response to an applied bending moment. The time course of this shape change can be calculated using a generalized form of the beam equation. These theories can be used to calculate the amplitude and the kinetics of bending and buckling. The predictions accord with the experimental measurements.

Flexible rods are also bent by thermal forces. The resulting thermal fluctuations in shape can be described using the persistence length, the length over which the tangent angle of the rod changes appreciably. The persistence length is proportional to the flexural rigidity. An extreme case of a flexible rod is the freely jointed chain in which short, rigid segments are connected by swivels. Interestingly, such a rod possesses stiffness because the elongation by an applied force decreases the entropy of the chain and therefore increases the free energy. This entropic stiffness is the basis for the stiffness of rubber-like materials.

Problems

6.1 Bacterial chemotaxis   Bacteria such as *E. coli* move up a gradient of attractant molecules by a process of biased diffusion. In the absence of attractant (or repellent), the bacteria alternate between periods of "running," during which time a bacterium moves in a constant direction for about 1 second, and "tumbling," during which time a bacterium stops translating and undergoes rotational diffusion for about 0.1 second. During the tumble, the bacterium's orientation is randomized so the next run is likely to be in a different direction (Berg, 1993).

(a) If the running speed is 30 μm/s and the direction is completely randomized during the tumble, calculate the effective translational "diffusion" coefficient of the bacterium. [Answer: 818 μm²/s, similar to that of a sodium ion.]

(b) Using the rotational drag coefficient for a sphere of radius 1 μm and a viscosity of 0.7 mPa/s appropriate for water at 37°C, calculate the root-mean-square mean change in angle during a tumble. [Answer: ~1 rad ≅ 57 degrees.]

6.2 ATP synthase The $F_1 - F_0$ ATP synthase is a rotary engine that is driven by the flux of protons down their electrochemical gradient (Boyer, 1997). The rotation of the F_1 component has been directly visualized by attaching a fluorescently labeled actin filament to the γ subunit (Noji et al., 1997). It is found that the speed depends on the third power of the length of the actin filament, and that the rotational speed is 0.85 revolutions per second when a 2-μm-long actin filament is attached at its end (Yasuda et al., 1998).

(a) Assuming that the average height of the filament is ~100 nm above the surface, that the diameter of the filament is 3 nm, and that the viscosity is 0.89 mPa/s, calculate the torque that the motor must generate against the viscous drag (see Figure 6.7). [Answer: 38 pN·nm.]

(b) Given that 3 ATP molecules are hydrolyzed per revolution, calculate the work done per ATP hydrolyzed. [Answer: 80 pN·nm, though more recent work suggests that the work done per ATP against a viscous load is only about half this value (Soong et al., 2000).]

6.3 Torsion of slender rods The torsional rigidity, C, of a slender rod of cross-sectional area, A, is defined by

$$\frac{T}{A} = C\theta$$

where T is the torque and θ is the induced torsion.

(a) Show that for a homogenous, isotropic rod with shear modulus G and a circular section, $C = 2GI$, where I is the second moment of the section.

(b) Show that if the section is elliptical with eccentricity e, then $C = 4G \cdot I_h(e + e^{-1})$ where I_h is the second moment of a helical rod, as in Table 6.1. (Landau and Lifshitz, 1986.)

Cytoskeleton

PART II

The cytoskeleton is a protein scaffold found inside most eukaryotic cells. It consists of three major classes of filaments—cable-like actin filaments, pipe-like microtubules, and rope-like intermediate filaments—and a large collection of accessory proteins that bind to the filaments and either crosslink them together or crosslink the filaments to other cellular structures, such as the plasma membrane, membrane-bound organelles, and chromosomes.

The mechanical properties and organization of this scaffold determine, in large part, the morphology and mechanical properties of cells. This relationship is especially true in the case of animal cells, which have no cell walls and whose fluid-like plasma membranes are unable, on their own, to support complex cellular morphologies. In addition to determining the overall structure of cells, the cytoskeleton is also crucial for the internal organization of cells. For example, motor proteins use actin filaments and microtubules as tracks to move vesicles and organelles from one part of the cell to another.

The assembly of the cytoskeleton is one of the outstanding problems in cell biology. The cytoskeletal proteins have molecular dimensions on the order of 10 nm. Yet they build structures, the cytoskeletal filaments, that have cellular dimensions on the order of 1 to 1000 μm. The cytoskeletal proteins are "smart bricks" that somehow know their place within the cell. What information is needed to build a cell, and where is this information stored?

The next five chapters explore the structures of the cytoskeletal filaments, their mechanical properties, and the mechanisms by which they assemble and disassemble as cells change shape and move.

CHAPTER 7

Structures of Cytoskeletal Filaments

Eukaryotic cells contain three major classes of cytoskeletal filaments (Figure 7.1). Actin filaments, also called microfilaments, have a cable-like structure with a diameter of ~6 nm. Intermediate filaments have a rope-like structure composed of several intertwined protein strands; their diameter is ~10 nm. And microtubules have a pipe-like structure with an outer diameter of ~25 nm and an inner diameter of ~18 nm. The properties of the filaments and their constituent proteins are summarized in Table 7.1.

The cytoskeletal filaments were initially identified by electron microscopy; a comprehensive review of the history of these discoveries can be found in Schliwa (1986). Since these initial discoveries, the constituent proteins—actin, intermediate filament proteins, and tubulin—have been chemically purified, and the proteins have been polymerized in the test tube to give filaments that are virtually indistinguishable from those observed in cells. The amino acid sequences of these proteins, as well as many relatives from many different organisms, have been determined. Furthermore, the actin and tubulin proteins have been crystallized and their structures solved at atomic resolution. These atomic structures have then been used to build atomic models of the entire filaments. Considering that actin filaments and microtubules grow to lengths of several tens of microns, these filaments are the largest known atomic structures of any biomolecules. The atomic structure of the coiled-coil protein motif, which forms the backbone of the intermediate filament protein, has also been elucidated, but the arrangement of the intermediate filament proteins within the filament is still uncertain. Detailed information on the biochemical properties of the filament proteins, their associated proteins, as well as some newly discovered cytoskeletal filament systems can be found in Kreis and Vale (1999).

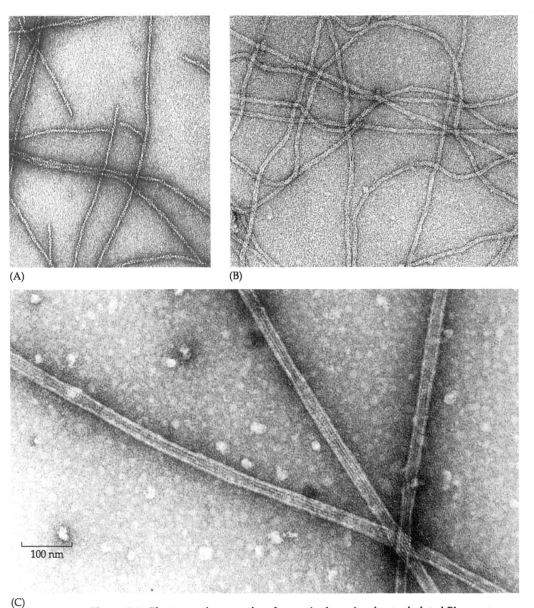

Figure 7.1 Electron micrographs of negatively stained cytoskeletal filaments
(A) Actin filaments have a diameter of ~6 nm and display a characteristic helical repeat of period 36 nm. (B) Intermediate filaments have a diameter of ~10 nm. Despite having a larger diameter than actin filaments, the intermediate filaments have greater curvature, indicating that they are more flexible. (C) Microtubules have a diameter of ~25 nm; the striations within each filament correspond to the individual protofilaments. (A courtesy of Jim Spudich; B courtesy of Peter Steinert; C courtesy of Manfred Schliwa.)

In this chapter I will first present the structures of the major cytoskeletal filaments. Then I will show how the key functional properties of these filaments—their strength, length, straightness, and polarity—are determined by the structures of the subunits and the packing of these subunits into the filaments.

Table 7.1 Properties of the cytoskeletal proteins and filaments

Material	Molecular mass of subunits (kDa)	Number of protofilaments Average	Range[a]	Diameter (nm)	Cross-sectional area (nm^2)[b]
Actin	45	2	2	6	19
Tubulin	50	1–3	9–17	25	200
Intermediate filaments	40–180	8	6–10	~10	~60
Coiled coil	—	2	2	2	1.9

[a] Tubulin (Wade and Chretien, 1993); intermediate filaments (Heins et al., 1993).
[b] Assuming a density of 1.2 nm^3 per 1 kDa of protein, or a volume per amino acid of 0.14 nm^3.

Structures of the Subunits

The building blocks of actin filaments, microtubules, and intermediate filaments are, respectively, the actin monomer, the αβ tubulin heterodimer, and the intermediate filament dimer. The structures of actin and tubulin are shown in Figure 7.2.

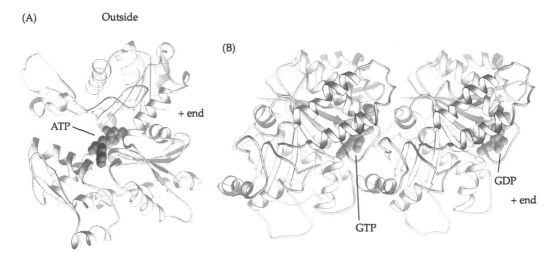

Figure 7.2 Structures of actin and tubulin
(A) The structure of the actin monomer from skeletal muscle, elucidated by X-ray crystallography. It has a deep cleft in which the ATP and a divalent ion are bound (Protein Data Bank ID: 1ATN; Kabsch et al., 1990). View with the outside surface up. (B) The structure of the αβ-tubulin dimer from brain determined by electron microscopy of two-dimensional sheets of tubulin (Protein Data Bank I.D.: 1TUB; Nogales et al., 1998). View looking at the outer surface of the microtubule; note the two surface helices. See Figure 7.4 for a different view. (Courtesy of Ken Downing.)

Actin is a globular protein that contains a deep cleft in which the nucleotide, ATP or ADP, and the Mg^{2+} ion bind (Kabsch et al., 1990). The role of the nucleotide is to regulate polymerization and depolymerization (Chapter 11). The tubulin dimer is also globular; the α- and β-subunits have similar atomic structures and are joined together in a head-to-tail fashion so that the dimer has translational symmetry rather than rotational symmetry. A filament or a molecule has **rotational symmetry** if rotation about an axis allows superposition on the original structure; it has **translational symmetry** if translation in one direction allows superposition on the original structure or on part of it. Each tubulin subunit contains a shallow cleft in which the nucleotide, in this case GTP or GDP, and Mg^{2+} ion bind (Nogales et al., 1998). The nucleotide associated with the α-subunit is trapped in the interface between the two subunits, and consequently it exchanges very slowly with the nucleotide in solution. This is known as the N site (nonexchangeable) and it is usually occupied by GTP. The nucleotide associated with the β-subunit is partially exposed to the solvent and exchanges more readily. This is known as the E site (exchangeable). GTP in the E site is hydrolyzed after the subunit is incorporated into a microtubule, whereupon it is trapped in the interface between two dimers.

By contrast to the globular actin and tubulin subunits, the intermediate filament proteins are highly elongated. Each monomer within the dimer is a long α-helix that slowly winds around its partner to form a coiled coil. Unlike actin and tubulin, intermediate filament proteins do not bind nucleotides; instead, their assembly and disassembly is regulated by phosphorylation. There are no atomic structures of intermediate filament proteins, though the structures of several coiled coils have been solved (see Figure 7.7).

Families of Cytoskeletal Proteins

Actin and tubulin are found in all eukaryotic organisms, where they define two large families of structurally and evolutionarily related proteins. The intermediate filament proteins are even more diverse. The large number of proteins in each of these families leads to some ambiguity about what we actually mean by "actin" or "tubulin." In this section I will define more precisely what I mean by actin, tubulin, and intermediate family proteins.

We usually say that two proteins are the same when they have similar amino acid sequences. One source of protein variation is genetic polymorphism of individuals within one species. Most of the individual organisms within a species contain the same set of genes in their chromosomes. However, the nucleotide sequence of a particular gene may vary from individual to individual, or may differ in the two copies of the gene that are found within the one individual. As a consequence, the encoded proteins may vary in amino acid sequence. Such **variant proteins** are often functionally identical, but sometimes they lead to disease. For example, a single nucleotide in the human keratin *K14* gene causes a single amino acid change of the corresponding K14 intermediate filament protein (arginine to cysteine at position 125); this structural change leads to a skin blistering disorder called epidermolysis bullose simplex (Coulombe et al., 1991).

A second source of protein variation is genetic variation between species. Individuals from closely related species have a similar organization of genes in their chromosomes; the genes that occupy corresponding chromosomal locations are said to be orthologous, and the encoded proteins are said to be **orthologues**. Orthologous proteins have similar sequences. For example, ACTS, the human skeletal muscle actin, has an identical amino acid sequence to the corresponding protein from mice and chickens. Orthologous proteins are thought to perform similar functions in the different organisms; they are sometimes said to be **homologues**.

A third source of protein variation, this time within individuals, is gene duplication and subsequent divergence of the gene sequences. Often, an organism has two or more different genes, located at different positions in their chromosomes, that are very similar to one another and that code for proteins that have sequence identity greater than ~80%. These proteins are called **isoforms**, or **paralogues**. For example, budding yeast has one actin, two α-tubulins, and one β-tubulin, whereas vertebrates have several actins, and several α- and β-tubulins; all these isoforms are encoded by different genes. It is noteworthy that different isoforms found in the one species are often more different in sequence than are the orthologous proteins found in different species.

There are two main reasons why an organism might have two or more isoforms of a protein. The first is that the different genes that encode the different isoforms can be regulated differently. For example, the two α-tubulins from baker's yeast are used at different stages in the life cycle and can substitute for one another provided that they are expressed at sufficient levels. This suggests that the isoforms are functionally identical, despite the 10% difference in amino acid sequence, and that the reason why there are two of them is so that their expression can be differentially regulated (Raff, 1994). Differential regulation might also explain the existence of multiple isoforms in multicellular organisms, where different isoforms are often found in different cell types. The second reason for having different protein isoforms is that the different structures can be used for different functions. For example, the nematode worm *C. elegans* has microtubules with 11, 13, and 15 protofilaments (see below for the definition of protofilament): 11-protofilament microtubules are found in most cells (this is a peculiarity of *C. elegans*), 13-protofilament microtubules are restricted to the doublets of cilia, and 15-protofilament microtubules are found only in the six mechanoreceptive cells (Chalfie and Thomson, 1982). Deletion of the *mec*-7 gene, which encodes one of the β-tubulin isoforms, leads to a specific deletion of the 15-protofilament microtubules and to loss of mechanosensitivity in the worm (Savage et al., 1989). How the MEC-7 tubulin isoform gives rise to the different microtubule structure and how that in turn leads to a different physiological function is not understood. Nevertheless, the tubulin isoforms in the nematode do illustrate how slightly different proteins can be used to build different structures with different functions.

Usually we will refer to variants, orthologues, and even paralogues as being the "same" protein or in the same class of protein. However, when two proteins differ at about half the positions in their amino acid sequences, they are no longer called orthologues or paralogues. Instead they are said to be different proteins. Even so, because this high level of sequence similarity implies that they evolved from a common ancestor, the proteins are said to be **relatives** that belong to the same **family** or **superfamily**. For example, α- and β-tubulin have about 50% amino acid sequence identity and are said to be different proteins, even though this level of amino acid identity ensures that the proteins have similar structures, as is the case (see Figure 7.2B). Indeed, 20% amino acid identity is sufficient to ensure that the protein folds are similar (Creighton, 1993).

Actin

The actin family contains over a dozen classes of proteins. In addition to actin (which forms the microfilaments), there are several classes of **actin-related proteins**, called Arps, that have approximately 50% amino acid sequence identity to conventional actin (Frankel, 1999). Arp1 forms the short, 35-nm-long filament at the core of the dynactin complex, which links dynein to membranes and microtubules (Schafer et al., 1994). Arp2 and Arp3 are components of a protein complex found in *Acanthamoeba* (Machesky et al., 1994) and human platelets (Welch et al., 1997) that nucleates actin filaments during polymerization-driven intracellular motility (see Figures 1.2 and 10.1; Chapter 10). Members of the same actin class from different species have highly conserved amino-acid sequences: For example, ACT-1, a conventional actin from the slime mold *Dictyostelium*, differs from ACTG, a human actin, in only 17 of the 375 amino acids (Sheterline et al., 1995).

Tubulin

The tubulin family contains five classes of proteins. In addition to the α- and β-tubulins which form microtubules, there are three other classes of tubulins: γ-tubulin (Oakley et al., 1990), δ-tubulin (Dutcher and Trabuco, 1998), and ε-tubulin (Chang and Stearns, 2000). γ-Tubulin is localized to the centrosome and has microtubule-nucleating activity (Zheng et al., 1995). δ-Tubulin is localized to the basal body or the centriole and is necessary for basal-body duplication in the green alga *Chlamydomonas reinhardtii*. ε-Tubulin is localized to the pericentriolar material. In most cases, members of the same tubulin class from different species have quite similar amino acid sequences. For example, β-tubulin from the yeast *S. cerevisiae* has 72% amino acid sequence identity to β-tubulin 2 from humans. However, γ-tubulins from yeast and humans have only ~40% sequence identity; nevertheless, they are considered to be in the same class based on similar localization and function.

Intermediate Filament Proteins

The intermediate filament family of proteins is much more structurally diverse than the actin or tubulin families. It includes keratins found in hair, nail, and epithelial cells; vimentin from fibroblasts, and its relatives desmin (from the Z-bands of muscle) and glial fibrillary acidic protein (from glial cells); the neurofilament proteins of neurons; and the lamins, which form the nuclear envelopes of all eukaryotic cells. The common structural unit of the intermediate filament proteins is a ~340 amino acid region which is ~90% α-helical and is predicted to form a coiled coil of length ~43 nm (see Figure 7.8A for the structure of a coiled coil). At the N- and C-terminals of this central rod are globular "head" and "tail" domains of various sizes. Diversity in sequence arises in the head, rod, and tail domains. Sequence comparisons of cytoplasmic intermediate filament proteins indicate that they arose from a mutated lamin early in the evolution of animals (Weber, 1999).

Filament Structures

The structures of actin filaments and microtubules have been solved. The structure of intermediate filaments is still uncertain.

Actin Filament

The lattice structure of the actin filament was deduced by electron microscopy (Figure 7.3A). An actin filament is a one-start, left-handed helix of actin monomers (Moore et al., 1970). The full period of the filament is 72 nm, though the repeat appears as 36 nm under the electron microscope, where the image of the filament is projected onto the plane. One full period contains 26 subunits, which make 12 complete turns—the rise is therefore 2.77 nm per sub-

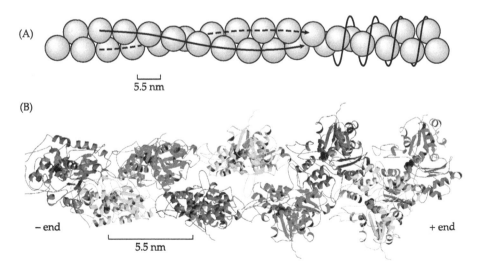

Figure 7.3 Structure of the actin filament
(A) Lattice structure showing the two-stranded (or 2-start) representation on the left and the one-stranded (or 1-start) representation on the right. (B) The atomic model of the actin filament. (From Geeves and Holmes, 1999.)

unit (Squire, 1981). Because the rotation per monomer is large (166°), and because there is extensive monomer–monomer contact between alternate monomers, the actin filament is more appropriately viewed as a two-stranded, right-handed helix, with the strands, called protofilaments, half staggered and wrapping slowing round each other with a repeat period of 72 nm.

The atomic model of the actin filament is shown in Figure 7.3B. The orientation of the individual actin monomers within the lattice was determined by comparing atomic models of the filament with electron microscopic images of filaments (Holmes et al., 1990) and with X-ray diffraction patterns generated by gels of oriented actin filaments (Lorenz et al., 1993). Only one orientation of the monomer within the helically symmetrical filament was found to be consistent with the electron microscope and X-ray data. The atomic model confirms that there are extensive contacts between the two strands, which is necessary if the monomers are to assemble into long polymers (Chapter 9). Because the actin monomers are asymmetrical, the actin filament is polar and its ends are structurally different. One consequence of this **polarity** is that polymerization is faster at one end than the other. The fast-growing end is called the **plus (+) end**, whereas the slow-growing end is called the **minus (–) end**.

Microtubule

The lattice structure of the microtubule, deduced from electron microscopy, is shown in Figure 7.4A. The building block, the $\alpha\beta$ heterodimer, is very stable and can be dissociated only by harsh treatments, such as with the ionic detergent SDS. The dimers associate head-to-tail to form a protofilament. The protofilaments then associate laterally to form a sheet that closes to form the cylindrical tube that is the microtubule. There is a small offset, ~0.92 nm, between dimers of neighboring protofilaments; after 13 protofilaments, the accumulated offset is exactly equal to the length of three monomers, 12 nm ($= 13 \times 0.92$ nm). This ensures that the protofilaments run parallel to the axis of the microtubule, giving rise to the **3-start helical structure** (meaning that they can be thought of as comprising three helices of monomers; see Figure 7.4A) first described by Amos and Klug (1974) and Erickson (1974).

The microtubule, like the actin filament, is polar. This is a consequence of the asymmetry of the dimers and the facts that (1) the dimers associate head-to-tail to make a polar protofilament, and (2) all the protofilaments are parallel. As in the case of actin, the fast-growing end is called the plus end and the slow-growing end is called the minus end. Because GTP, which exchanges only at the terminal β-subunit, binds preferentially to the plus end (Mitchison, 1993), and because antibodies specific for the α-subunit bind only to the minus end (Fan et al., 1996), the polarity of the microtubule is such that the β-subunit is at the plus end (Hirose et al., 1997), as shown in Figure 7.4.

Most cellular microtubules have 13 protofilaments (Tilney et al., 1973) and 3 starts. However, microtubules are highly polymorphic in structure. Microtubules with as few as 8 protofilaments and as many as 19 protofilaments have been observed in vivo and in vitro (see Chretien and Wade, 1991, for references). The number of starts can range from 2 to 4 (Ray et al., 1993; Hirose et al., 1995).

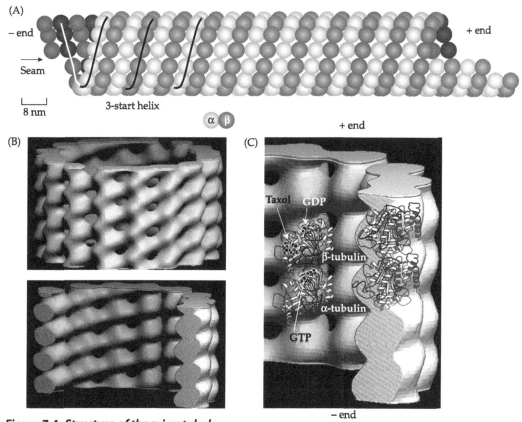

Figure 7.4 Structure of the microtubule
(A) Lattice structure of a 3-start, 13-protofilament microtubule. The solid line traces one of the 3-start helices. (B) Low-resolution structure of the microtubule deduced by cryoelectron microscopy. (C) An atomic model of the microtubule showing the orientation of the dimer in the low-resolution structure. (B and C courtesy of Linda Amos; see also Nogales et al., 1999; Amos, 2000.)

The structural basis for the polymorphism of microtubule structure is **lattice rotation**. A sheet containing more than or fewer than 13 protofilaments can still close despite the misregistration at the join if the whole lattice rotates slightly (Figure 7.5). This requires little or no distortion within the sheet, as can be demonstrated using a sheet of ruled paper: The sheet can be rolled into a closed tube with the ruled lines out of register provided that the sheet is rotated about the axis of the tube. The consequence of lattice rotation is that the protofilaments do not run parallel to the axis of the microtubule, but rather follow shallow helical paths around the surface of the microtubule (Chretien and Wade, 1991). The pitches of the resulting **supertwists** for 12-, 14-, and 15-protofilament microtubules have been measured by electron microscopy to be $+4.5 \pm 0.3$ µm, -5.8 ± 0.3 µm, and -3.3 ± 0.3 µm, where the minus signs indicate left-handed helices (Chretien and Wade, 1991; Ray et al., 1993). These pitches correspond closely to the predictions of the lattice rotation model, +4.1,

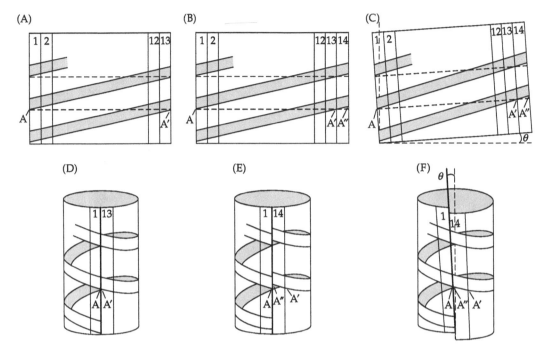

Figure 7.5 Closure of 13- and 14-protofilament microtubules according to the lattice rotation model
Each band corresponds to one 3-start helix of monomers (see Figure 7.4), and it slopes up to the right (when looked at from the inside) due to the offset between neighboring protofilaments. After 13 protofilaments (A), the accumulated offset exactly equals 3 monomer lengths (12 nm), so the 13-protofilament microtubule (D) closes with its protofilaments running parallel to the axis of the microtubule. The accumulated offset after 14 protofilaments is greater than 3 monomers (B), so the 14-mer cannot close (E) unless the lattice rotates (C), giving rise to protofilaments that supertwist about the axis of the microtubule (F). (After Wade and Chretien, 1993.)

–5.6, and –3.2 μm, respectively, where the different signs arise from different directions of lattice rotation. It is remarkable that a minute misregistration, less than 1 nm in the case of a 14-protofilament microtubule, leads to the formation of a structure—the supertwist—with a dimension of several microns. The lattice rotation model also accounts for the 2-start and 4-start microtubules. The model predicts that less rotation is required to close a 16-protofilament microtubule if the lattice rotates clockwise instead of counterclockwise and that the microtubule closes to form a 4-start helix. Likewise, the model predicts that 10-protofilament microtubules should have only 2 starts. These microtubules are indeed observed, and their supertwists closely match the predictions of the lattice rotation model (Ray et al., 1993).

The close correspondence between measured and predicted supertwists indicates that little distortion of the individual subunits takes place. This means that *locally* all the microtubules are identical, independent of the number of protofilaments and starts (just as a small region of a sheet of paper rolled up into a tube is not stressed even if a supertwist is introduced into the tube). Thus we say that microtubules with different numbers of protofilaments have the same quaternary structure, but different **quinary** structures.

Because the building block is a dimer, there are two ways that the neighboring protofilaments of a microtubule can join. Either they are nearly aligned (with only the 0.93 nm offset), or they are staggered by approximately one monomer length. When the protofilaments are nearly aligned, the join is said to be of the A-type, and when they are staggered, the join is said to be of the B-type. The names arise from electron microscopic studies on the doublet microtubules of axonemes (see Figure 8.3B): Amos and Klug (1974) postulated that the complete, 13-protofilament microtubule (called the A-tubule) has all A-type joins and forms what they called an A-lattice; and they postulated that the incomplete, 11-protofilament microtubule (called the B-tubule) has B-type joins and therefore forms what they called a B-lattice (Figure 7.6). Although it is clear that the B-tubule has a B-lattice, whether the A-tubule has an A- or a B-lattice is still controversial (Amos, 1995; Mandelkow et al., 1995). Indeed, there are no confirmed examples of A-lattices. For example, decorating the outer singlet microtubules of cricket sperm with motor domains shows that these microtubules have B-lattices (Hirose et al., 1996). Most microtubules grown in vitro have B-lattices (Mandelkow et al., 1986). Furthermore, decoration of cytoplasmic microtubules with motor proteins shows almost all the joins are of the B-type (Kikkawa et al., 1994). However, a 13-protofilament, 3-start microtubule cannot be made up of only B-type joins. There must be an odd number of A-type joins, because, for all microtubules, the number of A-type joins plus the number of starts must be divisible by two (this is a consequence

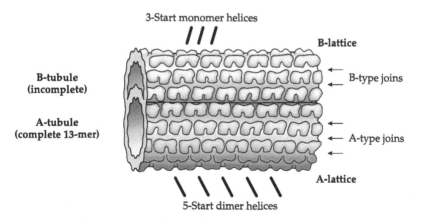

Figure 7.6 The Amos and Klug model of the doublet microtubule
According to the model, the complete, 13-protofilament A-tubule has the A-lattice, while the incomplete, 11-protofilament B-tubule has the B-lattice. However, the model is not proven and it is possible that the A-tubule actually has a B-lattice, in which case it would not be helically summetrical. (After Amos and Klug, 1974.)

of the building block being a dimer). In particular, a 13-protofilament, 3-start microtubule cannot be helically symmetrical. (A filament is **helically symmetrical** if, following a rotation about its axis, an axial translation allows superposition on the original structure.) Consistent with these geometric considerations, seams corresponding to A-type joins have been observed by deep-etch electron microscopy of cytoplasmic microtubules (Kikkawa et al., 1994). Evidently the structures of the α- and β-subunits must be similar enough that they can form both types of joins; the B-type join must either be more stable or form more quickly than the A-type join.

The issues of lattice type and protofilament number are not esoteric ones, but rather cut to the crucial question of how microtubules grow and shrink. If a microtubule were helically symmetrical, then one might expect that polymerization would also be helical, with growth occurring on each of the 5-start helices of dimers (e.g., Bayley et al., 1990). Instead, the lack of helical symmetry of 13-start B-lattice microtubules, the polymorphism in the number of starts and protofilaments, and the finding that both the protofilament and start numbers can change while a microtubule is growing (Chretien et al., 1992), all suggest that growth occurs via sheets that subsequently close to form the microtubule, like a sheet of paper rolling up to form a tube (as depicted in Figure 7.5) (Erickson, 1974). The visualization of sheets at the growing ends of rapidly growing microtubules by cryoEM (Chretien et al., 1995; Arnal et al., 2000) lends further support to the sheet-closure model. We will return to the structure of growing microtubules in Chapter 11.

Coiled Coils

The building block of intermediate filaments is the coiled coil, a parallel dimer of α-helices. The coiled coil is a widespread protein domain found in several other classes of proteins including motor proteins, tropomyosins, viral fusion proteins, and transcription factors (where the coiled coil is called a leucine zipper). There are a number of computer programs that estimate the likelihood that a novel sequence will form a coiled coil based on the sequences of known coiled-coil proteins (Berger et al., 1995; Lupas, 1996).

Coiled coils are composed of α-helices that are hydrophobic on one side and hydrophilic on the other, and whose hydrophobic surfaces pack together to avoid exposure to the solvent (Figure 7.7). The snug, "knobs-in-holes" packing of two parallel α-helices to give a coiled coil was postulated by Crick (1953) and Pauling and Corey (1953), and both groups correctly predicted that keratin is a coiled-coil protein. Indeed, Crick even predicted the leucine zipper motif: "One might expect [valine, leucine, and isoleucine] to make up about two-sevenths of the residues" of coiled coils (Crick, 1953, p. 696). The atomic structures of coiled coils with two, three, and four strands have been solved both in proteins (the catabolite-activating protein was the first; Weber and Steitz, 1987) and synthetic peptides (GCN4 leucine zipper; O'Shea et al., 1991).

The twist of a coiled coil arises because the periodicity of the hydrophobicity does not exactly match the pitch of the α-helix. Coiled-coil proteins have a char-

Figure 7.7 The structure of the coiled-coil dimer
(A) Packing of the hydrophobic amino acids in the *a* and *d* positions of the heptad repeats. The two α-helices are parallel and in register when they come together to form the coiled coil. (B) Ribbon diagram of a coiled coil in which two α-helices wrap slowly around each other. (After Lupas, 1996; B courtesy of Andrei Lupas.)

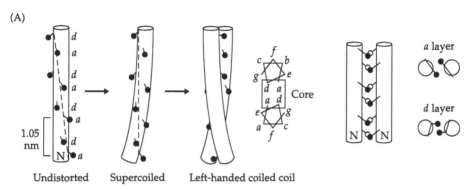

acteristic heptad repeat of amino acids in which hydrophobic residues usually occupy the *a* and *d* positions, and hydrophilic residues usually occupy the other five positions. In leucine zippers, leucine often resides in the *a* position. Because the hydrophobic period ($7/2 = 3.5$ amino acids) differs from the α-helix period (3.64 ± 0.09 amino acids in globular proteins [Phillips, 1992]) the two helices coil gently around each other. The pitch is

$$P \approx \frac{\delta}{1/3.64 - 1/3.5} = -14 \text{ nm}$$

using $\delta = +0.15$ nm for the axial rise per amino acid in the α-helix. The negative sign means that the coiled coil is left-handed, which is a consequence of the α-helix being right-handed. This pitch agrees closely with that measured from the known atomic structures of coiled coils (Phillips, 1992). The agreement is remarkable: It shows that the geometry of the hydrogen bond in the backbone of the polypeptide sets the geometry of the coiled coil (the hydrophobic period is constrained to be an integer). The disparity of only 0.021 nm between the hydrophobic and amino acid periodicities (($3.64 - 3.5$) × 0.15 nm) gives rise to a structure with a periodicity of 14 nm, nearly one thousand times greater. Thus, as in the case of supertwisted microtubules, small disparities give rise to large pitches. For the aficionados, this large spatial period corresponds to the low-frequency beating of two closely matched high frequencies.

Intermediate Filaments

The structural organization of the coiled-coil dimers in the intermediate filament is not known for sure, and it is possible that the arrangement differs between classes of filaments. One model is that homodimers form antiparallel

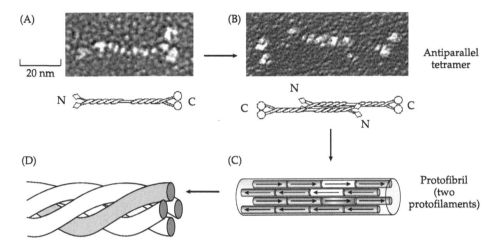

Figure 7.8 Hypothetical structure of an intermediate filament
Two coiled-coil dimers (one shown in A) form an antiparallel tetramer (B), which is the building block of the protofilament (From Geisler et al., 1998). Two protofilaments form a protofibril (C), and four protofibrils wrap around each other to form the intermediate filament protein (D). (A, B from Geisler et al., 1998. The model is after Heins et al. [1993], Parry and Steinert [1995], and Geisler et al. [1998]).

tetramers, which associate head-to-tail to form a protofilament. A pair of protofilaments forms a protofibril, and 4 protofibrils wrap around each other to form a 10 nm filament with an organization like a piece of string (Figure 7.8). The model predicts 16 coiled coils in each section, in agreement with a mass per unit length of 47 kDa/nm for neurofilaments composed of the NF-L protein (Heins et al., 1993). The average cross-sectional area for these particular intermediate filaments is ~60 nm^2; in general, the cross-sectional area will depend on the sizes of the head and tail domains, which vary considerably between different intermediate filament proteins.

Summary: Structural Basis for the Length, Strength, Straightness, and Polarity of Filaments

The length, strength, straightness, and polarity of a cytoskeletal filament are consequences of the packing of the subunits into the polymers.

Length

All three filaments are multistranded, and high resolution electron micrographs show that there are extensive contacts between the protofilaments in actin filaments and microtubules. As will be shown in Chapter 9, intrastrand contact explains why filaments are so long, and why polymerization and depolymerization occur by monomer addition and subtraction at the ends rather than by annealing and breakage of filaments.

Strength

The actin and tubulin subunits fit snugly into their respective polymers such that the protein–protein interfaces in the filaments are well packed. As a result, the filaments are rigid, with Young's moduli that approach the theoretical limit for materials held together by van der Waals bonds (Chapters 3 and 8). The large, tubular cross-section of a microtubule suggests that this filament will have a much greater resistance to bending than the others. Measurements described in the next chapter show that this is indeed the case. The tubular arrangement also makes microtubules rather tolerant of lattice defects; an analogy can be made to a steel tube into which holes can be drilled without seriously weakening the structure.

Straightness

The filaments are straight because they are helically symmetrical (actin filaments, probably intermediate filaments) or nearly so (microtubules). Without symmetry, filaments made of helical protofilaments would also be helical. However, with helical symmetry any intrinsic curvature in the individual protofilaments (i.e., any tendency to adopt a helical shape) is canceled out. Although microtubules are not helically symmetrical due to the existence of seams, the α- and β-subunits evidently have similar enough structures that the A-type and B-type joins have similar mechanical properties and thus the filament can be regarded as quasi-symmetrical. Intermediate filaments may be helically symmetrical, though their structure is still uncertain.

Polarity

Actin filaments and microtubules are polar because the subunits are arranged head-to-tail in the protofilaments, and the protofilaments are parallel (rather than antiparallel). Polarity has two important consequences. First, polymerization is expected to be kinetically faster at one end than the other: The end with the faster kinetics is called the plus end, whereas the other end is called the minus end. Second, the surfaces of the filaments are asymmetrical. An analogy can be made to the coat of a short-haired cat which is smooth to stroke in one direction but rough in the other. The asymmetry of the actin filament is revealed by decoration with myosin head fragments in the absence of ATP (see Figure 12.8). The "arrows" point away from the "barbed" plus end toward the "pointed" minus end. Asymmetry is essential for the unidirectional movement of motor proteins along actin filaments and microtubules.

Intermediate filaments may not be polar. Indeed, the homopolymeric intermediate filaments such as desmin and vimentin are almost certainly not polar because the antiparallel arrangement of the coiled-coil dimers in the tetramer produces a symmetrical protofilament, which in turn will lead to a symmetrical, nonpolar filament. By the same argument, the heteromeric keratin intermediate filaments are also likely to be nonpolar. The acidic and basic chains form a heterodimeric coiled coil but, like vimentin, the two coiled coils in the tetramer are arranged in an antiparallel fashion. On the other hand, neurofilaments might be polar. They are composed of three classes of neurofilament proteins, which could form polar tetramers (Amos and Amos, 1991). No intermediate filament motors are known, and this may be due to their lack of polarity.

In summary, the structures of the cytoskeletal polymers reveal filaments that resemble cables (microfilaments), pipes (microtubules), and rope (intermediate filaments). In the next chapter we consider the mechanical properties of these filaments and find that these analogies are quite appropriate.

CHAPTER

8

Mechanics of the Cytoskeleton

The mechanical properties of the cytoskeleton and its organization determine, in large part, the morphology and mechanical properties of cells. This is especially true in the case of animal cells, which have no cell walls and whose fluid-like plasma membranes are unable, on their own, to support complex cellular morphologies. Direct evidence that the cytoskeleton is necessary for the establishment and maintenance of cell morphology comes from experiments in which the cytoskeletal filaments are disrupted pharmacologically or genetically. For example, when actin filaments depolymerize following exposure to the drug cytochalasin B, the cleavage furrow of a dividing cell retracts (Schroeder, 1969), and the axon of a cultured neuron ceases to elongate as the growth cone rounds up (Wessells et al., 1971). When microtubules in the marginal zone of a platelet are depolymerized by the drug colchicine, these flattened, discoid cells become spherical (White and Rao, 1998). When mutated keratin intermediate-filament genes are introduced into mice, the morphology and mechanical integrity of skin epidermal cells become grossly altered, leading to a blistering skin disorder (Vassar et al., 1991).

In this chapter, I show that in a number of cases the cytoskeleton is also sufficient to account for the mechanical properties of cells. Indeed, the mechanical properties of intact muscle cells, sensory hair cells, and sperm can all be quantitatively accounted for by the mechanical properties of the constituent cytoskeletal filaments. Thus knowledge of the mechanics of purified cytoskeletal filaments, together with their location within the cytoplasm and the geometry of their crosslinking, should provide a solid base for predicting the mechanical properties of cells.

Study of the mechanical properties of the cytoskeleton also provides a unique insight into the mechanical properties of proteins in general. Because of their large size, the cytoskeletal polymers are amenable to direct, single-molecule mechanical manipulation. Such experiments provide information about the mechanical properties of the individual protein subunits that make up the polymers, information that is not readily obtained from small proteins in isolation. These studies give support to the notion that proteins, despite their complex atomic structures, can be thought of as having material properties such as elasticity.

Rigidity of Filaments in Vivo

The mechanical properties of some specialized cells, such as muscle cells and sperm, are determined by the mechanical properties of a single type of cytoskeletal filament. Measurements on such specialized cells (and their organelles) have provided estimates of the rigidity of actin filaments, microtubules, and intermediate filaments in the context of the living cell.

Longitudinal Stiffness of Actin in Muscle

The highly ordered structure of skeletal muscle fibers has made it possible to estimate the stiffness of actin filaments. Skeletal muscle fibers have a banded or striated appearance under the light microscope due to the regular repeating of the structural unit, the sarcomere. The sarcomere is composed of (1) interdigitating thin filaments, primarily made up of actin with a smaller contribution from tropomyosin and other proteins, and (2) thick filaments, which contain myosin (see Figure 1.1). When a muscle fiber is depleted of ATP, such as occurs after death, it goes into rigor mortis; but even in this relatively rigid condition, the muscle still has significant compliance (compared to much more rigid materials such as wood or steel). This compliance could arise from several structures—the Z-lines that hold the sarcomeres together, the overlap region where the myosin heads protruding from the thick filaments bind to the actin in the thin filaments, or the regions that contain only thin filaments or thick filaments.

A clever experiment by Higuchi et al. (1995) shows that about half of the compliance of muscle arises from the thin filament (Figure 8.1). These workers compared the stiffness of a muscle fiber at two different sarcomere lengths, 1.8 μm and 2.4 μm. These sarcomere lengths were chosen so that the only difference was that the region containing only actin filaments was increased in length by 0.6 μm in the longer muscle, with the other regions being unchanged in length in the two conditions. In this way the compliance contributed by the Z-lines and the thick filaments should be the same in both cases. They found that the shorter muscle was significantly less compliant than the longer one. From these measurements, they estimated that a 1-μm-long thin filament has a longitudinal stiffness of 53 pN/nm. Taking into account the cross-sectional area of the actin and tropomyosin, the major components of the thin filament, this stiffness corresponds to a Young's modulus of 2.3 GPa (Table 8.1).

The significant compliance of the thin filaments is confirmed by X-ray studies on intact, active muscle. When the muscle is shortened by 6 nm per half sar-

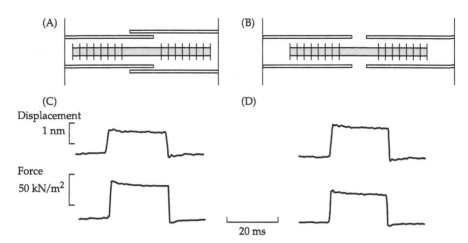

Figure 8.1 Measurement of the stiffness of the thin filament in rabbit skeletal muscle A rigor muscle is held at two different sarcomere lengths, 1.8 μm (A) and 2.4 μm (B). The shorter muscle is stiffer: An increase in tension of 50 kPa in the shorter muscle increases the length by only 1.2 nm per sarcomere (C), whereas a somewhat smaller increase in tension in the longer muscle increases the length considerably more (the same 50 kPa increase in tension would increase the length in the longer muscle by 1.8 nm). Assuming that the additional compliance of the longer muscle is due to the additional 0.6 μm length of thin filament in the nonoverlap region of each sarcomere (2.4 μm–1.8 μm), it is calculated that the stiffness of a 1-μm-long thin filament is 53 pN/nm. The calculation follows by noting that a muscle tension of 50 kPa corresponds to 53 pN per thin filament (there are 0.95×10^{15} m^{-2} thin filaments per unit cross-sectional area), and the change in the tension in the 0.6 μm of excess thin filament in the longer muscle increases its length by 0.1% (1.8 nm–1.2 nm). Other proteins such as titin contribute negligible stiffness at these sarcomere lengths. (After Higuchi et al., 1995.)

Table 8.1 Young's moduli (GPa) of actin and thin-filament proteins

Type of measurement	Thin filament	Actin-tropomyosin	Actin alone
Longitudinal stiffness	2.3[a]	2.8[b]	2.3[b]
Flexural rigidity	—	2.0[c]	1.3[c], 2.6[d]
Torsional rigidity	—	—	1.5[e]

[a]Skeletal muscle in rigor (no ATP or Ca^{2+}). The Young's modulus is calculated using the stiffness of 53 pN/nm per μm length (Higuchi et al., 1995) and a cross-sectional area of 23 nm^2 (19 nm^2 for actin plus 2×2 nm^2 for tropomyosin). The contribution of other thin-filament proteins is ignored.
[b]Actin filaments stabilized by the drug phalloidin. The stiffnesses of 44 ± 5 and 65 ± 6 pN/nm per μm length for filaments without and with tropomyosin (Kojima et al., 1994) were converted to Young's moduli using the cross-sectional areas from [a].
[c]Unstabilized actin filaments. Persistence lengths of 20 μm and 9 μm for filaments with and without tropomyosin (Isambert et al., 1995).
[d]Phalloidin increases the persistence length by a factor of two (Gittes et al., 1993).
[e]Phalloidin-stabilized filaments in Mg^{2+}. The Young's modulus was calculated from the torsional rigidity (Yasuda et al., 1996) assuming that the filaments are homogenous and incompressible (see Problems 3.6 and 6.3).

comere to reduce its tension to zero, the positions of the actin layer lines change, indicating that the average length of the ~1-μm-long actin filaments has decreased by ~3 nm (Huxley et al., 1994; Wakabayashi et al., 1994). Given that each actin filament supports a tension of about 200 pN in an active muscle (see Figure 16.5), this length decrease indicates a stiffness of about 70pN/nm, consistent with the measurement of Higuchi et al. (1995). Thus the mechanical and X-ray studies both show that the myosin heads, which form the crossbridges in the overlap region, can contribute no more than one-half (and probably only about one-third) of the total compliance of the sarcomere. These results contradict earlier mechanical experiments that had incorrectly shown that the thin filament was nearly inextensible and that the crossbridges contributed at least 80% of the compliance of muscle fibers (Ford et al., 1981). The considerable compliance of the thin filament means that a large body of mechanical experiments on muscle must be reinterpreted. We will return to this important point in Chapter 16 when we consider the mechanical properties of motors.

Bending Stiffness of Actin in Stereocilia

The highly regular structure of the hair bundle, the mechanoreceptive organelle of sensory hair cells of the vertebrate inner ear, has made it possible to estimate the flexural rigidity of actin. A hair bundle consists of some 20 to 300 stereocilia, each of which is composed of a cylindrical core of crosslinked actin filaments ensheathed by and connected to a sleeve of plasma membrane (Figure 8.2A,B). Direct microscopic observation of hair cells from frog sacculus, an organ whose function is to detect low-frequency, groundborne vibrations (Lewis et al., 1985), reveals that when a force is applied to the tip of a hair bundle, the rigid stereocilia pivot about their basal insertion points, and adjacent stereocilia shear with respect to each other (Figure 8.2 C,D). Consistent with the stereocilia pivoting about their bases, the stiffness of a hair bundle is inversely proportional to the square of the height at which the force is applied (Crawford and Fettiplace, 1985; Howard and Ashmore, 1986; Appendix 8.1). Also consistent with shearing between adjacent stereocilia is the finding that the stiffness is proportional to the number of stereocilia (Howard and Ashmore, 1986). The frog saccular hair bundles also contain a true cilium (called the kinocilium) composed of microtubules; however, this does not contribute significantly to the mechanical properties of the bundle (Howard and Ashmore, 1986).

When a force is applied at a height of $L = 6.8$ μm, as shown in Figure 8.2C, the stiffness attributed to the stereocilial pivots is $\kappa = 600$ μN/m (Howard and Hudspeth, 1987b). Because the actin filaments in the stereocilia are crosslinked (Tilney et al., 1980), we expect the stiffness of a hair bundle to be

$$\kappa = N \frac{nEI}{lL^2} \frac{a^2}{r^2} \tag{8.1}$$

(Appendix 8.2), where $N = 50$ is the number of stereocilia, $n = 30$ is the number of actin filaments in the basal region of each stereocilium, $a = 50$ nm is the

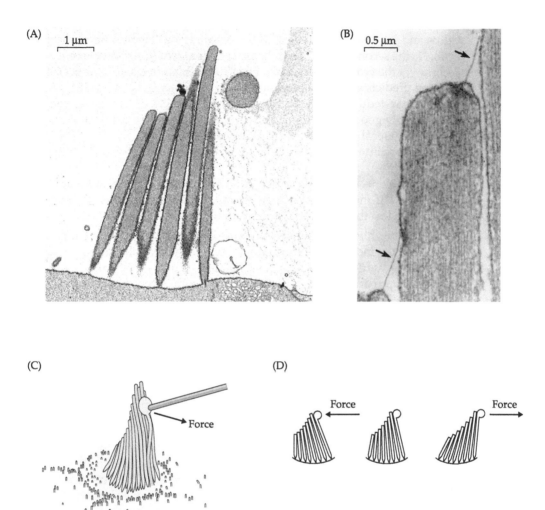

Figure 8.2 The sensory hair bundle
(A) Electron micrograph of a hair bundle showing the stereocilia, which taper at their bases where they insert into the apical surface of the hair cell. (B) Additional magnification shows that each stereocilium is composed of 6-nm-diameter filaments that have been identified as actin filaments based on antibody staining and other criteria. The arrows point to the "tip links," fine filaments of unknown composition that run from the tip of each stereocilium to the side of its tallest neighbor. These tip links are thought to pull on and open ion channels that are located at one or both of the ends where the tip link inserts into the membrane (Example 5.1). (C) Illustration of a bullfrog hair bundle being deflected by an attached glass fiber. (D) Applied forces cause the stereocilia to pivot about their tapered bases and shear with respect to their neighbors; the shear will tense the tip links, and this is thought to be the proximate stimulus that opens the mechanically sensitive ion channels.
(A,B from Jacobs and Hudspeth, 1990; C after Howard and Hudspeth, 1987b; D after Howard et al., 1988.)

radius and $l = 1$ μm is the length of the basal region (Jacobs and Hudspeth, 1990). I is the second moment of a single actin filament ($= A^2/4\pi$, where $A = \pi r^2 = 19$ nm^2 is the cross-sectional area of the actin filament, see Table 6.1). This equation predicts a Young's modulus for actin filaments, E, of 1.6 GPa, nearly equal to that deduced from the stiffness of muscle (2.3 GPa). The agreement is remarkably good considering that the deduced rigidity depends critically on the mechanical model of the hair bundle.

The stereocilium illustrates a number of interesting mechanical points.

1. Actin filaments must be tightly crosslinked together in order to resist bending forces. If there were no crosslinking, the stiffness of the hair bundle would be less than 1% that in the crosslinked case (Appendix 8.2).

2. Even a large, highly crosslinked structure like a hair bundle is still quite soft: The stiffness measured at the tip of the bundle, where physiological forces act during auditory or vestibular stimulation, is only ~1 mN/m = 1 pN/nm. This is approximately equal to the stiffness of the head domain of a single motor molecule (Chapter 16). Considering that the hair bundle is approximately 1000 times larger than a motor domain, ~10 μm compared to ~10 nm, it is clear that stiffness is not scaling with linear dimension as expected for a solid structure (we expect $\kappa = EA/L \propto L$, Chapter 3).

3. Owing to its large size, the motion of the frog saccular hair bundle is expected to be highly overdamped, with a relaxation time on the order of $\gamma/\kappa = 3\pi\eta L/\kappa \cong 100$ μs. This accords with experimental measurements showing that the motion of these hair bundles is overdamped, with a relaxation time constant of ~300 μs (Howard and Hudspeth, 1987a; Denk et al., 1989), consistent with the arguments made in Chapter 3 that the motion of the cytoskeleton and cells are expected to be overdamped. The relaxation time is sufficiently short so as not to impede the physiological function of these cells, which is to detect low-frequency vibrations (<100 Hz). But what about hair cells tuned to high sound frequencies: Could their hair bundles follow vibrations at auditory frequencies? In Appendix 8.3 I show that the very short hair bundles found in the high-frequency regions of mammalian cochleas might be underdamped and capable of mechanical resonance. This may be an exception to the rule that cell motions are overdamped; alternatively, it is possible that the mechanical resonances measured in the ear are due to active processes within hair cells (Camelet et al., 2000; Eguiluz et al., 2000).

Bending Stiffness of Microtubules in Sperm

The flexural rigidity of microtubules has been inferred from the bending stiffness of sperm. The mechanical arrangement is shown in Figure 8.3A. The stiffness depends on the physiological state of the sperm. A sperm is much more rigid when it is in rigor (in the absence of ATP) than when it is relaxed by the addition of ADP and vanadate or in the presence of CO_2 (Table 8.2). Presumably, the higher rigidity of sperm in rigor is due to crosslinking between micro-

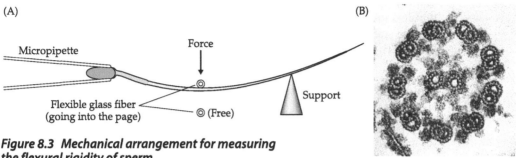

Figure 8.3 Mechanical arrangement for measuring the flexural rigidity of sperm
(A) Apparatus for measuring the bending stiffness of a sperm. While the head and tail are held fixed, the tip of a flexible glass fiber, whose axis is orthogonal to that of the sperm, is positioned against the middle section while the base of the fiber is moved downwards. This causes the sperm to bend. Withdrawing the fiber slightly allows it to slip past the sperm and assume its unflexed, free position. The force is calculated from the flexure and calibrated stiffness of the fiber according to Figure 15.3. (B) Electron micrograph showing a cross-section through a sperm; nine doublet microtubules (whose detailed structure is modeled in Figure 7.6) surround a pair of single microtubules. Protruding from the complete tubule in each doublet are fuzzy projections toward the incomplete tubule of the adjacent doublet; these projections correspond to the dynein motor proteins that power the sliding of adjacent doublets and, as a consequence, the serpentine motion of the sperm. (A after Okuno and Hiramoto, 1979; B from Tilney et al., 1973.)

Table 8.2 Flexural rigidity of sperm and microtubules

Organism	Flexural rigidity (sperm) (10^{-21} N·m^2)	Flexural rigidity (microtubule) (10^{-24} N·m^2)	Young's modulus[a] (microtubule) (GPa)
Sea urchin sperm[b]			
Rigor	10.9 ± 1.0	—	—
ATP	1.9 ± 0.2	—	—
ADP + V_i	0.9 ± 0.3	30 ± 5[c]	2.7
CO_2	0.7 ± 0.1	24 ± 3[c]	2.1
Bovine brain microtubule[d]	—	26 ± 2	1.9

[a]Using Table 6.1, assuming a protofilament cross-sectional area $A = 15$ nm^2, and a contact radius $R = nz/2\pi$, where n is the number of protofilaments (13 in vivo and 14 in vitro) and $z = 5.16$ nm is the width of the protofilament (Chretien and Wade, 1991).
[b]From the setup illustrated in Figure 8.3. Measurements were performed on demembranated sperm, except in the case of CO_2 treatment, in which the sperm was intact; 20°C (Okuno and Hiramoto, 1979; Okuno, 1980).
[c]Derived from the whole-sperm data assuming that there are no crosslinks between the microtubules of the most relaxed sperm. The rigidity of the 9 + 2 sperm (see Figure 8.3) is expected to be ~30 times that of a single microtubule because there are 9 doublets and 2 singlet microtubules per axoneme. (The second moment of inertia of each doublet microtubule is expected to be $(23/13)^2 = $ ~3 times greater than that of a singlet microtubule, because the doublet has 23 protofilaments to the singlet's 13, see Table 6.1.) The error corresponds to a calibration uncertainty of ± 30%, which is interpreted as two standard errors.
[d]37°C (Mickey and Howard, 1995).

tubule doublets by dynein crossbridges (Figure 8.3B). Assuming that in the most relaxed sperm there is no crosslinking between doublet microtubules, then the flexural rigidity of a single microtubule is 20 to 30×10^{-24} N·m^2 (see Table 8.2).

Rigidity of Keratin-Containing Materials

Because many materials composed of intermediate filament proteins are of commercial interest (e.g., hair and wool), the mechanical properties of intermediate filaments have been studied extensively (e.g., Fraser and Macrae, 1980). Hair and wool, nails and hooves, and feathers, scales, quills, and stratum corneum (the outer layer of skin) are composed primarily of keratins that are embedded in a matrix of disulfide-linked proteins (Alexander and Earland, 1950). Examples of the elastic moduli of these proteinaceous materials are found in Table 8.3. A number of interesting points emerge.

1. Dry hair is mechanically isotropic. The Young's moduli of horsehair in the longitudinal and transverse directions are equal (within a factor of two), and the shear modulus, G, of wool is approximately ⅓ the Young's modulus, consistent with an isotropic, incompressible material (i.e., Poisson ratio equal to 0.5, see Problem 3.6). Thus the dehydrated matrix that crosslinks the keratins is able to resist relative sliding in the transverse direction.

2. The keratin-containing materials plasticize when hydrated. However, whereas the longitudinal Young's modulus decreases only about 3-fold when hair is fully hydrated, the transverse Young's modulus decreases about 30-fold. Likewise, the shear modulus of wool decreases 16-fold when hydrated. The molecular explanation is that hydration of the intermediate filament proteins has little effect on their longitudinal stiffness, but hydration softens the connecting matrix, allowing the intermediate filament fibers to shear and slide past one another (Fraser and Macrae, 1980). The resulting limpness is the hallmark of a bad hair day!

Table 8.3 Young's moduli (E) and shear moduli (G) of keratin-containing materials

Material	Type	Direction	Modulus (GPa) (dry)	Modulus (GPa) (hydrated)
Horsehair	E	Longitudinal	6.8	2.4
Horsehair	E	Transverse	3.5	0.1
Sheepswool	E	Longitudinal	5.6	2.0
Sheepswool	G	Torsional	1.8	0.11
Nail	E	Longitudinal	2.3	1.5
Skin	E	Longitudinal	0.19	0.13

Source: Fraser and Macrae, 1980.

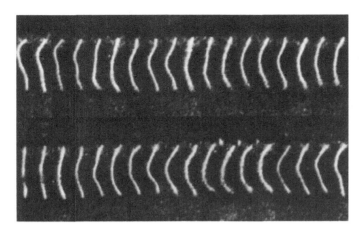

Figure 8.4 Thermal fluctuations of an actin filament
The filament, 11.5-μm-long, was decorated with tropomyosin and viewed by darkfield microscopy. To enhance the brightness, the filament was also decorated with myosin HMM. The sequential micrographs were taken at 1/12 s intervals. (From Nagashima and Asakura, 1980.)

Rigidity of Filaments in Vitro

In general, the rigidities of purified cytoskeletal filaments measured in vitro are in excellent agreement with those values measured in vivo.

Actin Filaments

The rigidity of actin filaments, measured using several different techniques and geometries, is summarized in Table 12.1. The flexural rigidity was first measured from observations of thermal fluctuations in curvature of actin filaments by Nagashima and Asakura (1980) (Figure 8.4). These were the first "real-time" single-molecule mechanical measurements, and they came four years after the first electrical measurements from single ion channels (Neher and Sakmann, 1976). The measured persistence length of 12 μm corresponds to a Young's modulus of 1.7 GPa (see Equation 6.11). Subsequently, the longitudinal stiffness of actin filaments has been measured by micromanipulation of individual filaments using glass fibers: A 1 μm length of actin filament stabilized by phalloidin has a stiffness of 44 pN/nm, corresponding to a Young's modulus of 2.3 GPa (Kojima et al., 1994). More recently, the torsional rigidity of individual actin filaments has been deduced from thermally driven torsional fluctuations (Yasuda et al., 1996). The torsional rigidity of 28×10^{-27} N·m² corresponds to a shear modulus of 0.5 GPa (see Problem 6.5): for an isotropic material with Poisson ratio 0.5, this corresponds to a Young's modulus of 1.5 GPa, similar to the Young's modulus of 2.1 GPa estimated from the flexural rigidity of the same phalloidin-stabilized filaments.

The rigidity of the thin filament is well accounted for by the combined rigidities of actin and tropomyosin, the two principle components. The thin filament is composed of two molecules of the coiled-coil protein tropomyosin wrapped around an actin filament (plus some less abundant proteins). Tropomyosin increases the mass of the filament (and cross-sectional area) by ~22%. If it has the same Young's modulus as actin (as expected, see the section on coiled coils below) and has no effect on the rigidity of actin, it should increase the longitudinal stiffness by 22% and the flexural rigidity by ~50% (the flexural rigidity depends on the square of the cross-sectional area, see Table 6.1). To a first approximation, experiments bear this out (Kojima et al., 1994, Isambert et al., 1995). This leads to a close agreement between the Young's modulus of the thin-filament protein calculated from the measurements on muscle (2.3 GPa) and that calculated from the measurements on reconstituted actin-tropomyosin filaments (2.8 Gpa by Kojima et al., 1994; 2.0 GPa by Isambert et al., 1995).

Microtubules

The rigidity of microtubules polymerized from purified brain tubulin agrees with rigidity estimated from measurements on sperm (see Table 8.2). The most accurate in vitro measurements have been made by observing the very small thermal fluctuations in curvature of long microtubules (Gittes et al., 1993; Figure 8.5). The flexural rigidity of "naked" microtubules, those made from pure tubulin, at 37°C is $26 \pm 2 \times 10^{-24}$ N·m², corresponding to a Young's modulus of 1.9 GPa (see Table 8.2).

Because the rigidity of naked, purified microtubules is sufficiently high to explain the rigidity of microtubules in sperm, there is no physiological require-

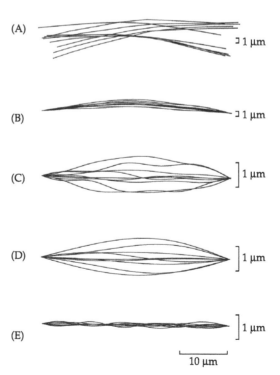

Figure 8.5 Measurement of the flexural rigidity of a microtubule from analysis of its thermal fluctuations in curvature
(A) Ten digitized images of a microtubule at 10 s intervals diffusing laterally and rotationally in a shallow experimental chamber. The 2–3 μm depth of the chamber prevents axial rotation of the intrinsically curved microtubules, but does not interfere with the microtubules' other degrees of freedom (e.g., the lateral diffusion shows that the microtubule is free from the surface). The ends of the images are superimposed (B), the mean curvature is subtracted, and the horizontal scale expanded by a factor of three to reveal the fluctuations in shape (C). The filtered image shows that the fluctuation in the first two modes (D) is much greater than the experimental noise (E), which is the difference between the raw image (C) and the sum of the first two modes (D). (From Mickey and Howard, 1995.)

ment for an agent or additional proteins to make microtubules more rigid in vivo. Furthermore, microtubule-associated proteins (MAPs) add only a small additional mass to the microtubule (~10%; Cleveland et al., 1977; Kurz and Williams, 1995); thus even if the MAPs were as rigid as the tubulin protein and were securely buttressed to the microtubule surface, they would be expected to increase the longitudinal stiffness by only 10% and the flexural rigidity by ~20%. Consistent with this argument, neither tau (Mickey and Howard, 1995) nor bovine brain MAPs (Kurz and Williams, 1995) have an appreciable effect on microtubule rigidity. However, a contradictory report was subsequently published: Felgner et al. (1996) reported that MAPs increased the flexural rigidity of naked microtubules approximately 4-fold to 16×10^{-24} N·m². Multiplying these values by a factor of 2, to correct an error in calculating the hydrodynamic drag (see Appendix 6.2), brings the higher value into agreement with the previously mentioned measurements made with or without MAPs. One possibility is that the lower value measured by Felgner et al. in the absence of MAPs is due to structural defects in the walls of their unstabilized microtubules, which were observed in a nonphysiological buffer. By stabilizing these microtubules against such defects, the MAPs might have increased the flexural rigidity to a value in agreement with the other values.

Interestingly, the nonhydrolyzable nucleotide analogue GMP-CPP, which is thought to trap tubulin in its GTP state, increases the rigidity of microtubules by about a factor of two (Mickey and Howard, 1995). This might be important for regulating the structure and GTPase of the growing microtubule end (Chapter 11). A related effect is seen for actin: The phosphate analogue BeF_3^- increases the stiffness of actin filaments by 50% (Isambert et al., 1995), suggesting that the kinetics of the ATPase cycle might influence the mechanical properties of the filament (or vice versa). This idea will be explored in Chapter 11.

Coiled Coils

The elastic modulus of coiled coils measured in vitro can account for the longitudinal stiffness of keratin-containing materials. The persistence length of the coiled-coil protein tropomyosin is 50 to 200 nm, based on crystallographic and solution studies (Phillips and Chacko, 1996). The persistence length of the coiled-coil tail of myosin and paramyosin is estimated from hydrodynamic studies to be 130 nm (Hvidt et al., 1982). A lower bound for the persistence length of the coiled coil can be obtained from the shapes of myosin tails adsorbed to EM grids. The curvature fluctuations are consistent with a persistence length of ~100 nm (Howard and Spudich 1996; Howard, unpublished); this is a lower limit because adhesive forces might bend the molecules as they bind to the EM grid. Finally, theoretical calculations for polyglycine and polyalanine α-helices predict persistence lengths of 40 and 65 nm, respectively (Suezaki and Go, 1976); the dimeric coiled coil, which has twice the cross-sectional area, is expected to have a persistence length four times greater (see Table 6.1), or some 160 to 260 nm. Taken together, we conclude that the persistence length of a hydrated coiled coil is 100 to 200 nm, and the corresponding Young's modulus is ~2 GPa (using a cross-sectional area of 2 nm^2). This is sufficiently high to account for the rigidity of the hydrated keratin-containing materials listed in Table 8.3.

Intermediate Filaments

The flexural rigidity of pure, hydrated intermediate filaments is surprisingly low, suggesting that there is slippage between the constituent coiled coils. Electron micrographs of intermediate filaments show an apparent persistence length of ~1 μm for neurofilaments (Heins et al., 1993) and ~3 μm for vimentin (Inagaki et al., 1989; Howard, unpublished). These are likely to be upper bounds for the true persistence length because forces associated with adsorption to the EM grids could cause significant nonthermal bending. On the other hand, light-scattering experiments on desmin, which is in the vimentin class of intermediate filaments, give an upper bound on the persistence length of ~1 μm (Hohenadl et al., 1999). Taken together, these experiments indicate that these intermediate filaments have persistence lengths of ~1 μm. If the 16 coiled coils in each cross-section of an intermediate filament were tightly crosslinked, we would expect the persistence length to be 16^2 times greater than that of an individual coiled coil (i.e., ~25 to 50 μm), whereas if they were not crosslinked, we would expect the persistence length to be only 16 times greater (i.e., 1.6 to 3 μm). Thus it appears that in hydrated intermediate filaments there is little crosslinking and therefore slippage between neighboring coiled coils: The result is a mechanical structure which, while still highly resistant to elongation, is much less resistant to bending and twisting. In this sense, the intermediate filament has similar mechanical properties to string or rope. Presumably slippage accounts for the low torsional stiffness of hydrated wool. But when the material is dry, it becomes much more isotropic as slippage is prevented by rigid crosslinks.

Bacterial Flagellum

The bacterial flagellum (Example 2.2) is a helical tube of length ~5 μm composed of 11 protofilaments made of the 40 kDa flagellin protein. Due to an unusual deviation from helical symmetry (Calladine, 1983), the flagellum adopts a coiled shape, like a tiny spring. Fujime et al. (1972) used light scattering to measure the flexural rigidity of isolated repolymerized bacterial flagella: the value of 2.2 to 4×10^{-24} N·m^2 corresponds to a Young's modulus of 0.5 to 0.9 GPa, assuming inner and outer radii of 8 and 10 nm, respectively. Hoshikawa and Kamiya (1985) used a hydrodynamic method to measure the stiffness of the flagellar spring (Example 3.3). The shear modulus $G = 0.5$ GPa corresponds to a Young's modulus of 1.5 GPa, assuming the protein is isotropic and incompressible. The rigidity of the flagellar protein is similar to that of the cytoskeletal proteins, and the similarity of values for the Young's modulus measured using the different approaches is consistent with the flagellum being mechanically isotropic.

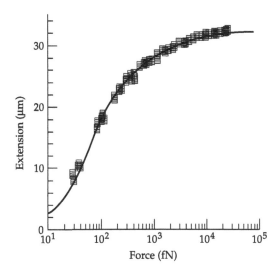

Figure 8.6 The force–extension curve for DNA
The solid curve is the prediction of the worm-like chain model with a persistence length of 53 nm. (After Bustamante et al., 1994a.)

DNA and Titin

Single-molecule techniques have been used to measure the rigidity of other biological polymers such as DNA (Smith et al., 1992; Bustamante et al., 1994b; Figure 8.6) and the very long muscle protein titin (Kellermayer et al., 1997; Rief et al., 1997; Tskhovrebova et al., 1997; Carrion-Vazquez et al., 1999). The extension curve resembles the Langevin function, which describes a freely jointed chain (see Figure 6.14).

Summary: Material Properties of Cytoskeletal Proteins

A number of general conclusions can be drawn about the material properties of the cytoskeletal proteins, whose Young's moduli are compared to those of other biological and nonbiological materials in Table 8.4.

1. The rigidity of purified actin filaments and microtubules measured in vitro accounts well for the rigidity of the corresponding cytoskeletal filaments in cells. Therefore actin-binding proteins and microtubule-associated proteins have a comparatively small effect on rigidity, as expected, given that they contribute only a small fraction of the total mass of the filaments.

2. To a first approximation, the cytoskeletal proteins behave as isotropic, homogenous materials. This is supported by experiments on actin filaments showing that the stretching stiffness, the bending stiffness, and the torsional stiffness all yield similar values for the Young's modulus. Likewise the flexural and torsional rigidities of bacterial flagella give similar Young's moduli. Furthermore, the rigidity of dry hair is similar in the longitudinal and transverse directions, and the flexural and torsional rigidities of wool are also consistent with isotropy. The main exception to this general statement is that

Table 8.4 Young's moduli of biological materials

Material	Young's modulus (GPa)	Notes
Proteins		
Actin	2	Filament persistence length, L_p = 15 µm
Tubulin	2	L_p = 6 mm
Coiled coil	2	L_p = 100–200 nm
IF protein	2 (longitudinal)	Hydrated keratins and filaments. L_p ~1 µm
Flagellin	1	L_p = 0.5–1 mm
Silk	~5	Silk-moth cocoon (depends on humidity)
Collagen	~2	Tendons and ligaments
Abductin	0.004	Hinge ligament of molluscan shell
Resilin	0.002	Hinge of insect flight system
Elastin	0.002	Vertebrate smooth muscle and ligaments
Sugars		
Cellulose	80	Dry ramie fibers
Cellulose	20–40	Wet fibers
Chitin	45	Dry beetle exoskeleton
DNA	~1	L_p = 50 nm, A = 2 nm^2
Composites		
Teeth	75	Human
Shell	68	Conus
Bone	19	Human femur
Wood	16	Douglas fir, along the grain
Muscle	0.040	Active frog sartorius muscle
Cartilage	0.015	Human costal cartilage
Muscle	0.00025	Relaxed frog sartorius muscle
Others		
Rubber	0.001	Polyisoprene
Plastic	2.4	Polycarbonate
Concrete	24	Unreinforced
Glass	71	Crown
Steel	215	Stainless (18/12)

Source: Wainwright et al., 1976, Kaye and Laby, 1986; this chapter.

hydrated keratinaceous materials as well as individual intermediate filaments are not isotropic; instead, there appears to be considerable shear between the constituent coiled coils.

3. Hydrated actin, tubulin, flagellin, and coiled-coil proteins all have Young's moduli in the range 1 to 3 GPa, similar to that of hard plastics, collagen (the major constituent of tendons), and silk (see Table 8.4). The finding that the

cytoskeletal proteins are among the most rigid proteins known supports the hypothesis that these proteins have evolved to serve structural roles in cells. Because the proteins are unrelated in sequence and structure, high protein rigidity has probably evolved independently a number of times.

4. The absolute value of the Young's modulus of these rigid proteins is within a factor of 2 of the Young's modulus of a van der Waals solid (Chapter 3). Thus we infer that the compliance of these proteins is primarily determined by the compliance of the van der Waals bonds that hold the polypeptide chains in their folded conformation. This is expected given that van der Waals forces are the weak links in protein structures and will therefore tend to dominate the compliance.

CHAPTER

9

Polymerization of Cytoskeletal Filaments

The most striking feature of cytoskeletal filaments is their length. Actin filaments range in length from ~35 nm in the cortex (just underneath the plasma membrane) of erythrocytes and other cells (Byers and Branton, 1985), to 10 to 100 µm in the stereocilia of hair cells, the sensory receptors of the vertebrate inner ear (Tilney and Tilney, 1988). Microtubules range in length from 1 µm or less in the mitotic spindle of the yeast *S. pombe* (Ding et al., 1993), through 100 µm in the axons of rat neurons (Bray and Bunge, 1981), to greater than 1 mm in insect sperm (Ashburner, 1989) and the comb plates of ctenophores (Tamm, 1973). The lengths of intermediate filaments are also in the micron range (Quinlan et al., 1995). Thus the cytoskeletal filaments span the scale from molecular to cellular. Because the individual proteins that make up these filaments are only a few to a few tens of nanometers long, it follows that cytoskeletal filaments contain tens to tens of thousands of protein subunits. One of the major points made in this chapter is that the existence of such long filaments requires a special structural adaptation: Filaments must be multistranded because single-stranded filaments are generally very short due to their tendency to break in the middle.

A second striking feature of actin filaments and microtubules is that their assembly and disassembly occur primarily by the addition and subtraction of subunits at one or both ends. This is clearly demonstrated in vitro where polymerization onto stable seeds can be visualized for actin by electron microscopy (Pollard, 1986), and for tubulin by video-enhanced light microscopy (Walker et al., 1988). Microtubules in vivo also grow by monomer addition onto their ends, as can be demonstrated by injection of fluorescently labeled tubulin into cells (Soltys and Borisy, 1985). In the case of actin, there are many observations that are consistent with polymerization also occurring via end addition in vivo, though there is no direct experimental support due to the difficulty of visualizing individual actin filaments under the light microscope. *Taken together, these in vitro and in vivo results demonstrate that end-polymerization is the major growth and shrinkage mechanism for actin filaments and microtubules.* By contrast, the spontaneous breakage and annealing of actin filaments and microtubules in vitro is very slow, though there are proteins that sever actin filaments (such as gelsolin; Yin and Stossel, 1979) and others that sever microtubules (such as katanin; McNally and Vale, 1993). We will see in this chapter that end-polymerization is another consequence of actin and microtubules being multistranded—because of the contacts between the strands, it is energetically less favorable to break the filament in the middle than to remove a subunit from the end. The polymerization of intermediate filaments is not well understood and will not be discussed.

Passive Polymerization: The Equilibrium Polymer

As a first step to understanding polymerization, we will consider some very simple models. These models ignore the important fact that both actin and tubulin are nucleotide triphosphatases. The free energy derived from this hydrolysis reaction endows cytoskeletal polymerization with unusual and unexpected properties not found in crystallization or other forms of "passive" self-assembly. Some of these properties will be discussed in Chapter 11. But in order to appreciate the added richness provided by nucleotide hydrolysis, it is necessary to understand first the physical constraints that exist when a free energy source is not tapped during polymerization. For this reason we consider the passive polymerization of the so-called **equilibrium polymer**.

The simplest equilibrium polymer is the single-stranded filament shown in Figure 9.1. This is also known as the Einstein polymer (Hill, 1987). Whereas the single-stranded model has the advantage that its properties can be derived quite easily, it predicts that the average polymer is only a few subunits long. This fundamental weakness will be corrected later by considering the more realistic two-stranded model of Figure 9.2.

To solve the single-stranded model, we assume that all the individual subunit-addition reactions have the same **dissociation constant**, K. That is, we write

$$A_n + A_1 \underset{k_{\text{off}}}{\overset{k_{\text{on}}}{\rightleftharpoons}} A_{n+1} \qquad \frac{[A_n] \cdot [A_1]}{[A_{n+1}]} = K = \frac{k_{\text{off}}}{k_{\text{on}}} \qquad n \geq 1 \qquad (9.1)$$

where A_1 denotes the monomer and A_n denotes the n-mer. The dissociation constant has the same unit as concentration, namely M, and is the reciprocal of the equilibrium constant (Chapter 5). It is more convenient to use the dissociation constant than the equilibrium constant because the latter has the awkward unit M^{-1}. Another reason to use the dissociation constant is that it is equal to the critical concentration, an important parameter for describing polymerization (see below). Because we are going to discuss polymerization of both actin filaments and microtubules, we need to adopt a potentially ambiguous terminology. We will refer to the building block of the filament as a **subunit**, which can be in either a **monomeric** or **polymeric** form. This definition is

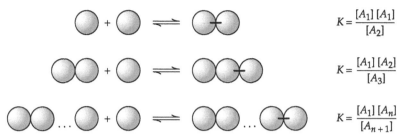

Figure 9.1 Single-stranded filament (Einstein polymer)

unambiguous in the case of actin because the building block is the 45 kDa actin monomer. However, in the case of tubulin, there is a potential ambiguity because tubulin is a dimer of α- and β-tubulin subunits. But because the dimer cannot be split into its two constituent subunits without irreversibly denaturing the protein, the dimer really is the building block and, for the purposes of polymerization, we will refer to it as a "subunit" that can even be in a "monomeric" form. This ought not create confusion. The dissociation constant is defined in terms of the ratios of concentrations. Even though it is often useful to write it as a ratio of the dissociation rate constant (k_{off}) and the second-order association rate constant (k_{on}) as shown in Equation 9.1, it is important to realize that the equilibrium concentration of the n-mers does *not* depend on the individual rates or the details of the mechanism by which the subunits come off and go on. Instead, the equilibrium concentration depends only on the equilibrium constant, which in turn depends only on the differences in energy of the subunits when they are in their various states of aggregation. This is a consequence of Boltzmann's law (Chapter 5).

The Einstein polymer is a crude model. It is a useful starting point for understanding polymerization because it is simple, but it is important to realize that the model is inexact. For example, the dissociation constants for each of the addition reactions shown in Figure 9.1 are assumed to be equal. However, in reality, the equilibrium constants are expected to depend on the length of the polymer, though the differences are fairly small. The reason for this length dependence can be appreciated by considering the physical meaning of the dissociation constant. The dissociation constant is associated with a "standard free energy change," ΔG^0, via

$$K = \exp\left(\frac{\Delta G^0}{kT}\right) \text{moles/liter} \tag{9.2}$$

(Chapter 5). ΔG^0 is the free energy change when the reaction occurs under standard conditions. In the molecular spirit of this book, we consider ΔG^0 as an energy per molecule rather than per mole and so use kT, rather than RT, as the denominator in the exponential. The standard state is 1 M concentration, by chemical convention. We can interpret ΔG^0 as the sum of a potential energy, usually negative, associated with formation of the bond, and an entropic energy, usually positive, which includes a component associated with the loss of translational and rotational entropy that occurs when the subunit goes from

1 M concentration to being bound to other subunits in a polymer. This interpretation allows us to see why the dissociation constants are not all the same: When a monomer binds to a long polymer it loses more translational and rotational entropy than when it binds to another monomer, because a dimer moves and rotates more freely than a long polymer. Thus the association between two monomers will be stronger than the association between a monomer and a long polymer, and so the dissociation constants will be smaller for the former reaction. Using quantum mechanical calculations for objects the size of proteins and assuming that the polymers are rigid cylinders, Hill (1987) estimated that the dissociation constant is about ten times smaller for the monomer–monomer reaction compared to the monomer–long polymer reaction. These calculations also show that once a polymer is more than about 10 subunits long it can be considered to be a "long polymer" in the sense that the dissociation constant for binding to a 10-mer is within a factor of 2 of that for an infinitely long polymer. Thus we can think of this entropic effect as influencing nucleation but not growth. Despite this length-dependence of the dissociation constants, the conclusion I am going to draw regarding the equilibrium lengths of polymers is not qualitatively affected. Indeed, because the length dependence implies that smaller polymers are even more favorable, the conclusion that single-stranded filaments are short is reinforced.

Single-Stranded Filaments Are Short

The crucial question is: How does the average length of the polymers depend on the total concentration of subunits available for polymerization?

It is relatively easy to show that at equilibrium the lengths of the polymers are exponentially distributed

$$[A_n] = K \exp\left(-\frac{n}{n_0}\right) \tag{9.3}$$

(Appendix 9.1). In other words, there are always fewer $(n + 1)$-mers than n-mers, and the ratio of $(n + 1)$-mers to n-mers is a constant. As the total concentration of subunits, $[A_t]$, is increased, the average length of the polymers increases. However, even when $[A_t]$ is much greater than the dissociation constant K, the average length of a single-stranded filament is still modest:

$$n_{av} \cong \sqrt{\frac{[A_t]}{K}} \tag{9.4}$$

This equation illustrates the fundamental shortcoming of the single-stranded model if it is to be applied to cytoskeletal filaments. It predicts that even when the total subunit concentration is one hundred times greater than the dissociation constant, the average polymer would contain only about 10 subunits. Yet cytoskeletal filaments must be designed to contain thousands of subunits. For a single-stranded polymer this would require total concentrations of subunits that are millions of times the dissociation constant. But for actin the total concentration (~200 μM; Bray, 1992) is only ~1000 times the dissociation constant

(~0.1 µM; Pollard, 1986). The discrepancy is even greater for tubulin, where the total concentration (~20 µM) is only about twice the monomer concentration (~10 µM) (Zhai and Borisy, 1994).

The reason why the single-stranded model predicts short filaments is that the ends of the polymers are not very unfavorable from an energetic point of view—a monomer can decrease the free energy of the system equally well by associating with another monomer as it can by associating with a long polymer.

Multistranded Filaments Are Long

It could be argued that the reason why real filaments are much longer than predicted by the single-stranded model is that they are not at equilibrium, and that, if one waited long enough, the filaments would eventually shorten to the predicted lengths. Alternatively, it is possible that the free energy derived from nucleotide hydrolysis could somehow be used to overcome the shortcomings of the single-stranded model. However, these possibilities are not consistent with the observation that long filaments can be grown with ADP-actin, which has a relatively high dissociation constant (Pollard, 1986) and for which there is no source of free energy.

Instead, there is a much simpler explanation: Actin filaments and microtubules are multistranded, and multistranded filaments are inherently longer. The two-stranded model (Figure 9.2) differs from the one-stranded model because it has two different classes of bonds, one within the strands and one

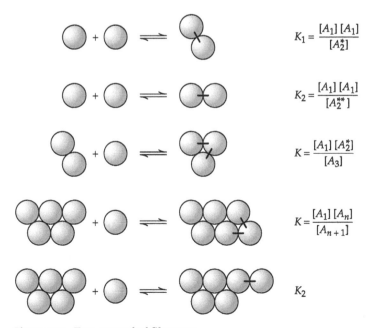

Figure 9.2 Two-stranded filament

between the strands. As a result, there are two different nuclei, A_2^* and A_2^{**}, and three different dissociation constants, K, K_1, and K_2. The model is solved in Appendix 9.2. Like the single-stranded model, it predicts that the lengths of the polymers are exponentially distributed, but the average length is much greater

$$n_{av} \cong \sqrt{\frac{K_1}{K}} \sqrt{\frac{[A_t]}{K}} \tag{9.5}$$

(assuming $[A_t] \gg K$). For example, the dissociation constants for actin are $K \sim 1\ \mu M$ (see Table 11.1), and $K_1 \cong K_2 \cong 0.1\ M$ (see below). If the total subunit concentration, $[A_t]$, is $10\ \mu M$, then $n_{av} \cong 1000$, corresponding to a 2.75-μm-long actin filament. Thus the two-stranded model predicts filament lengths that are consistent with the polymer lengths found in cells. The reason why the two-stranded filaments are much longer than single-stranded filaments is that the ends of a two-stranded filament are energetically unfavorable, so there will be only a low concentration of them at equilibrium.

Multistranded Filaments Grow and Shrink at Their Ends

A second crucial difference between single-stranded and multistranded polymerization is that the lengthening and shortening of multistranded filaments occurs almost exclusively by monomer addition and subtraction at the ends. This is expected to hold even when the polymer is not at equilibrium, provided that the monomer concentration is not vanishingly small (in which case only annealing could take place). That growth and shrinkage of multistranded polymers occur at the ends is the key to the regulation of polymerization of actin filaments and microtubules. By providing stable nuclei it is possible to control the location of filament polymerization. Furthermore, it is possible to stabilize a filament by simply capping it, and it is not necessary to bind stabilizing proteins all along its length. These issues of nucleation and capping will be considered in more detail in Chapter 11.

The rate of elongation of multistranded filaments, in subunits per second, is given by

$$\frac{dn}{dt} = k_{on}[A_1] - k_{off} \tag{9.6}$$

where k_{on} and k_{off} are the on- and off-rates and $[A_1]$ is the monomer concentration. In general, the on- and off-rates will differ at the two ends (Chapter 11), but for the moment we will consider only one end and assume that the other end is capped.

The proof that growth and shrinkage occur exclusively by monomer addition and subtraction is quite subtle, but worth going through because it illustrates a number of important points about the kinetic properties of large molecules. Consider first the annealing reactions that lead to an increase in length of an n-mer:

$$\frac{dn}{dt} = k_{on}[A_1] + 2k_{on,2}[A_2] + \cdots + mk_{on,m}[A_m] + \cdots$$

where the first term on the right-hand side corresponds to addition of monomers, the second term corresponds to addition of dimers, and so on. If all the second-order annealing rate constants ($k_{on,m}$) were equal to the second-order monomer association rate constant (k_{on}), then the total growth rate would be

$$\frac{dn}{dt} = k_{on} \sum_{m=1}^{\infty} m[A_m] = k_{on}[A_t]$$

where $[A_t]$ is the total concentration of subunits. This shows that elongation is not necessarily an end property because, in general, $[A_t] > [A_1]$, and in some cases $[A_t] \gg [A_1]$. However, when m is large, the translational and rotational diffusion of n-mers become very slow. The diffusion coefficients scale as $\ln(m)/m$ for translation, $1/m$ for axial rotation, and $\ln(m)/m^3$ for rotation about an axis perpendicular to the filament axis (Chapter 6). As a result, the second-order annealing rates will become diffusion limited for large m and will be much slower than the monomer association rate. Because the second order on-rate constant for the monomer, $k_{on} \sim 10^6$ $M^{-1} \cdot s^{-1}$ (see Tables 11.1 and 11.2), is already close to the diffusion-limited rate (Berg and von Hippel, 1985; Northrup and Erickson, 1992), it is likely that the annealing rate of even 10-mers will be significantly less than k_{on}. The strong length dependence of the perpendicular rotation term ($\ln(m)/m^3$) means that we can consider the annealing of all but the shortest polymers to be "frozen out," so that only monomer and short polymers will contribute significantly to the growth rate. But the concentration of short polymers is much less than that of the monomers (by a ratio $[A_n]/[A_1] \cong K/K_1 \cong 10^{-5}$, see Appendix 9.2). As a consequence, the total contribution of annealing to growth will be minor. In other words, growth is by monomer addition, with rate $k_{on}[A_1]$.

Now consider breakage. The change in length due to breakage is

$$\frac{dn}{dt} = -k_{off} - 2k_{off,2} + \cdots - mk_{off,m} - \cdots$$

where $k_{off,m}$ is the rate constant for the breaking off of an m-mer. Because the dissociation constant for annealing/breakage of an m-mer is independent of m (i.e., independent of the length of the fragment), the freezing out of the annealing of long fragments as discussed in the previous paragraph means that the breakage into long fragments must also be frozen out. This unexpected result is an example of the Principle of Microscopic Reversibility: At the molecular level, the breakage rate is frozen out because long polymer fragments diffuse away from each other so slowly that they have a high chance of re-annealing before they escape. Thus only the dissociation of monomers and the breaking off of smaller oligomers will contribute significantly to shortening. However, the total contribution from small oligomers is small because breaking a two-stranded filament requires severing three bonds, whereas removing a subunit from the end requires severing only two bonds (Figure 9.3A). As a consequence, $k_{off,m}/k_{off} \leq K/K_1 \cong 10^{-5}$ for $m > 1$ (Appendix 9.3). This means that shrinkage is by monomer subtraction, with rate constant k_{off} as written in Equation 9.6. In contrast to multistranded filaments, single-stranded filaments will

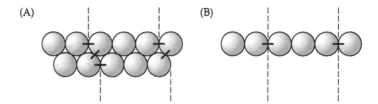

Figure 9.3 Breaking a two-stranded filament is difficult
Dissociation of the terminal subunit from a two-stranded filament (A) is more likely to occur than breakage in the middle because it involves breaking only two bonds rather than three. By contrast, breakage of and dissociation from a one-stranded filament involves breaking only one bond (B).

shorten primarily by breakage because the breakage and dissociation both require only one bond to be severed (Figure 9.3B).

Other Properties of Multistranded Filaments

Multistranded filaments have a number of other interesting and important properties. Though our discussion has centered on 2-stranded filaments, qualitatively similar results hold for 3-stranded, 4-stranded (thin filaments comprising two actin filaments and two tropomyosin filaments), and even 13-stranded filaments (microtubules).

1. There exists a **critical concentration**, K_c, such that when the total subunit concentration ($[A_t]$), is less than K_c, there are hardly any polymers, whereas when the total subunit concentration exceeds K_c, almost all the excess subunits go into filaments. When the monomer concentration is above the critical concentration, the polymers will grow, whereas below the critical concentration the polymers will shrink. Thus the critical concentration is the monomer concentration at which the elongation rate is zero; comparison of Equations 9.6 with Equation 9.1 therefore shows that

$$K_c = \frac{k_{off}}{k_{on}} = K \qquad (9.7)$$

These properties are shown graphically in Figure 9.4A,B. Note that when the total subunit concentration is greatly in excess of the critical concentration, the monomer concentration approaches the critical concentration.

2. The concentrations of the nuclei, A_2^* and A_2^{**} are very small, equal to $\sim K^2/K_1$ and K^2/K_2, respectively. In our numerical example, $K = 1\ \mu M$, $K_1 = K_2 = 0.1\ M$. Thus the concentrations of the nuclei are 0.1 nM. If the cell can provide more nuclei than this, then little growth from spontaneously formed nuclei will occur. The Arp2/3 complex (Machesky et al., 1994, Welch et al., 1997) nucleates actin polymerization and is one mechanism that permits cells to regulate their rate of actin polymerization in response to extracellular or intracellular signals (Machesky et al., 1999).

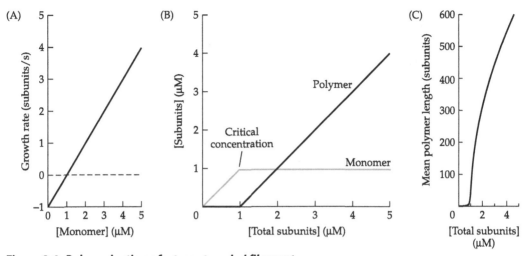

Figure 9.4 Polymerization of a two-stranded filament
(A) Growth rate for a multistranded filament according to Equation 9.6. (B) The concentration of subunits in their monomeric and polymeric forms as a function of the total subunit concentration. Note the sharp transition in polymer concentration about the critical concentration. (C) Mean length as a function of the total subunit concentration. In these examples, the off-rate constant is 1 s^{-1}, the on-rate constant is 1 μM^{-1}s^{-1}, and the dissociation constant (equal to the critical concentration) is 1 μM.

3. The ends of filaments will be predominantly blunt rather than having long protofilaments growing from them. This follows from the extra stability afforded by a subunit filling the "snug" site (see Figure 9.2).
4. The mean lengths of filaments increase very steeply about the critical concentration (see Figure 9.4C). Thus slight changes in the free monomer concentration, mediated by proteins that bind to monomers and sequester them (e.g., thymosin-β4, which binds to actin; stathmin, which binds to tubulin; Safer et al., 1991; Gigant et al., 2000), are expected to have a profound effect on the polymer length.

Binding Energies and the Loss of Entropy

I need to elaborate on the binding constants used in our example. For the two-stranded filaments there are two bonds, one between protofilaments, with equilibrium constant K_1, and one within protofilaments, with dissociation constant K_2 (see Figure 9.2). Thus we might think that when both bonds are formed, the "binding energy" is simply the sum of the two separate contributions, and that the dissociation constant K is simply the product of K_1 and K_2. However, this is not the case. The reason is that, as first mentioned in the discussion of the single-stranded model, the standard free energy associated with the individual bond, ΔG_i^0, is the sum of an "intrinsic binding energy", ΔG_i, which is generally negative, and an entropic energy, ΔG_S, which is generally positive due to the loss of translational and rotational entropy as a result of bond formation (Jencks, 1981). That is, $\Delta G_i^0 = \Delta G_i + \Delta G_S$ ($i = 1, 2$). Now, when a monomer binds to a snug site to form two bonds, the entropic cost is taken only once because the translational and rotational entropy is lost only once when the first bond is formed. According to this model, $\Delta G^0 = \Delta G_1 + \Delta G_2 + \Delta G_S \neq \Delta G_1^0 + \Delta G_2^0$. However, there is also another term: The formation of both bonds at the same time may require strain in the monomer or the polymer; furthermore, some of the entropy loss will be compensated for by vibrational entropy of the newly formed bonds. Thus a positive "interaction energy," ΔG_{12}, must also be included. Thus we write the standard free energy for formation of the two bonds together as

$$\Delta G^0 = \Delta G_1 + \Delta G_2 + \Delta G_S + \Delta G_{12} = \Delta G_1^0 + \Delta G_2^0 - \Delta G_S + \Delta G_{12} \qquad (9.8)$$

In practice it is difficult to separate the entropic term (ΔG_S) from the interaction-energy term (ΔG_{12}), so the reader may wonder why this is a useful description of a protein–protein association. However, this approach seems preferable to the usual $\Delta H^0 - T\Delta S^0$ formulation because the standard enthalpy (ΔH^0) and entropy ($-T\Delta S^0$) are very difficult to interpret for complex structures such as proteins interacting in water. In contrast, our mechanical picture of a protein–protein bond provides many qualitative insights into the mechanical interactions between proteins and is especially useful when interpreting how changes in protein conformation could alter binding strengths by inducing strain in the subunits and therefore increasing the interaction energy (Chapters 5 and 11).

The dissociation constant associated with Equation 9.8 is

$$K = \exp\left(\frac{\Delta G^0}{kT}\right) = K_1 K_2 \exp\left(\frac{-\Delta G_S + \Delta G_{12}}{kT}\right) \qquad (9.9)$$

where $K_i = \exp[(\Delta G_i + \Delta G_S)]/kT]$. The magnitude of the entropy-interaction term ($\Delta G_S - \Delta G_{12}$) is uncertain. In many texts the term is simply ignored (e.g., Israelachvili, 1992). This is the same as setting it equal to zero and interpreting ΔG_i^0 as binding energies alone. On the other hand, quantum mechanical calculations that assume that the bonds between subunits are completely rigid predict an entropic term equal to ~40 kT (Chothia and Janin, 1975). The best experimental estimate for the entropy-interaction term is ~10 kT, based on the polymerization of tubulin (Erickson and Pantaloni, 1981), actin (Erickson, 1989), and sickle hemoglobin (Cao and Ferrone, 1997); on the binding of desthiobiotin to avidin (Jencks, 1981); and on the modeling of protein–protein association constants in terms of solvation energies (Horton and Lewis, 1992).

The entropy-interaction term for actin is likely to be about 10 kT in order to account for filament lengths in the micron range. Let us first consider an extreme case where $\Delta G_S - \Delta G_{12} = 0$, and the entropy is ignored. Then, for a dissociation constant $K \sim 1$ μM, we need $K_1 = K_2 \sim 1$ mM, which gives a mean length of only ~100 (Equation 13.6 with $A_t/K = 11$). This is significantly shorter than the 1000 that is typical in cells and in vitro. On the other hand, if we take the other extreme of a very large entropic term, $\Delta G_S - \Delta G_{12} = \Delta G_S = 40\ kT$, then

$K_1 = K_2 \sim 10^6$ M (!) and two extremely weak bonds give a strong bond. In this case, the mean length is $\sim 3 \times 10^6$, which is too long. If we take an intermediate value of $\Delta G_S - \Delta G_{12} = kT\ln(10{,}000) \cong 9\ kT$, then $K = 1\ \mu$M, $K_1 = K_2 = 0.1$ M, and if the total subunit concentration is 10 times the critical concentration, then the mean length is 1000. Erickson (1989) also argues that the entropy-interaction term must be about $10\ kT$ in order to account for the low frequency of breakage and annealing of actin filaments.

There is a simple mechanical explanation for why the entropic term should be only $\sim 10\ kT$. Since the Debye length at physiological ionic strength is ~ 1 nm, it is likely that within this distance electrostatic forces will bring proteins into precise alignment (Schurr, 1970a,b; Chapter 5). Thus most of the entropy loss has occurred by the time the protein's position and orientation have been restricted to within 0.5 to 1.0 nm (at which point intermolecular forces take over), rather than within the 0.05 to 0.1 nm characteristic of the stereospecific alignment that results when the final bond is formed. Thus the translational entropy lost going from 1 M concentration (corresponding to 1 molecule per $(1.2\ \text{nm})^3$) to the bound state is approximately equal to that lost as a result of being concentrated $\sim$2-fold in each of the three directions, a total of perhaps $(3 \cdot \ln 2)k \cong 2\ k$. The loss of rotational entropy will be greater because a 50 kDa protein has a circumference of ~ 16 nm: Alignment within 1 nm in each of three dimensions is associated with a total entropy loss of $8\ k$ ($\cong (3 \cdot \ln 16)k$). Thus a total entropy loss of $10\ k$ is reasonable. If concentration by another factor of ten in each of six dimensions were necessary, the entropy loss would increase by an additional $14\ k$ ($\cong 6 \cdot \ln 10k$), to $24\ k$, nearer to that predicted by the quantum mechanical calculations.

Structure and Dimensionality

A key feature of linear polymers is that, at equilibrium, the lengths of the filaments are exponentially distributed. In other words, the polymers are polydisperse: They have a wide range of sizes. This is quite different from polymerization in two and three dimensions, where, at equilibrium, all subunits in excess of the critical concentration have aggregated into a single oligomer (Israelachvili, 1991). In two dimensions the aggregate is a "raft"; in three it is a clump or a crystal. In both cases, we expect that at equilibrium there will be just one raft or one clump. The reason why there is polydispersion in one dimension but precipitation in higher dimensions can be understood by considering the "surfaces" of the various polymers. In one dimension the surface corresponds to the two ends of a filament. In two dimensions it is the perimeter of a raft. And in three it is the surface of a clump. We can think of polymerization as being driven by the minimization of the surface areas. Now in the case of filaments, the fusion of two filaments reduces the number of ends by two, but this reduction is independent of the initial lengths of the filaments. By contrast, the fusion of two rafts reduces the surface area by an amount that is greater, the greater the sizes of the individual rafts. The same is true for clumps. In other words, if there is a driving force for aggregation, then in two and three dimensions this driving force gets larger as the clumps get larger and this drives the system to precipitate.

Summary

Irrespective of the details of polymerization, two things are clear: Multistranded filaments are much longer than single-stranded filaments, and multistranded filaments polymerize and depolymerize by subunit addition and subtraction at their ends, whereas single-stranded filaments change length by annealing and breaking. These functional differences between single-stranded and multistranded filaments highlight the importance of the intrastrand contacts in actin filaments. If there were no intrastrand contacts (i.e., $K_2 = \infty$ in Figure 9.2), then the actin filament would be effectively single-stranded and would not have the observed functional properties (Erickson, 1989).

Problems

9.1 Average length Show that the average length of the polymers, n_{av}, (i.e., not counting the monomers and nuclei) is approximately equal to the characteristic length n_0.

9.2***Three-stranded model** Find the critical concentration and the mean polymer length for a three-stranded filament. (Difficult!)

9.3*Transition about the critical concentration Show that for a two-stranded filament the mean filament length, n_{av}, depends very sensitively on the total subunit concentration near the critical concentration (i.e., when $[A_t] = K$):

$$\frac{dn_{av}}{d([A_t]/K)} = \tfrac{1}{2}\left(\frac{K_1}{K}\right)^{2/3}$$

Show that the slope ~1000 in the numerical example used in the text ($K_1 = 0.1$ M and $K = 1$ μM).

---

*Asterisks denote more advanced problems.

9.4* Closure confers stability Consider the ring formed by five monomers shown in the figure below. Let $K = 1$ µM. If $\exp[(\Delta G_S - \Delta G_{12})/kT] = 10^4$, show that $K_1 = 10^{-16}$ M! Show that if the total subunit concentration is 5 µM, then ~99.8% of the subunits are contained in closed rings. This demonstrates the stabilizing influence that closure brings. The incredible stability of icosahedral viruses is a consequence of closure in three dimensions.

$$K = \frac{[A_1][A_1]}{[A_2]}$$

$$K = \frac{[A_1][A_2]}{[A_3]}$$

$$K = \frac{[A_1][A_3]}{[A_4]}$$

$$K_1 = \frac{[A_1][A_4]}{[A_5]}$$

CHAPTER

10

Force Generation by Cytoskeletal Filaments

If a polymer is at equilibrium, it is unable to do work. However, if the monomer concentration is in excess of the critical concentration, there will be a flux of monomers onto the polymers. In principle, it is possible to couple this flux to an external load in order to drive movement and generate force, and this has been confirmed by experiments using purified proteins. Furthermore, there is now good evidence that actin polymerization, and perhaps also tubulin polymerization and depolymerization, is used to do mechanical work in cells. This evidence is summarized in the first section of this chapter. These findings raise the following questions: Is polymerization fast enough, and can it generate enough force? By generalizing the equilibrium polymer to include the effect of an external force, I show that the answer is yes in both cases.

The discoveries that actin polymerization and microtubule depolymerization can generate force are noteworthy because they directly contradict the "tensegrity" model for cellular architecture (Ingber, 1997, 1998). According to this model, the structural integrity of the cell depends on tension in actin filaments being balanced by compression in microtubules: The actin filaments and microtubules are analogous to the wires and struts in structures such as the Needle Tower sculpture by Kenneth Snelson in the Hirshhorn Museum and Sculpture Garden, Washington, D.C. The term *tensegrity* was coined by Buckminster Fuller (e.g., Fuller and Applewhite, 1975) to describe these discontinuous compression structures invented by Snelson (U.S. Patent 3,169,611). However, as we will see, polymerizing actin filaments are actually in compression, while depolymerizing microtubules are in tension. Evidently, each type of cytoskeletal filament can be in compression or tension depending on its cellular location or the activity of the cell (actin is in tension in contracting muscle, for example), and a more sophisticated model for cell structure is needed. Such general models are beyond the scope of this book; instead we will continue our discussion of the nuts and bolts of building the cytoskeleton.

Generation of Force by Polymerization and Depolymerization in Vivo

There are several examples in which actin polymerization generates forces in cells. The clearest one is the motility of intracellular viruses and bacteria. Such viruses, including the vaccinia (or cowpox) virus, and bacteria, such as *Listeria monocytogenes* and *Shigella flexneri*, nucleate bundles of actin behind them. These "comet tails" appear to push the virus or bacterium through the cytoplasm of infected cells (Tilney and Portnoy, 1989; Theriot et al., 1992; Cudmore et al., 1995; Figure 10.1). This polymerization-based motor mechanism has been firmly established by the recent reconstitution of motility of the bacterium *Listeria* using a small number of purified proteins (actin, Arp2/3, ADF, and capping protein; Loisel et al., 1999). *Listeria* moves through the cytoplasm at speeds up to 1 µm/s (Dabiri et al., 1990). If the cytoplasmic viscosity were similar to that of water, a force of only ~0.01 pN would be needed to overcome the viscous drag force on a 1-µm-long bacterium moving at this speed. However, it has been shown that large particles are almost immobile in cytoplasm: Polysaccharides of diameter >50 nm injected into the cytoplasm of Swiss 3T3 cells have <3% of the mobility that they have in water (Luby-Phelps et al., 1987), and larger particles in newt lung cells have only ~0.3% of the mobility that they have in water (Alexander and Rieder, 1991). If the bacterially infected cells also have high effective viscosity, then actin polymerization would need to generate much larger forces, in the piconewton range.

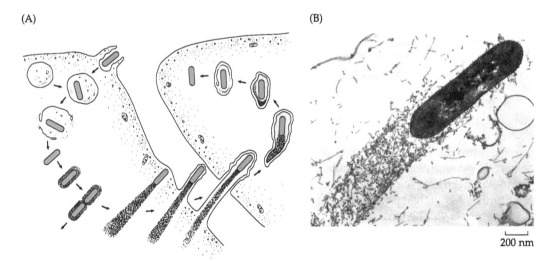

Figure 10.1 Polymerization-driven motility
(A) The bacterium *Listeria monocytogenes* propels itself through the cytoplasm of its host cell by inducing the polymerization of an actin filament–containing tail. Because it can move from one cell to another without being in direct contact with the extracellular space, *Listeria* can avoid the immune system. (B) Electron micrograph showing the actin filaments in the tail. (A from Tilney and Portnoy, 1989; B from Tilney et al., 1992.)

It is likely that an actin-polymerization motor mechanism also drives the protrusion of the **leading edge** of motile cells. (The leading edge is the lamellipodium in fibroblasts, macrophages, keratocytes, and motile white blood cells such as leukocytes; it is the growth cone in neurons.) The reasons for believing this are as follows: (1) the cytoplasm at the leading edge contains a zone adjacent to the plasma membrane that is enriched in polymerizing actin filaments (Wang, 1985; Forscher and Smith, 1988; Theriot and Mitchison, 1991); (2) the actin filaments form characteristic crisscrossed bundles (Small, 1988; Svitkina and Borisy, 1999; see Figure 1.2) similar to those found behind the motile bacteria; and (3) the crisscrossed pattern is due to branching of actin filaments mediated by the Arp2/3 complex (Blanchoin et al., 2000), an essential component required for *Listeria* motility (Loisel et al., 1999) that is also found at branch points in the leading edges of keratocytes and fibroblasts (Svitkina and Borisy, 1999). Thus it appears that the intracellular parasites have hijacked the cellular motile machinery by expressing proteins on their surfaces (such as ActA expressed on the surface of *Listeria*; Kocks et al., 1992) that bind directly to and activate the Arp2/3 complex (Welch et al., 1998), bypassing the cellular regulatory mechanisms (Machesky et al., 1999).

Another motile process driven by actin polymerization is the acrosomal reaction of some echinoderm sperm: When a sperm of the sea cucumber *Thyone* encounters an egg, the sperm produces a fine thread, the acrosomal process, that penetrates the jelly surrounding the egg and fuses with the egg membrane. The elongation of the acrosome is very fast, as much as 90 µm in only 10 seconds and is caused by the polymerization of a large pool of actin in the acrosomal cup at the apex of the sperm (see Example 10.3).

Microtubule polymerization generates force in cells. During the later phases of mitosis, anaphase B and telophase, there is an elongation of the spindle microtubules associated with the separation of the poles. Forces greater than 3 pN per microtubule are inferred from the occasional buckling of the spindle microtubules observed when a physical barrier impedes elongation (Aist and Bayles, 1991; Gittes et al., 1993). Polymerization could provide some or all of the force associated with pole separation, the rest being provided by microtubule-based motors. Microtubule polymerization probably underlies the "polar ejection" force that pushes the chromosome arms away from the poles during prophase (Rieder et al., 1986).

Microtubule depolymerization might also generate force in mitosis. During anaphase A, the newly separated sister chromatids move from the metaphase plate to the poles (Figure 10.2A). Associated with this movement is a shortening of the connecting kinetichore microtubules. The poleward forces are large: Micromanipulation with glass fibers indicates that anaphase A can still proceed against loads as high as 10 to 100 pN per microtubule (Nicklas, 1983, 1988). This force is much larger than the viscous force calculated for movement of a chromosome through the cytoplasm: The speed of ~1 µm/min corresponds to a drag force of ~0.2 pN for a chromosome of diameter 5 µm moving through a fluid of viscosity 300 times that of water (estimated from the diffusion of particles near the mitotic spindle; Alexander and Rieder, 1991).

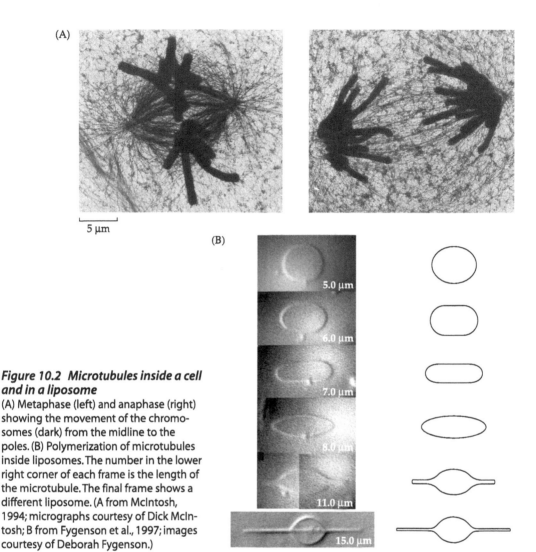

Figure 10.2 Microtubules inside a cell and in a liposome
(A) Metaphase (left) and anaphase (right) showing the movement of the chromosomes (dark) from the midline to the poles. (B) Polymerization of microtubules inside liposomes. The number in the lower right corner of each frame is the length of the microtubule. The final frame shows a different liposome. (A from McIntosh, 1994; micrographs courtesy of Dick McIntosh; B from Fygenson et al., 1997; images courtesy of Deborah Fygenson.)

It is not absolutely certain whether mitotic forces are generated by microtubule polymerization and depolymerization, by motors such as dynein (Steuer et al., 1990) or kinesin-related proteins (Yen et al., 1992) that pull directly on the microtubules, or by motors such as MCAK that modulate microtubule dynamics (Wordeman and Mitchison, 1995; Walczak et al., 1996; reviewed in Hunter and Wordeman, 2000). But it is clear that the growth and shrinkage of the microtubules are at least necessary for mitotic movements, and it is expected that the large forces involved in mitosis will influence the kinetics of microtubule polymerization and depolymerization (see the next section).

Generation of Force by Polymerization and Depolymerization in Vitro

Polymerization-driven force generation has been demonstrated in vitro in experiments in which actin or tubulin subunits are loaded into a membrane vesicle under conditions in which the monomeric state is favored. Polymerization is then initiated by warming (tubulin) or by increasing the ionic strength (actin). The growing polymers are observed to deform the vesicle as they do work against the membrane's elasticity (Cortese et al., 1989; Hotani and Miyamoto, 1990; Fygenson et al., 1997; see Figure 10.2). In the case of microtubules, forces as large as 2 pN have been inferred from instances when a microtubule inside a vesicle buckles (Fygenson et al., 1997). Polymerization forces up to 4 pN have also been inferred from the buckling of polymerizing microtubules that strikes a barrier (Dogterom and Yurke, 1997). Depolymerization-driven forces have been demonstrated in experiments in which a bead is attached to one end of a microtubule and the monomeric tubulin is washed away to induce depolymerization: The bead remains attached near the end of the microtubule as it shortens. By opposing the depolymerization-driven motion with fluid flow, forces as high as 1 pN have been inferred (Coue et al., 1991).

An important point established by these experiments is that *force generation can take place without additional proteins, even though in the cell other proteins are involved.*

Equilibrium Force

How much force can be generated by polymerization or depolymerization? The equilibrium polymer model developed in the last chapter is readily generalized to include the effect of an external force. If polymerization is countered by an opposing, compressive force of magnitude F, then the mechanical energy of the $(n + 1)$-mer exceeds that of the n-mer by $F \cdot \delta$, where δ is the increase in length due to incorporation of one monomer (Figure 10.3). The distance δ equals the monomer length divided by the number of strands. For actin this is 5.5 nm ÷ 2 = 2.75 nm, while for a microtubule it is 8 nm ÷ 13 ≅ 0.6 nm. Thus, at equilibrium in the presence of the force, application of Boltzmann's law gives a dissociation constant, $K(F)$, equal to

$$K(F) = K_c \exp\left(\frac{F \cdot \delta}{kT}\right) \tag{10.1}$$

where K_c is the critical concentration in the absence of an external force. $K(F)$ is the concentration at which there is no net elongation of the filament. If the force is positive (compressive), the dissociation constant is increased; but provided that the monomer concentration exceeds $K(F)$, polymerization will still proceed and can do work against the force. Conversely, if the force is negative, the dissociation constant is decreased; but if the monomer concentration is sufficiently low, then depolymerization will still occur.

Figure 10.3 A compressive force opposing the polymerization of a polymer

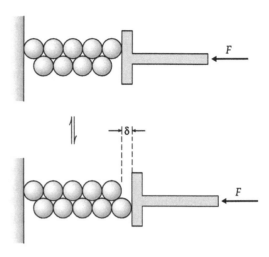

We can turn this equation around to say that if there is no net polymerization at monomer concentration $[A_1]$, then the **equilibrium force** is:

$$F_{eq} = \frac{kT}{\delta} \ln \frac{[A_1]}{K_c} \qquad (10.2)$$

For a given $[A_1]$, the system is at equilibrium when $F = F_{eq}$, in which case the chemical potential corresponding to the "concentration gradient" ($kT \cdot \ln([A_1]/K_c)$) is exactly balanced by the mechanical potential ($F_{eq} \cdot \delta$). The **chemical-force equation** (Equation 10.2) is analogous to the Nernst equation of electrochemistry, which relates the equilibrium electrical potential difference across a semipermeable membrane to the concentration difference of the permeant ion (Hille, 1992). Another way of interpreting this equation is that F_{eq} is the **reversal force**. For a fixed monomer concentration, $[A_1]$, increasing the force through F_{eq} will change the direction of polymerization from net growth to net shrinkage. This is analogous to the reversal potential of electrophysiology. Finally, if the monomer concentration differs from the critical concentration, work up to $F_{eq} \cdot \delta = kT\ln([A_1]/K_c)$ can be done for each monomer added.

The forces associated with polymerization can be large. For an actin filament in which the monomer concentration is 100 times the critical concentration (discussed later in this chapter), the equilibrium force is $kT\ln(100)/\delta \cong 7$ pN, larger than the average force exerted by a myosin molecule in muscle (Chapter 16). For a microtubule, the equilibrium force is even larger, ~30 pN, owing to the smaller length increment. This force exceeds the polymerization forces measured in vitro as well as the forces measured during anaphase B and telophase.

It is natural to ask whether the cytoskeletal filaments are strong enough to exert forces while polymerizing. The problem is that polymerization-based motility requires that the filaments be in compression, and there is only so much compressive force that a filament can withstand before it buckles (Chapter 6). Buckling is expected to be a more serious problem for actin filaments because they have a much lower flexural rigidity than microtubules. And, after all, in the well-known actin-based motile systems such as muscle, the actin filaments are in tension and can withstand very large forces, up to ~500 piconewtons, before they break (Tsuda et al., 1996). However, it turns out that buckling is not a problem provided that the filaments are not too long. Application of the buckling equation (Equation 6.6) shows that an actin filament can exert 7 pN of compressive force provided that it is less than ~300 nm long (using $EI = 60 \times 10^{-27}$ N·m^2 from Table 6.4 and Equation 6.11). This is shorter than the individual filaments seen in the comet tail behind a bacterium (see Figure 10.2B) and in the leading edge of a cell (see Figure 1.2), so buckling should not be a problem. Another potential problem arises from the fact that the actin filaments are angled with respect to the bacterial or the cell membrane and will tend to bend when the force is large; however, it can be shown that the maximum force in this geometry is of similar magnitude to the buckling force (Gittes et al., 1996), therefore polymerization of short actin filaments (<300 nm long) should be able to generate forces in the pN range even if the filaments are angled. Because they are about 500 times more rigid than actin filaments ($EI = $ ~30×10^{-24} N·m^2 from Table 8.2), microtubules can support compressive forces in the piconewton range even when they are several microns long (see Example 6.4), similar to their length in the mitotic spindle (see Figure 10.2A). Thus we conclude that both actin filaments and microtubules are rigid enough to support the polymerization forces that are observed in cells.

When comparing the behavior of a real cytoskeletal filament to that of a model equilibrium polymer, a number of points emerge:

1. The equilibrium force is independent of the polymerization mechanism. It is equally valid whether the polymer pushes directly on the particle, or whether the polymer is coupled to the particle via an elaborate set of accessory proteins, as is the case both for actin filaments (Arp2/3 complex) and microtubules (e.g., the kinetichore).

2. The maximum force exerted by a real polymer will be less than or equal to the equilibrium force. In other words, polymerization may be so slow at higher forces that growth is effectively stalled at forces significantly smaller than the equilibrium force. This appears to be the case for microtubules, whose polymerization stalls at about 4 pN (van Doorn et al., 2000).

3. At a fixed monomer concentration, an equilibrium polymer can either do work while polymerizing or depolymerizing, but not both. You can't have it both ways! Yet we have seen that during mitosis, work might be done by *both* growing *and* shrinking microtubules! This cannot be accounted for by an equilibrium model unless the monomer concentration is different at different phases of mitosis. This could be achieved by regulating proteins (such as stathmin, which binds to free tubulin subunits) and preventing them from polymerizing (Gigant et al., 2000). Alternatively, it could be explained if polymerization is not an equilibrium process but is coupled to nucleotide hydrolysis, as we will discuss in the next chapter.

Brownian Ratchet Model

A polymerization mechanism can, in principle, generate sufficiently large forces to account for a variety of in vivo and in vitro processes. The question remains: Is a polymerization mechanism fast enough? To answer this question one must consider specific kinetic mechanisms by which polymerization is coupled to force generation. In this section we will consider one such mechanism, the **Brownian ratchet** model (Peskin et al., 1993), since this model can be made explicit and solved under a variety of different conditions. In the next section we will consider other models that circumvent some of the problems associated with the Brownian ratchet model.

It is difficult to envision how the end of a growing or shrinking filament could push or pull on a particle. In the former case, one might think that the particle would block the end and so prevent polymerization. In the latter case, it is difficult to understand how a depolymerizing polymer could maintain contact with the particle being pulled. Although it is possible that these difficulties are overcome by proteins that connect the end of a filament to the surface of the particle, the in vitro experiments mentioned above demonstrate that additional protein machinery is not absolutely required for force generation. This has led to the idea that in the case of a growing polymer pushing against an opposing force, the particle being pushed must be able to undergo thermal motions sufficiently large to unblock the adjacent filament end and permit subunit addition. This mechanism, termed a Brownian ratchet (Peskin et al., 1993), in which diffusive motion is an essential component of a chemical reaction, goes back to Kramers (1940) who first solved the diffusive motion of a damped spring (Chapters 4 and 5).

The Brownian ratchet model is shown in Figure 10.4. According to the model, a filament of length n monomers is anchored at one end, while the other end abuts the particle. It is hypothesized that the particle diffuses so that the gap, x, between the end of the polymer and the particle's surface fluctuates. If the gap exceeds the distance required to incorporate another monomer (δ)—that is, if the gap is "open"—then there is a chance that a monomer can bind to the end (with on-rate k_{on}) to create an $(n + 1)$-mer. Alternatively, a subunit could unbind (with off-rate k_{off}) to create an $(n - 1)$-mer. Because the transition state has the gap fully open (like the final state) this model is an example of a Kramers-like kinetic mechanism (Chapter 5).

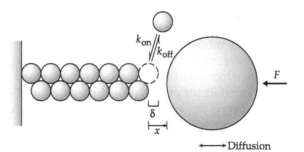

Figure 10.4 The Brownian ratchet mechanism for force generation by polymerization
The particle is free to diffuse horizontally as shown. It is assumed that additional constraints (not shown) prevent vertical diffusion.

The model is formulated precisely in Appendix 10.1 as a reaction–diffusion equation. The solution depends on the rate of the diffusive motion of the particle compared to the rate of the polymerization reaction. If diffusion is very fast, we say that the process is **reaction limited**: In this case, the creation of a sufficiently large gap occurs frequently, but the probability that the gap will be filled is low because the on-rate for subunit addition is low. On the other hand, if diffusion is slow, we say that the process is **diffusion limited**: In this case, the actual creation of a gap is the rate-limiting step, and once a gap is created, a subunit will always drop in. Note that at high-enough forces, polymerization will become reaction limited because the lifetime of the "open" state will be very short. In both cases, we assume that the transition rate constants, k_{on} and k_{off}, do not depend on the gap size, as implicitly assumed in this discussion.

Reaction-Limited Polymerization

In the reaction-limited case, the diffusion coefficient is very large, and the probability, $p(x)$, that the gap distance equals x satisfies Boltzmann's law: $p(x) = (F/kT) \exp(-Fx/kT)$, where $F > 0$, is the magnitude of the compressive force directed toward the left. We can think of the gap exploring all possible positions as though the particle were at equilibrium, and only very occasionally does a monomer drop in. In this case the elongation rate is

$$v = \delta \frac{dn_{av}}{dt} = \delta \left(k_{on}[A_1] e^{-\frac{F\delta}{kT}} - k_{off} \right) \qquad (10.3)$$

(Appendix 10.1). The interpretation of this equation is that the on-rate is reduced by a factor equal to the probability of the gap being larger than δ. At equilibrium, the elongation rate is zero, and Equation 10.3 reduces to the equilibrium force equation (Equation 10.2) because $K_c = k_{off}/k_{on}$. (See Example 10.1 for an application of Equation 10.3 to actin polymerization.) For microtubules, $kT/\delta = 7$ pN, but the experiments indicate that the speed decreases e-fold every 2 pN (Dogterom and Yurke, 1997). This suggests that the rate-limiting step is associated with a gap of ~2nm, perhaps associated with growth on several protofilaments (van Doorn et al., 2000).

Diffusion-Limited Polymerization

In the diffusion-limited case we assume that the monomer addition reaction is so fast that almost as soon as a gap opens up, a monomer will drop in. In Appendix 10.1 I show that the polymerization rate is approximately equal to the rate of gap creation. The rate of gap creation is simply the time it takes the particle to diffuse a distance δ against the force: This is the mean passage problem of Chapter 4 whose solution leads to the elongation rate

$$v = \delta \frac{dn_{av}}{dt} \cong \frac{2D}{\delta} \frac{(F\delta/kT)^2/2}{e^{F\delta/kT} - 1 - F\delta/kT} \qquad (10.4)$$

where D is the diffusion coefficient of the particle. This equation is plotted in Figure 10.5. In the case of very low external force, $F \cdot \delta \ll kT$, the "diffusion-limited"

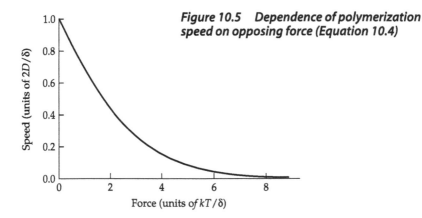

Figure 10.5 Dependence of polymerization speed on opposing force (Equation 10.4)

elongation rate approaches $2D/\delta$. As the opposing force increases, the speed then drops approximately exponentially, with an e-fold decrease per force of kT/δ. This equation for the force dependence of a polymer elongation rate is similar, though not identical, to the Goldman-Hodgkin-Katz (GHK) current equation of electrophysiology. The GHK equation gives the current flow across a semipermeable membrane for which the transmembrane potential and the ionic concentrations on each side of the membrane are not at equilibrium (Hille, 1992).

If, in the case of zero external force, the elongation rate is diffusion limited, the drag force against which the polymerization mechanism acts is

$$F_{\text{drag}} = \gamma \cdot v = \gamma \cdot \frac{dn_{\text{av}}}{dt} \cdot \delta = \frac{kT}{D} \frac{2D}{\delta^2} \delta = \frac{2kT}{\delta} \quad (10.5)$$

where we have used the Einstein relation $\gamma = kT/D$. For an actin filament this force is ~3 pN and for a microtubule it is ~13 pN. This force is of similar magnitude to the equilibrium force (see Equation 10.2). The force is independent of the viscosity because F_{drag} corresponds to the maximum force inherent to the polymerization mechanism. (See Example 10.2 for an application of Equation 10.5.)

Examples of Motility Driven by Actin Polymerization

The following examples show that the Brownian ratchet model can account for the speed of two actin-based motile processes, the intracellular movement of the bacterium *Listeria* and the extension of the acrosomal process in sea cucumber sperm. Because the actin-based amoeboid movement of cells is slower than that of *Listeria* (~0.1 μm/s for keratocytes and ~0.01 μm/s for fibroblasts; Svitkina and Borisy, 1999), the Brownian ratchet model can also account for these motions as well. However, the fact that a model can account for the data does not prove the model. Indeed, there are a number of conceptual problems with the Brownian ratchet model; these problems, discussed below, require that the model be modified.

> **Example 10.1 Actin polymerization is fast enough and powerful enough to drive the movement of Listeria** The fastest movements of *Listeria* are ~1 μm/s, corresponding to a growth rate of 360 monomers/s. This is consistent with a polymerization mechanism because the on-rate for actin binding to the end of a filament is 12 $\mu M^{-1} \cdot s^{-1}$ (see Table 11.1), and the total actin concentration in nonmuscle cell is ~200 μM (Bray, 1992); if 15% of this actin is free to polymerize (i.e., 30 μM), a speed of 1 μm/s can be attained (see Equation 10.3). 30 μM is about three hundred times higher than the critical concentration for ATP-actin measured in vitro (see Table 11.1); a maximum (equilibrium) force of ~8 pN per actin filament is therefore possible (Equations 10.2 and 10.3). Because the actin filaments in the tails are crisscrossed, rather than being in line with the direction of movement, even high forces are possible (the factor is ~$1/\cos\theta$, where θ is the angle of the filament with respect to the direction of motion). Given that a transverse section through the tail contains several filaments (see Figure 10.1B), total forces up to 100 pN are therefore possible. It should be noted that in order for the bacterium to be propelled forward, the actin-containing tail must somehow be anchored within the cell, perhaps via crosslinking to the existing cytoskeleton.

> **Example 10.2 The diffusion-limited speed of Listeria** For a 1-μm-long bacterium being pushed by actin polymerization through water, the diffusion-limited maximum speed is 360 μm/s (= $2D/\delta$, where using D = 0.5 $\mu m^2/s$, δ = 2.75 nm), well in excess of the measured speeds of 0.1 to 1 μm/s. This suggests that *Listeria*'s motion may not be diffusion limited. However, if the mobility of a bacterium in cytoplasm is only 1/100 to 1/1000 that in water, the diffusion-limited speed is reduced by a corresponding factor to a value more nearly equal to the observed speeds. If the motion were diffusion limited, the force would equal ~3 pN per filament, by Equation 10.5. This is of the same order of magnitude as the maximum reaction-limited force (Example 10.1). It is interesting that *Listeria* moves at different speeds in different cell types (Dabiri et al., 1990): The three-fold difference could arise from differences in the monomeric actin concentration if the motion were reaction limited, or to differences in "viscosity" (i.e., different mobilities of the bacteria) if the motion were diffusion limited.

Example 10.3 Actin polymerization is fast enough in vitro to account for the speed of the acrosomal reaction in sea cucumber sperm The highest rate of elongation of the acrosomal process (Figure A) is ~10 μm/s (solid circles in Figure C). For an actin filament with $\delta = 2.75$ nm, the monomer addition rate must be 3600 monomers/s. With an on-rate of 12 $\mu M^{-1} \cdot s^{-1}$ (see Table 11.1), this elongation rate requires a cytoplasm-free ATP–actin monomer concentration of 300 μM. The actin concentration in the acrosomal cup (at the base of the acrosome) is actually ~3 mM, sufficiently high to account for the elongation rate. In Appendix 10.2 I show that, contrary to the view of some workers (e.g., Oster and Perelson, 1987), the diffusion of actin down the acrosome (as depicted in Figure B) is indeed fast enough to feed the polymerization taking place at the tip of the acrosome (solid curve in Figure C).

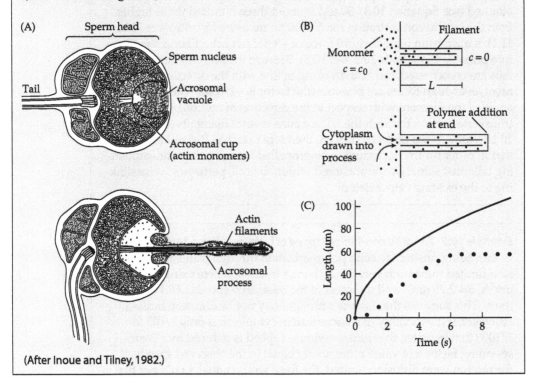

(After Inoue and Tilney, 1982.)

Other Kinetic Models

In the Brownian ratchet model it is assumed that the particle (or membrane against which the filament is pushing) must diffuse away from the tip of the growing filament in order to permit the addition of another monomer. As a consequence, a compressive force decreases the polymerization rate (Equations 10.3 and 10.4), rather than increasing the depolymerization rate. However, it is possible to conceive of other models in which the force does not affect the growth rate, but instead accelerates the shrinkage rate (Figure 10.6, lower path-

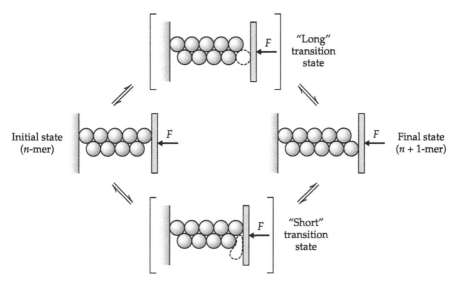

Figure 10.6 Two possible transition states for the polymerization reaction
In the top pathway, the piston must move to the right before another subunit can be added. Thus the transition state looks like the $(n + 1)$-mer (i.e., it is long). In the bottom pathway, the piston is still at its left-hand position when the next subunit binds. Thus the transition state looks like the n-mer (i.e., it is short). The idea in the latter case is that the rate-limiting step is the insertion of a wedge and the ensuing induced fit that quickly pushes the piston to the right. In the top pathway, the force slows polymerization, while in the bottom pathway, the force accelerates the depolymerization. The top pathway (the Brownian ratchet) is an example of a Kramers-like mechanism, whereas the lower pathway (induced fit) is an example of an Eyring-like mechanism.

way). In this case, the rate-limiting step is the binding of the next monomer, which then quickly undergoes a conformational change—an induced fit—that pushes the particle to the right. The monomer acts as a wedge that uses some of the binding energy to push the particle and thus create a gap. In the upper pathway, the force slows polymerization, while in the lower pathway, the force accelerates the depolymerization. Thus the Brownian ratchet model is an example of a Kramers-like mechanism, in which the reaction must diffuse to its final position, whereas the induced-fit pathway is like the Eyring-like mechanism, in which the reaction is pushed into its final position (see Chapter 5). Depending on the geometry of the reaction pathway, one can conceive of any combination of effects on the growth and shrinkage rates depending on the position of the transition state in the force field. The only constraint is that the ratio of the on- and off-rates must be consistent with the equilibrium force equation (Equation 10.2) where $K_c = k_{off}/k_{on}$.

The attractive feature of the Brownian ratchet model is that it is explicit. Furthermore, it is hard to imagine that any other mechanism could lead to microtubule polymerization inside membrane vesicles (see Figure 10.2B), and indeed rough calculations show that diffusive fluctuations in curvature of membranes is sufficiently fast to account for the observed microtubule growth rate (Fygenson et al., 1997). However, a purely diffusive mechanism seems less likely in the case of bacterial motility. The main argument against it is that the bacterium is firmly attached to the tail (Gerbal et al., 1999), so there is never a gap between the tail and the surface of the bacterium. Also, what would stop the bacterium diffusing laterally and falling off the tail?

One can envisage a number of ways that the problems associated with a Brownian ratchet model for bacterial motility could be circumvented. Because there are many actin filaments in the tail, some could be free while others are bound. It is possible that thermal fluctuations in the membrane of the bacterium open up gaps that are sufficiently large to allow an actin subunit to bind to the end of a free filament. Alternatively, the gap could be created by a thermal bending of an actin filament whose end is free, giving a brush-type mechanism, termed an elastic Brownian ratchet model (Mogilner and Oster, 1996). Finally, it is possible that the accessory proteins (Arp2/3, ActA, etc.) catalyze the formation of a gap of appropriate size.

Summary

Polymerization and depolymerization can generate forces when the monomer concentration differs from the critical concentration. The forces are in the range of a few piconewtons per filament, high enough to account for in vivo and in vitro observations. Furthermore, consideration of some specific mechanisms of polymerization-mediated motility indicate that polymerization is fast enough, even against physiological forces, to account for polymerization-driven processes such as the movement of *Listeria* within the cytoplasm and the amoeboid movement of crawling cells.

CHAPTER

11

Active Polymerization

In cells, the cytoskeletal proteins are not at equilibrium. If they were, then the random addition and subtraction of subunits at the polymer's ends would be much too slow to account for the dynamic properties of microtubules and actin filaments in vivo, as argued in the following paragraph.

The length of an equilibrium polymer fluctuates in a diffusive manner. The diffusion coefficient is equal to $k_{on}[A_1]\delta^2$, where k_{on} is the on-rate, $[A_1]$ is the monomer concentration, and δ is the increase in length due to incorporation of one monomer (Kasai and Oosawa, 1969; Appendix 11.1). This leads to very small length changes, even over long times. For example, for a microtubule with $k_{on} = 5 \times 10^6 \, M^{-1} s^{-1}$, $[A_1] = 10 \, \mu M$, and $\delta = 0.6$ nm, the standard deviation of the length over 1 minute is 46 nm, and over 1 week it is only 4.7 μm. This is not what is seen in cells. Fluorescently labeled tubulin dimers incorporate throughout microtubules in interphase cells within an hour (Saxton et al., 1984), and labeled actin monomers injected into cultured cells exchange into actin filaments in stress fibers (Kreis et al., 1982) and lamellipodia (Theriot and Mitchison, 1991) with half-times of several minutes. These studies are supported by observations of the "dynamic instability" of individual microtubules in cells: Microtubules can switch between a growing phase (with rate ~1 μm/min) and a shrinking phase (with rate ~10 μm/min) on a timescale of minutes (Cassimeris et al., 1988; Sammak and Borisy, 1988; Hayden et al., 1990). Furthermore, nearby microtubules can be in different phases, indicating that the length fluctuations are not due to fluctuations in the monomer concentration. Such a rapid turnover of filaments is inconsistent with an equilibrium mechanism. Conversely, in some fully differentiated cells, the lengths of filaments are stabler than expected for equilibrium. For example, the length of a cilium is fixed within a fraction of a micron for a particular cell type (Rosenbaum and Child, 1967), and the thin filaments in muscle are fixed in length within a few tens of nanometers over the week-long half-life of the actin protein (Littlefield and Fowler, 1998).

Having a cytoskeleton that is not in equilibrium gives the cell morphological flexibility. It can build structures where it needs them, stabilize them by capping the ends of the filaments, then take them apart by removing the caps. And all this can be done quickly without having to synthesize new protein or degrade old protein. In this chapter, I show how the energy derived from nucleotide hydrolysis frees polymerization from the constraints imposed by the equilibrium mechanisms considered in the previous chapters. Such "active" polymers are capable of much richer behavior than passive equilibrium polymers: Nucleotide hydrolysis accounts for dynamic instability and treadmilling, and it provides energy for filaments to do mechanical work while polymerizing or depolymerizing.

Actin and Tubulin Hydrolysis Cycles

Nonequilibrium behavior requires an energy source. For both actin and tubulin, the energy is derived from the hydrolysis of nucleotides—ATP in the case of actin and GTP in the case of tubulin. Straub and Feuer discovered that actin is an ATPase whose activity is coupled to polymerization: ATP is bound to monomeric actin, but, following assembly, only ADP is retained in the filament; when the filament is disassembled, the ATP is not resynthesized (Straub and Feuer, 1950). Similar results are found for tubulin (Weisenberg et al., 1976). These studies show that depolymerization is not the chemical reverse of polymerization.

The model for how polymerization and hydrolysis are coupled is shown in Figure 11.1. During the growing phase, monomers with nucleotide triphosphate (NTP) add on to the end of a filament and later get trapped in the interior of the filament, where they hydrolyze their NTP. The lag between polymerization and hydrolysis creates a "cap" of ATP or GTP subunits at the end. The hydrolysis of nucleotides in this cap converts the filament to a shrinking phase, during which the nucleotide diphosphate (NDP)-subunits dissociate from the end. Alternatively, in a process called treadmilling (described later in this chapter), dissociation of NDP-subunits occurs at the other, nongrowing end of the filament (Wegner, 1976; Margolis and Wilson, 1978). The cycle is completed with the exchange of NTP for NDP on the monomer, and it is ultimately driven by metabolic processes that convert NDP back to NTP.

There are two crucial features that assure that polymerization and hydrolysis are tightly coupled. First, hydrolysis of NTP is catalyzed by polymerization (hydrolysis of ATP and GTP is very slow when they are free in solution or when bound to monomers). Thus, for example, the microtubule is a GTPase-activating protein for tubulin (Erickson and O'Brien, 1992). And second, the exchange of NTP for NDP is catalyzed by depolymerization. This is because exchange cannot happen in the polymer and can only happen after the subunit has dissociated (Martonosi et al. 1960; Weisenberg et al., 1976; Margolis and

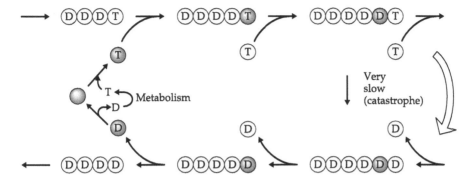

Figure 11.1 The polymerization cycle for actin and tubulin
In the growing phase (top row), the addition of subunits carrying ATP or GTP creates a nucleotide triphosphate (T) "cap" at the end, while accelerating nucleotide hydrolysis at internal subunits. A cap consisting of just a single NTP-subunit is shown in the top row. For both actin and tubulin, any lag between polymerization and hydrolysis will lead to an increase in cap size with the rate of polymerization. Hydrolysis of the ATP or GTP in the cap converts the filament to a shrinking phase (bottom row) during which the ADP- or GDP-containing subunits dissociate. In an alternative mechanism described in the text, diphosphate-containing subunits dissociate from the other, nongrowing end, leading to treadmilling.

Wilson, 1978). In the case of actin, the exchange of ATP for ADP on the monomer is regulated by other proteins: It is accelerated by profilin (Mockrin and Korn, 1980), which acts as an exchange factor, and it is inhibited by thymosin (Goldschmidt-Clermont et al., 1992). In the case of tubulin, nucleotide exchange is rapid (~1 s; Brylawski and Caplow, 1983) so there is no need for an exchange factor.

The model is based on the following experimental evidence (in addition to the studies referred to in the preceding paragraph):

1. The majority of the monomers have nucleotide triphosphate bound to them. This follows because even though the monomers have similar affinities for NDP and NTP (actin, K_{ATP} = 1 nM, K_{ADP} = 0.3 nM; Kinosian et al., 1993; tubulin, K_{GTP} = 20 nM, K_{GDP} = 60 nM; Zeeberg and Caplow, 1979), the cytoplasmic concentrations of ATP and GTP (~ 1 mM) are much higher than those of ADP and GDP (~ 10 μM). The Ks are dissociation constants.

2. There is a strong driving force for polymerization because the cytoplasmic concentrations of NTP-monomers are well above the critical concentrations ([ATP-actin] is ~30 μM, from discussion for *Listeria* in Chapter 10 >> critical concentration for ATP-actin, which is ~0.1 μM from Table 11.1; [GTP-tubulin] is ~10 μM [Zeeberg and Caplow, 1979] >> critical concentration for GTP-tubulin, which is ~0.03 μM, Table 11.2).

3. Following polymerization, the nucleotide triphosphates are hydrolyzed and the phosphate is released. Biochemical assays indicate that the incorporation of actin into a filament increases the ATP hydrolysis rate to ~0.05 s^{-1}

Table 11.1 Rate constants for actin polymerization and depolymerization

Rate constant or equilibrium constant	Plus end		Minus end	
	ATP-actin	ADP-actin	ATP-actin	ADP-actin
k_{on} ($\mu M^{-1} \cdot s^{-1}$)	11.6	3.8	1.3	0.16
k_{off} (s^{-1})	1.4	7.2	0.8	0.27
K_c	0.12	1.9	0.6	1.7

Note: In vitro measurements were made in a fairly low ionic strength buffer (50 mM KCl, 1 mM MgCl, 0.2 mM ATP or ADP, 1 mM EGTA; Pollard, 1986). Similar critical concentrations have been measured in 100 mM KCl (Wegner and Isenberg, 1983).

(Carlier et al., 1984; Pollard and Weeds, 1984), and the incorporation of tubulin into a microtubule increases the GTP hydrolysis rate to >0.2 s^{-1} (Erickson and O'Brien, 1992). Thus actin filaments older than a few minutes and microtubules older than several seconds have only NDP-subunits.

4. Hydrolysis destabilizes the polymer. The diphosphate monomers have a higher critical concentration than the triphosphate monomers (the critical concentration for ADP-actin is 2 µM >> critical concentration for ATP-actin, which is ~0.1 µM, see Table 11.1; the critical concentration for GDP-tubulin is ~100 µM >> critical concentration for GTP-tubulin, which is ~0.03 µM, see Table 11.2).

Table 11.2 Rate constants for tubulin polymerization and depolymerization

Rate constant or equilibrium constant	GTP-tubulin		GDP-tubulin	
	GTP	GMP-CPP	GDP	GMP-CP
k_{on} ($\mu M^{-1} \cdot s^{-1}$)	3.2	5.4	—	2.0
k_{off} (s^{-1})	[a]	0.1	290	128
K_c (µM)	0.03	0.02	90	64

Note: Concentrations refer to the dimers and all parameters are measured for the plus end. The rate constants were measured using video microscopy to monitor the lengths of individual microtubules: The rate of change of length is multiplied by 1600 dimers per µm to obtain the molecular rate. GMP-CPP, a GTP analogue that is very slowly hydrolyzed by tubulin, is assumed to behave like GTP-tubulin. In support of this, the on-rate for GMP-CPP-tubulin is similar to that for GTP-tubulin (top row, columns 2 and 3) and the off-rate for GMP-CP-tubulin is similar to that of GDP-tubulin (middle row, columns 4 and 5). GTP and GDP data from Drechsel et al. (1992, 34°C). The critical concentration for GDP-tubulin assumes that the on-rate equals that for GTP-tubulin, and the critical concentration for GTP assumes that the off-rate equals that for GMP-CPP-tubulin. GMP-CPP data from Hyman et al. (1992, 37°C). GMP-CP data from Caplow et al. (1994, 23°C).

[a] A value of 44 s^{-1} measured by Walker et al. (1988) depends on a particular interpretation of the experimental data.

Filament Polarity, Treadmilling, and Nucleation

The energy derived from nucleotide hydrolysis removes several of the constraints inherent to the equilibrium polymer model.

Filament Polarity

An important constraint on an equilibrium polymer is that the critical concentration must be the same at both its ends. The reason is that at equilibrium the addition of a monomer onto one end of an n-mer creates an $(n + 1)$-mer that is structurally identical to that created by monomer addition onto the other end (Figure 11.2). By Boltzmann's law, the equilibrium constants must be the same. However, the kinetics of assembly need not be equal at the two ends. According to the definition in Chapter 7, the end with the faster kinetics is referred to as the plus end, while the other end is called the minus end. For both actin filaments and microtubules, the plus end is often associated with membranes; for example, microvilli and stereocilia are filled with actin filaments whose plus ends are at the tips, and cilia and flagella are composed of microtubules whose plus ends are distal to the cell body or sperm head.

Treadmilling

Nucleotide hydrolysis adds an irreversible step to the assembly reaction and this removes the constraint of the equilibrium model that the critical concentration be the same at both ends (Wegner, 1976). For both actin and tubulin, the plus ends have lower critical concentrations (see Table 11.1 for actin data). At monomer concentrations intermediate between the critical concentrations for the two ends, the plus end is expected to grow while the minus end is expected to shrink (Figure 11.3). This behavior, demonstrated in vitro for both actin filaments (Wegner, 1976) and microtubules (Margolis and Wilson, 1978), has been termed **treadmilling**.

Treadmilling of tubulin occurs in vivo: During metaphase there is a poleward flux of photoactivated fluorescent tubulin through the mitotic spindle as subunits incorporate at the kinetichore and leave at the poles (Mitchison, 1989); treadmilling is also seen in cells in interphase (i.e., cells not in mitosis; Rodio-

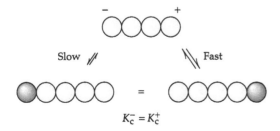

Figure 11.2 Polymerization at the two ends of an equilibrium polymer
Although the rates at the two ends might be different, the critical concentrations must be the same.

Figure 11.3 Treadmilling
The growth rates at the two ends of an actin filament are plotted against the ATP-actin concentration (on-rates are for ATP-actin, off-rates are for ATP-actin on the plus end and ADP-actin on the minus end). Because polymerization is coupled to nucleotide hydrolysis, the critical concentrations at the two ends are not the same. If the actin monomer concentration is precisely set between these concentrations, then the rate of growth at one end will balance the shrinkage at the other, so that total length of the polymer will remain constant as subunits flux through it. This is called treadmilling.

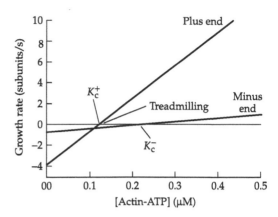

nov and Borisy, 1997). Treadmilling of actin was first seen in the leading edge of fibroblasts (Wang, 1985), and has subsequently been seen in the leading edge of keratocytes (Theriot and Mitchison, 1991) and in the tails behind moving bacteria (Theriot et al., 1992). However, the treadmilling of actin observed in cells is not in quantitative agreement with the treadmilling observed in vitro with purified actin (Wegner, 1976). The problem is that treadmilling of pure actin is expected to occur only at monomer concentrations of ~0.13 µM, at which the growth rate of the filaments would only be ~0.1 monomers per second, or 0.3 nm/s (see Figure 11.3). But this is one to three orders of magnitude lower than the rate of treadmilling (and movement) in these cells. The resolution of this problem is that other proteins, such as Arp2/3 complex, capping protein, profilin, and actin depolymerizing factor (ADF), profoundly affect the growth and depolymerization of actin. In particular, the depolymerizing and severing activity of ADF (and perhaps other proteins) increases the depolymerization rate by up to two orders of magnitude or more (Loisel et al., 1999; Pollard et al., 2000). This ensures that in cells the pool of monomers is at a high enough concentration (~30 µM; Example 10.1) to produce the very high growth rates associated with polymerization-driven motility. The detailed mechanism by which these (and other) actin-binding proteins influence actin dynamics and motility is just now beginning to be understood (e.g., Pollard et al., 2000).

Nucleation

Another constraint of an equilibrium polymer is that the critical concentration for nucleation is the same as the critical concentration for growth. What this means is that if the total subunit concentration is above the critical concentration, then nuclei will form spontaneously (if one waits long enough) and filaments will grow on them. A nonequilibrium polymer is not bound by this constraint. There are monomer concentrations that are high enough to allow growth off existing polymers, but too low for the spontaneous formation of new polymers. One way that this could occur is if the hydrolysis rate greatly exceeds the nucleation rate so that the prenuclei hydrolyze their NTP and depolymerize before a stable nucleus is formed; the critical concentration for nucleation would therefore equal the critical concentration for the NDP-subunit (K_D). Now, if the concentration of NTP-monomers is greater than the critical concentration for growth on existing polymers (K_T), but the concentration of NDP-monomers is less than K_D, growth will occur only on existing polymers. In this way, spontaneous nucleation is inhibited.

The finding that the nonhydrolyzable GTP analogue GMP-CPP promotes spontaneous assembly of microtubules at much lower tubulin concentrations than does GTP suggests that hydrolysis inhibits nucleation (Hyman et al., 1992). But whether GTP hydrolysis is the primary mechanism by which cells prevent spontaneous microtubule nucleation is not known, because spontaneous nucleation might be slow for kinetic reasons (Erickson and Pantaloni, 1981). Indeed, the nucleation of microtubules is dependent on the tubulin concentration raised to the power of 12 (Fygenson et al., 1994), which means that small changes in concentration have a large effect on nucleation rate. Thus tubulin in cytoplasm might be a super-saturated solution with an extremely low rate of spontaneous nucleation.

Dynamic Instability

An equilibrium polymer is either growing, shrinking, or, when it is at equilibrium, undergoing small, diffusive fluctuations in length. In contrast, microtubules in vitro (Mitchison and Kirschner, 1984a,b; Horio and Hotani, 1986; Mitchison, 1989) and in vivo (Cassimeris et al., 1988; Sammak and Borisy, 1988; Hayden et al., 1990) are observed to switch between phases of growth and shrinkage (Figure 11.4). Such behavior has been termed **dynamic instability** (Mitchison and Kirschner, 1984b).

One crucial cellular function of dynamic instability is the "capture" of the kinetichore during the prophase of mitosis. High-resolution video microscopy (Hayden et al., 1990) has shown that during mitosis, microtubules grow radically from each spindle pole in random directions. If the plus end of the growing microtubule makes contact with the kinetichore, then it is stabilized; if it misses the kinetichore, then the microtubule undergoes catastrophe (discussed in the next section), shrinks back to the pole, and another microtubule grows out in another randomly chosen direction. Other mechanisms to connect microtubules and kinetichores are difficult to envisage, given the large size and high rigidity of microtubules. Dynamic instability has been proposed as a general mechanism by which the orientation of microtubules, and therefore of cells, can be regulated: Microtubules grow in random directions and shrink back unless stabilized by capping proteins that select the correctly oriented microtubules (Kirschner and Mitchison, 1986). Such a selection principle for cell morphology is analogous to the selection mechanisms underlying evolution and immunology.

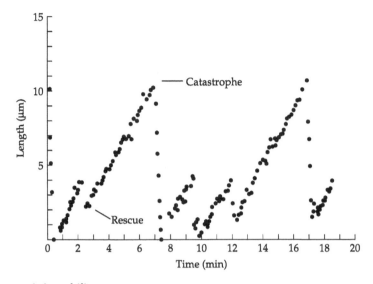

Figure 11.4 Dynamic instability
The length of a microtubule undergoing dynamic instability as a function of time. (From Fygenson et al., 1994.)

Switching between Growth and Shrinkage

As expected for the behavior shown in Figure 11.4, dynamic instability is described empirically by a four-parameter model. The parameters are the elongation rate in the growing phase (v_+), the shortening rate in the shrinking phase (v_-), the rate of the transition between growing and shrinking phases (the **catastrophe rate**, f_{+-}), and the rate of the transition between shrinking and growing phases (the **rescue rate**, f_{-+}). The model predicts a sharp conversion between bounded growth and unbounded growth depending on whether $f_{+-}v_- - f_{-+}v_+$ is greater than or less than zero (Dogterom and Leibler, 1993; see Appendix 11.2). The conversion to unbounded growth occurs when the rescue rate becomes sufficiently high that the average increase in length during the growth phase exceeds the average decrease in length during the shrinkage phase. Under conditions that lead to bounded growth, the lengths of the microtubules are distributed exponentially with mean length:

$$n_{av} \cong \frac{v_- v_+}{v_- f_{+-} - v_+ f_{-+}} \tag{11.1}$$

This equation has a simple interpretation: Ignoring rescue (i.e., setting $f_{-+} = 0$), it says that the mean length equals the elongation rate (v_+) times the average duration of the growth phase ($1/f_{+-}$).

The dynamic parameters have been measured in vitro (Walker et al., 1988) and in vivo (Cassimeris et al., 1988), and they predict, via Equation 11.1, mean lengths in the micron range, consistent with microtubule lengths in vivo (Soltys and Borisy, 1985). Interestingly, addition of cyclins to frog oocyte cytoplasm (to mimic mitosis) increases the catastrophe rate (Verde et al., 1992), resulting in a decrease in microtubule length such as that observed in cells that enter mitosis (Murray and Hunt, 1993). Such a decrease in microtubule length makes biological sense. Holy and Leibler (1994) have shown that the efficiency of dynamic instability as a strategy for searching for kinetichores is optimized when there is no rescue ($f_{-+} = 0$) and the average length of the microtubules ($n_{av}\delta$) is approximately equal to the distance between the pole and the target. In other words, in the optimal search, the microtubules must be long enough to reach the kinetichore, but the average length should not greatly exceed this distance, since continued growth after a miss will slow down the search. Thus by increasing the catastrophe rate, the mean microtubule length is shortened to more nearly equal the average pole–kinetichore distance, leading to a more efficient capturing of kinetichores.

GTP-Cap Model

Dynamic instability requires that a microtubule switch between different structural states. The simplest model for the structural transition is that the end of a growing microtubule has a cap of GTP-subunits and that the catastrophe and rescue transitions are related to the loss and reformation of this GTP-cap (Mitchison and Kirschner, 1984a; Figure 11.5). The cap has not been directly

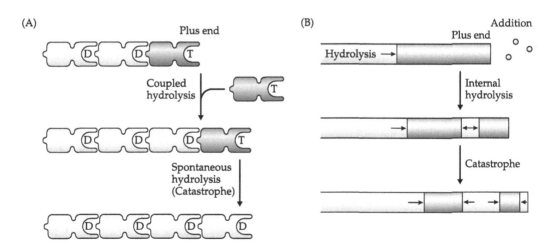

Figure 11.5 GTP-cap models for dynamic instability
(A) The incoming subunit interacts with the nucleotide at the plus end of the microtubule and catalyzes hydrolysis. Catastrophe is caused by occasional spontaneous hydrolysis or dissociation of the terminal subunit. Only one protofilament is shown: The microtubule actually has 13 subunits. (B) There is competition between internal hydrolysis and subunit addition (top). The cap shortens abruptly when a GTP on an interior subunit spontaneously hydrolyzes and this zone of hydrolysis spreads. Catastrophe occurs when fluctuations in end growth and interior hydrolysis cause loss of the cap.

visualized and its size is therefore uncertain. However, addition of as few as ~40 fluorescently labeled GMP-CPP-tubulin subunits onto the end of a microtubule is sufficient to stabilize it (Drechsel and Kirschner, 1994); this suggests that the minimum cap is no more than a few tubulin-GTP subunits deep (keeping in mind that there are 13 protofilaments). This is confirmed by kinetic experiments, which suggest that the minimum cap is a ring of 13 or 14 GTP-subunits, and that loss of any one of these subunits is sufficient to trigger catastrophe (Caplow and Shanks, 1996).

It is expected that the size of the cap should increase as the growth rate increases and addition of new GTP-subunits outpaces hydrolysis. However, the delay before catastrophe following rapid dilution is independent of the growth rate (Walker et al., 1991), which is unexpected because longer caps are expected to take more time to be lost.

This last observation can be explained by two different models. In one, the GTP in the terminal β-subunit on the plus end of the microtubule is hydrolyzed very slowly (as GTP is hydrolyzed only very slowly by the dimer in solution). When another dimer binds to the end, residues in the α-subunit of the incoming dimer interact with this GTP and catalyze its hydrolysis (Nogales et al., 1999). Consequently, there is very tight coupling between polymerization and hydrolysis, and the GTP-cap is only one dimer deep (see Figure 11.5A). Catastrophe is then due to the spontaneous hydrolysis of GTP in the cap or the dissociation of a terminal GTP-subunit. The precise relationship between the biochemical parameters (k_{on}^T, k_{off}^T, k_{on}^D, k_{off}^D, and hydrolysis rate k_{hyd}) and the dynamic parameters (v_+, v_-, f_{+-}, f_{-+}) will depend on the detailed interactions between subunits and their neighbors. The other model envisages a larger GTP-cap. The hydrolysis of GTP in one dimer within the cap (at a random location) induces hydrolysis within neighboring dimers, effectively cutting down the size of the cap. Catastrophe occurs when the internal hydrolysis overtakes addition of new subunits (Flyvbjerg et al., 1994). In this model the size of the cap increases with the growth rate, but, following dilution, the lifetime of the longer caps is similar to that of shorter caps because spontaneous hydrolysis of subunits near the end is equally likely for the shorter and longer caps. This model has been solved in detail and can account for most of the experimental data (Flyvbjerg et al., 1996).

The structure of the GTP-cap is an open question. Growing microtubules often have two-dimensional sheets of tubulin at their ends (Erickson, 1974; Chretien et al., 1995; Arnal et al., 2000), whereas shrinking microtubules have curved protofilaments peeling out from their ends (Mandelkow et al., 1991). Microtubules with blunt ends are also seen. Thus the cap could be a sheet, or it could be the terminal subunits in the sheet (Arnal et al., 2000); the answer will require distinguishing the GDP-dimers from the GTP-dimers in the electron microscope.

Work during Polymerization and Depolymerization

The hydrolysis of GTP provides a mechanism by which a large amount of mechanical energy can be stored within the microtubule lattice. The critical concentration for GDP-tubulin is much higher than that for GTP-tubulin (see Table 11.2); presumably this is due to the GDP-tubulin having a more strained conformation that does not sit well in the microtubule lattice (Caplow et al., 1994). This strain energy is, in principle, available for doing mechanical work, as I now show.

Consider a model in which an external compressive force, F, opposes microtubule polymerization (see Figure 10.2). If the association rates of the GTP- and GDP-dimers (k_{on}^T and k_{on}^D, respectively) do not depend on the relative amounts of GTP- and GDP-subunits already at the end, then the total subunit association rate will be $\{k_{on}^T[A^T]+k_{on}^D[A^D]\}\exp(-F\delta/kT)$ where $[A^T]$ and $[A^D]$ are the cytoplasmic GTP-tubulin and GDP-tubulin concentrations, respectively. The total dissociation rate is $p^T k_{off}^T + p^D k_{off}^D$, where p^T and p^D are the proportion of GTP- and GDP-subunits in the cap (in general, p^T and p^D will depend on whether the microtubule is in a growing or shrinking phase). The maximum force, or the reversal force (the force at which there is no net growth) occurs when association and dissociation are balanced, so that

$$F = \frac{kT}{\delta}\ln\left[\frac{k_{on}^T[A^T]+k_{on}^D[A^D]}{k_{off}^T p^T + k_{off}^D p^D}\right] \quad (11.2)$$

The force reduces to the chemical-force equation of the last chapter if there is only one species of subunit. Equation 11.2 is analogous to the Goldman-Hodgkin-Katz (GHK) voltage equation, which describes how the reversal potential depends on the concentration and permeabilities of the ions on each side of a membrane (Hille 1992). The GHK equation assumes that the ions permeate by independent pathways; the analogous assumption here is that the association rate of a monomer does not depend on the subunit composition at the growing end.

During periods of rapid growth, the cap will have exclusively GTP-subunits and so $p^T = 1$ and $p^D = 0$. If this remains the case at high loads, then the maximum compressive force the microtubule can push against is

$$F_{polym} \cong \frac{kT}{\delta}\ln\frac{k_{on}^T[A^T]}{k_{off}^T} = \frac{kT}{\delta}\ln\frac{[A^T]}{K_c^T} \quad (11.3)$$

where K_c^T is the critical concentration for GTP-tubulin. This is an upper limit on the polymerization force because high loads may accelerate the dissociation of subunits. On the other hand, during rapid shrinkage, there is no GTP in the cap so $p^T = 0$ and $p^D = 1$. In this case, the maximum force is

$$F_{depolym} \cong \frac{kT}{\delta}\ln\frac{k_{on}^T[A^T]}{k_{off}^D} \cong \frac{kT}{\delta}\ln\frac{[A^T]}{K_c^D} \quad (11.4)$$

where K_c^D is the critical concentration for GDP-tubulin. Because $[A^T] < K_c^D$, the sign of the force is negative, which means that it is a tensile force: A depolymerizing microtubule pulls, whereas a polymerizing one pushes. The last equation is analogous to the equation that describes the "bi-ionic potential" setup across a membrane that has one permeant ion on one side and another permeant ion on the other (Hille, 1992). In our case, $F_{depolym}$ is the reversal force such that the flux of GTP-subunits onto the polymer equals the flux of GDP-subunits off the polymer. Using $\delta \sim 0.6$ nm, $[A^T] = 10$ μM, $K_c^T = 0.02$ μM, and $K_c^D = 100$ μM, we obtain maximum polymerization and depolymerization forces for microtubules of 40 pN and −15 pN, respectively. These forces are considerably larger than those generated by the motor protein kinesin because the subunit distance δ is much smaller than kinesin's step size. The forces can readily account for the magnitude of the forces acting during mitosis (Chapter 10). Using $\delta \sim 2.75$ nm, $[A^T] = 30$ μM, and $K_c^T = 0.12$ μM, we obtain a maximum polymerization force for actin of 8 pN.

The total work that can be done during a cycle of polymerization and depolymerization is

$$w_{total} = w_{polym} + w_{depolym} = F_{polym}\delta - F_{depolym}\delta \leq kT \ln\left[\frac{k_{off}^D}{k_{off}^T}\right] \cong kT \ln\left[\frac{K_c^D}{K_c^T}\right] \quad (11.5)$$

where the last approximate equality holds because $k_{on}^D \cong k_{on}^T$. It is important to point out that these expressions (Equations 11.3, 11.4, and 11.5) are upper limits on the force and work because the cap may be a mixture of GTP- and GDP-subunits. Because the GDP-tubulin off-rate is about 5000 times greater than the GTP-tubulin off-rate (see Table 11.2), the maximum work for a cycle of GTP hydrolysis is $kT\ln 5000 = 8.5\ kT$, about 35% of the maximum free energy available under cellular GTP, GDP, and P_i concentrations, and similar to the maximum work of the motor proteins kinesin and myosin (Howard, 1996; see also Chapter 16). For actin, the ratio of the off-rates at the plus end is only 5, which means that the maximum work in a cycle is $kT\ln 5 = 1.6\ kT$. However, as argued above, the presence of actin-binding proteins profoundly alters the dynamics of actin, and this could lead to an increase in the efficiency.

There is a close analogy between the thermodynamics of cytoskeletal polymers and those of membrane pumps and channels. I have commented in this chapter and the previous one that expressions for the forces and kinetics of polymerization and depolymerization are analogous to expressions governing the electrical potentials and the flow of ions across membranes. The underlying reason for the analogy is that an end-polymer, a polymer for which growth and shrinkage occur only at the ends, is analogous to an impermeable membrane-enclosed compartment with an ion channel in it (or two ion channels, if both ends are dynamic). If the monomer concentration differs from the equilibrium concentration, then polymerization can be coupled to the generation of a mechanical force, analogous to the electrical potential set up across a membrane when the impermeant ion is not at equilibrium. Furthermore, if the subunits are not at equilibrium and the monomer concentration is below the critical concentration, a very considerable mechanical potential energy can be stored in the polymer lattice. This energy is analogous to the electrochemical or electro-osmotic energy stored across a membrane. The mechanical energy can be considerable: A 60-μm-long microtubule contains roughly 100,000 subunits and could store an energy equal to $\sim 1{,}000{,}000\ kT$, equal to the free energy of some 40,000 ATP molecules. Nucleotide hydrolysis acts like the pump that transfers subunits against their "concentration gradient": The NDP-monomers are pumped into the polymer by the favorable association of the NTP-monomer, followed by hydrolysis to the unstable NDP form, and capping to stabilize the high-energy polymer. The cap plays a role analogous to that of an ion channel by regulating the flux of NDP-subunits down their chemomechanical gradient and out of the polymer in response to gating signals.

Structural Changes Attending Nucleotide Hydrolysis

What structural changes occur within the subunits that allow actin filaments and microtubules to store mechanical energy? The simplest idea is that the NTP-subunit fits snugly into a filament, whereas the NDP-subunit does not. In this case, the NTP-subunit will have a lower critical concentration than the NDP-subunit. If this structural change takes place after the subunit has been incorporated into the filament (due to hydrolysis), then the NDP-subunit will be forced to adopt a mechanically strained conformation. It is the mechanical potential energy associated with this strain that is stored in the filament and which is the driving force for depolymerization. Do actin and tubulin undergo such structural changes?

Comparison of the structure of actin in crystals with actin in thin filaments in muscle (Lorenz et al., 1993) indicates that when actin polymerizes, its nucleotide-binding cleft closes a little. A similar closure of the cleft is observed in actin-profilin crystals in different buffers (Chik et al., 1996). These observations can account for the lower critical concentration of ATP-actin if we assume that ATP stabilizes the cleft in a more closed state (Figure 11.6A). Following hydrolysis, the ADP-subunit then finds itself trapped in an unfavorable closed (and strained) conformation due to contacts with the surrounding subunits (Figure 11.6B). If cleft closure is to regulate polymerization, the protein must be rigid enough that the distortion could store enough mechanical energy. A back of the envelope calculation shows that a protein like actin with the Young's modulus of Plexiglas (Chapter 8) is indeed rigid enough that even small cleft closures on the order of a few angstroms would be associated with large enough energy changes (a few kT) to significantly alter the equilibrium con-

(A)
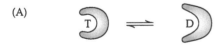

(B)
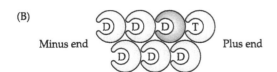
Minus end   Plus end

Figure 11.6 Structural model for coupling hydrolysis to actin polymerization
(A) The actin subunit in solution is in rapid equilibrium between open and closed states. ATP binds more strongly to the closed state. (B) To incorporate into the filament, the subunit must be even more closed. If the transition state for hydrolysis is more closed still, then hydrolysis will be accelerated by polymerization.

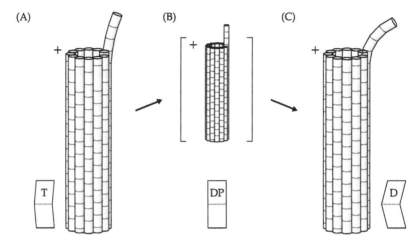

Figure 11.7 Structural changes attending GTP hydrolysis in microtubules
If the GTP-subunit is only slightly curved, the resulting protofilament need only be slightly deformed when it is in the straight wall of the microtubule (A). If, following polymerization, the subunit changes to a very curved GDP-state, then it will be very unstable in the microtubule wall, and so the GDP-microtubule will tend to depolymerize (B). If the GDP · P_i-subunit is straighter still, then it will be stabilized by being in the wall of the microtubule (C). Thus the transition GTP → GDP · P_i will be accelerated after closure. In this way polymerization catalyzes hydrolysis, as required for a coupled mechanism.

stant and therefore the critical concentration. An extension of this idea can also account for the coupling between polymerization and ATP hydrolysis: If incorporation of the subunit into the filament requires that the cleft close even more than in the ATP structure, and if the transition state for hydrolysis is also more closed, then assimilation of the subunit into the filament will accelerate hydrolysis as required to ensure coupling.

A similar idea can be applied to microtubules. Electron microscopy shows that the protofilaments at the ends of a growing GTP microtubule are fairly straight (Chretien et al., 1995), while those at the ends of a shrinking GDP microtubule curve outward (Mandelkow et al., 1991). The simplest interpretation is that whereas the GTP dimer is fairly "straight," the GDP dimer is "curved" (Figure 11.7). Thus the GTP-subunit fits more comfortably into the *straight* wall of the microtubule, and so has a lower critical concentration. But after GTP hydrolysis, the GDP-subunit is highly strained and mechanical energy is stored in the lattice. If the transition state for the hydrolysis reaction is straighter still than the GTP state (Figure 11.7B), then the straightening of the protofilament attending closure of the microtubule will catalyze hydrolysis. This is another way that hydrolysis could be coupled to polymerization (the first way is shown in Figure 11.5A). Just like the case with actin, association with other subunits provides some of the energy required to enter the transition state. Also, like the case of actin, calculations indicate the microtubule protein is rigid enough that such mechanical deformations can give rise to sufficiently large energy differences (Mickey and Howard, 1995).

Summary

Nucleotide hydrolysis endows actin filaments and microtubules with many properties not possible for equilibrium polymers. It removes the constraint that the critical concentration of the two ends be the same, permitting the simultaneous growth on one end and shrinkage at the other. It means that the critical concentration for nucleation might be less than that for elongation so that growth only occurs on preformed nuclei. In addition, it means that polymers need not always be in a growing or a shrinking phase, but can switch between them. Such dynamic behavior allows the actin and microtubule cytoskeleton to respond quickly to signals, and to do mechanical work to alter cell morphology during growth and movement.

Motor Proteins

PART

III

Motor proteins are enzymes that convert the chemical energy derived from the hydrolysis of ATP into mechanical work used to drive cell motility. The widely accepted framework for understanding this chemomechanical transduction process is the rotating crossbridge model (see Figure 12.1). The model contains two key ideas. First, the motor cycles between attached and detached states. Second, while attached, the motor undergoes a conformational change (the working stroke) that moves the load-bearing region of the motor in a specific direction along the filament. If recovery takes place during the detached phase of the cycle, there will be a net displacement of the motor toward its next binding site on the filament.

Each chapter in Part III examines the rotating crossbridge model from a different perspective. Chapter 12 describes the structures of the crossbridges, Chapter 13 the movement of the motors, Chapter 14 the mechanisms of ATP hydrolysis, Chapter 15 the behavior of single motor molecules, and Chapter 16 presents models that account for the macroscopic properties of motile systems—the contraction of muscle and the processive movement of kinesin—in terms of the structural, mechanical, and chemical properties of individual crossbridges.

CHAPTER

12

Structures of Motor Proteins

The discovery of the myosin crossbridges by H. E. Huxley in 1957 provided a molecular basis for the contraction of muscle: The bending or rotation of these crossbridges causes the actin-containing thin filaments to slide relative to the myosin-containing thick filaments, and the sliding of these filaments, in turn, leads to the shortening of the muscle (as had been demonstrated a few years earlier by H.E. Huxley and Hanson [1954] and A. F. Huxley and Niedergerke [1954]; and see Figure 1.1). Since the initial discovery, the rotating crossbridge has proven to be a general mechanism for cell motility that is responsible for the beating of cilia and flagella, the movement of organelles, and the segregation of the chromosomes.

The rotating crossbridge model continues to provide the framework within which the structural, biochemical, and mechanical properties of motor molecules, not just myosin, can be understood. This framework incorporates the following findings and hypotheses: (1) the Lymn–Taylor scheme, which describes how nucleotides regulate the attachment and detachment of myosin from the filament (Lymn and Taylor, 1971); (2) the swinging lever arm hypothesis, which provides a mechanism by which small structural changes around the nucleotide-binding pocket are amplified into much larger conformational changes of the crossbridge (Rayment et al., 1993a); and (3) the powerstroke model, which accounts for how the motor generates force by making the additional assumption that there is an elastic element within the crossbridge that is strained during the power stroke (A. F. Huxley, 1957). The rotating crossbridge model has been successfully applied to dynein (e.g., Johnson, 1985) and kinesin (Howard, 1996, 1997).

This chapter describes the structures of the crossbridges that are formed by motor proteins and shows how these structures change during the nucleotide hydrolysis cycle. The structural changes are readily interpreted in terms of the rotating crossbridge model.

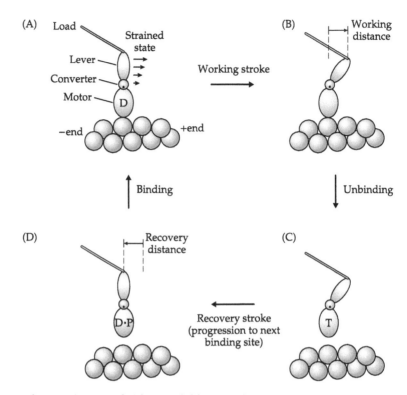

Figure 12.1 The rotating crossbridge model for myosin
(A) The binding of myosin to the actin filament catalyzes the release of phosphate from the motor domain and induces the formation of a highly strained ADP state. (B) The strain drives the rotation of the converter domain, which is connected to a lever domain that amplifies the motion, moving the load through the working distance. (C) Following ADP release, ATP binds to the motor domain and causes dissociation of myosin from the actin filament. (D) While dissociated, the crossbridge recovers to its initial conformation, and this recovery moves the motor toward its next binding site on the filament. T = ATP, D = ADP, P = Pi.

Crossbridges and the Domain Organization of Motor Proteins

According to the **rotating crossbridge model** (Figure 12.1), the movement of a motor protein along its track is driven by a directed conformational change within the crossbridge that the motor protein forms with the filament. A key goal, therefore, is to visualize this conformational change using structural techniques, and to understand how it is driven by the hydrolysis reaction. Of particular interest is how small structural changes associated with the chemistry of ATP hydrolysis are mechanically amplified into much larger conformational changes of protein domains. This amplification is at least 10-fold—from a few angstroms, the size of the phosphate ion, to several nanometers, the size of the working stroke deduced from physiological, biophysical, and biochemical experiments (Chapters 13 and 15). Over the last few years, the structures of the crossbridges have been resolved in enough detail to unambiguously detect these conformational changes. The myosin crossbridge has three distinct regions: One that creates the conformational change, one that converts it into directed motion, and one that amplifies it. Homologous regions have been identified in kinesin, and possibly in dynein.

The first glimpses of **crossbridges** were the linkages (seen by electron microscopy) that span between the thick and thin filaments of muscle cells (H. E. Huxley, 1953, 1957; Figure 12.2A). The chemical identity of these crossbridges

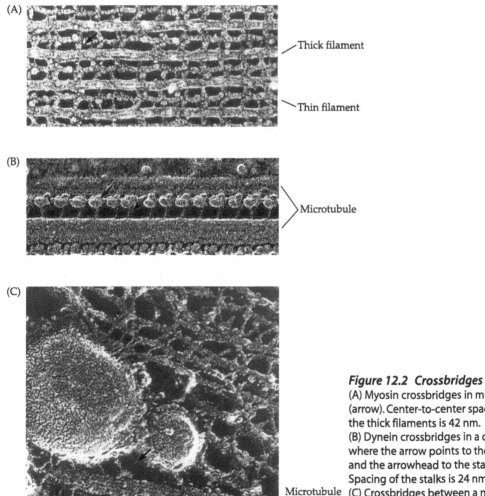

Figure 12.2 Crossbridges
(A) Myosin crossbridges in muscle (arrow). Center-to-center spacing of the thick filaments is 42 nm.
(B) Dynein crossbridges in a cilium where the arrow points to the head and the arrowhead to the stalk. Spacing of the stalks is 24 nm.
(C) Crossbridges between a membrane-bounded organelle and a microtubule. (A courtesy of Roger Cooke and John Heuser; B from Goodenough and Heuser, 1982; C from Hirokawa, 1998.)

was established by repolymerizing purified myosin in vitro: the resulting "synthetic" thick filaments have a similar size and shape to native thick filaments and have numerous projections that correspond to the crossbridges (Huxley, 1963). Crossbridges between the microtubule doublets of sperm (Afzelius, 1959) and between organelles and microtubules in neurons (Miller and Lasek, 1985) were discovered subsequently (Figure 12.2B,C). Dynein was shown to form the crossbridges in sperm by selective extraction with high salt; electron microscopy confirmed that the crossbridges had indeed been removed (Gibbons, 1963). The disorganized structure of organelles has made the molecular identification of their crossbridges less certain.

The crossbridges are formed by the globular **head** or **motor domains**, which have been defined by electron microscopy of isolated proteins (Figure 12.3). The evidence that the heads have motor activity is very strong. For example, limited proteolysis of purified skeletal muscle myosin cleaves the molecule into two fragments (Szent-Györgyi, 1953; Figure 12.3B). One of these, **HMM**, contains the globular head domains seen in individual myosin molecules and binds to actin (Huxley, 1963). The other, **LMM**, has no globular region and does not bind to actin; furthermore, though LMM forms synthetic filaments, these filaments do not have projections like the native filaments (Huxley, 1963), showing that the projections correspond to the globular domain of myosin. The actin-binding domain is further localized to **subfragment 1 (S1)**, which is produced by further proteolysis of HMM (Lowey et al., 1969). S1 has a dimension similar to that of a crossbridge, it hydrolyzes ATP at a rate that is greatly accelerated by actin (actin-activated ATPase), and it moves actin filaments in an in vitro motility assay (Toyoshima et al., 1987; Chapter 13). Thus the S1 head domain has the appropriate size and functional properties expected of a crossbridge. In the case of conventional kinesin, the globular head domain (Yang et al., 1989) has been identified as the motor domain: It has a microtubule-stimulated ATPase activity (Stewart et al., 1993), and it moves microtubules in in vitro motility assays (Scholey et al., 1989; Yang et al., 1990; Stewart et al., 1993). Dynein also has a microtubule-stimulated ATPase activity (Gibbons and Rowe, 1965); strangely, each head contains four nucleotide binding sites (Gibbons et al., 1991), though only one of them is active.

Figure 12.3 Domain organization of motor proteins ▶
(A) Electron micrographs of myosin from rabbit skeletal muscle. (B) Interpretation of the myosin EM structures, together with the definitions of the subfragments HMM, LMM, S1, and S2. S1 is the head or motor domain and forms the crossbridge. The long tail is a coiled coil. (C) Electron micrographs of three-headed, outer-arm dynein from *Tetrahymena* axonemes and two-headed cytoplasmic dynein. (D) Interpretation of the three-headed dynein structures, together with the definitions of the various domains. (E) Electron micrographs of conventional kinesin. (F) Interpretation of the kinesin EM structures, together with the definitions of the head and tail domains. The elongated domain that connects the head to the tail is a coiled coil which is broken approximately midway by a proline residue to form a hinge. The proteins were visualized by rotary shadowing. (A from Elliott and Offer, 1978; C from Goodenough and Heuser, 1984, micrographs courtesy of John Heuser; E from Hirokawa et al., 1989.)

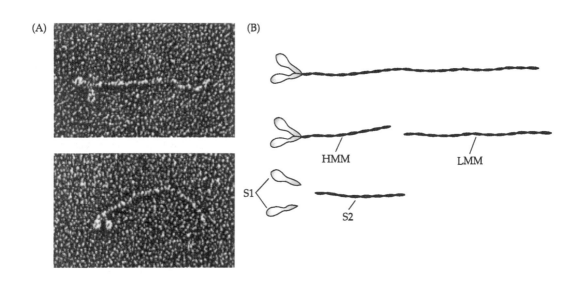

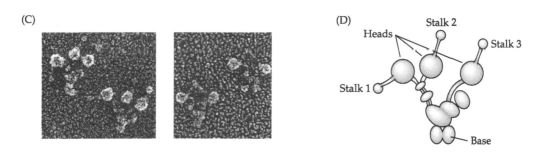

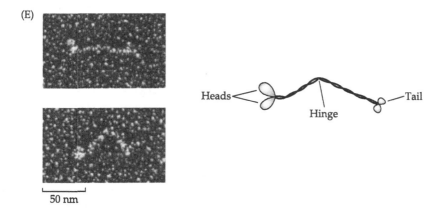

Muscle myosin, conventional kinesin, and axonemal dynein are all elongated molecules, though dynein is less so. They have a similar domain organization, with a head at one end and a **tail** (or a base) at the other. Each head is composed of a large subunit, called the α-subunit, or heavy chain. In addition, there are smaller subunits (β, γ, etc.) that associate with the heads in the case of muscle myosin, with the tail in the case of conventional kinesin (Cyr et al., 1991; Gauger and Goldstein, 1993), and with both in the case of dynein (Milisav, 1998). Both myosin and kinesin contain two identical heavy chains that dimerize via a coiled coil, and the motors can be written as $(\alpha\beta\gamma)_2$ and $(\alpha\beta)_2$, respectively. By contrast, the heavy chains of dynein have different, though similar, amino acid sequences, and thus the structure of this motor is not symmetrical.

The tails are association domains. As mentioned above, skeletal muscle myosin forms thick filaments (with lengths up to several microns) through the oligomerization of the LMM domains (Huxley, 1963). The tail of dynein binds to the A-tubule, while the head domain, and in particular a stalk-like subdomain, binds to the B-tubule (see Figures 7.6 and 8.3B; Goodenough and Heuser, 1984; Gee et al., 1997). The tail of kinesin is thought to bind to cellular cargo (Coy and Howard, 1994; Manning and Snyder, 2000). Interestingly, the kinesin tail also binds to the head (Stock et al., 1999); it is thought that in the absence of cargo the tail inhibits the motor and ATPase activities of the head, thereby switching the motor off when it is not carrying cargo (Hackney 1992; Coy et al., 1999a).

Motor Families

Skeletal muscle myosin, axonemal dynein, and conventional kinesin are the first-discovered members of the large families of myosin-related, kinesin-related, and dynein-related proteins. Part of the myosin family is shown in Figure 12.4. The families are defined by sequence similarity in the motor domains: Typically, proteins within each family share approximately 50% amino acid sequence identity within their globular motor domains. Because of this high

Figure 12.4 Myosin motor family ▶
(A) The distal end of each radial line corresponds to a myosin-related protein. The sequences of the proteins have been placed in categories based on the similarity of the motor domains, which are aligned in (B). Each category corresponds to a subfamily of myosin-related proteins (myosin I, II, etc.). Many of the subfamilies, such as myosins I, II, and V, contain proteins from species as distantly related as yeast and humans. The percentage of divergence of two sequences is proportional to the lengths of the branches connecting them. The particular alignment algorithm used in this figure attempts to identify hypothetical ancestral sequences corresponding to ancestral proteins from which the present-day proteins evolved; the ancestors (and the ancestors' ancestors) correspond to the nodes. The structures are based in some cases on experimental data and in other cases on sequence interpretation alone. A more complete version of this figure can be found at the web site http://www.mrc-lmb.cam.ac.uk/myosin/myosin.html. (Courtesy of Jake Kendrick-Jones.)

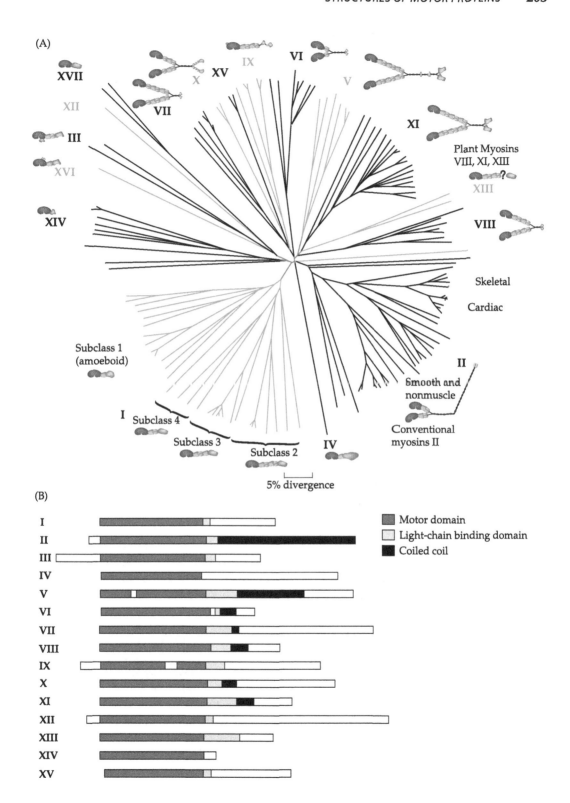

level of sequence similarity, the crossbridges formed by related proteins within each family are expected to have similar structures (protein domains with amino acid identity greater than ~20% have similar three-dimensional folds; Creighton, 1993; Brändén and Tooze, 1999).

The number of motor-related proteins that has been identified is growing rapidly, especially with the sequencing of entire genomes (e.g., the yeast *S. cerevisiae*, Goffeau et al., 1996; the nematode worm, *C. elegans* Sequencing Consortium, 1998; the fruit fly *D. melanogaster*, Adams et al., 2000), and the reader is referred to reviews (Bloom and Endow, 1995; Sellers and Goodson, 1995; Hirokawa, 1998) and databases (myosin home page: http://www.mrc-lmb.cam.ac.uk/myosin/myosin.html; kinesin home page: http://www.blocks.fhcrc.org/~kinesin/index.html). A great challenge for biologists is to elucidate the precise functions of all these proteins—for example, the baker's yeast has 6 genes encoding kinesin-related proteins and *Drosophila* has 23 genes. Are all of these proteins even motors?

The motor families contain proteins that were previously purified and found to have functional similarities to the founding proteins. For example, a protein purified from the amoeba *Acanthamoeba* was classified as a myosin based on its actin-activated ATPase activity (Pollard and Korn, 1973); subsequent sequencing showed that it was indeed a myosin-related protein (Jung et al., 1987). The families also contain proteins whose genes were identified based on mutant phenotypes. Finding that a gene codes for a motor-related protein then hints at the cause of the mutant phenotype and therefore to the cellular function of the protein. For example, the protein whose disruption gives rise to the *dilute* mouse, which has a light coat color, is a myosin-related protein (Mercer et al., 1991). Because the dilute phenotype is due to incorrect localization of pigment granules, the function of the dilute myosin, a myosin V (Cheney et al., 1993), may be to transport these granules along actin filaments. Disruption of myosin genes give defects in muscle development and contraction (Epstein et al., 1974), cell division (De Lozanne and Spudich, 1987), and the aggregation of surface receptor protein (Pasternak et al., 1989). In humans, mutations in cardiac muscle myosin II cause familial cardiac hypertrophic diseases (Geisterfer-Lowrance et al., 1990), and mutations in myosin VII cause Usher's syndrome, with its associated vision and hearing loss (Weil et al., 1995). Disruption of kinesin-related proteins in yeast, nematode worm, and fly give rise to defects in mitosis, meiosis, and the localization of synaptic vesicles (Bloom and Endow, 1995; Barton and Goldstein, 1996; Hirokawa, 1998). Interestingly, a kinesin-related protein is also involved in the development of cilia (Walther et al., 1994). Knocking out the murine homologue of this protein, KIF3A, causes loss of cilia and randomization of left–right symmetry of the internal organs (Nonaka et al., 1998). A similar phenotype is seen in a mouse carrying a mutant axonemal dynein (Supp et al., 1997). These murine conditions are similar to Kartagener's syndrome in humans, which is associated with defects in the structure of the cilia (Afzelius, 1976).

Comparisons of the sequences and the functions of related motor proteins provide important clues to the motor mechanism. For example, analysis of the sequences of myosin and kinesin confirm the domain organization established biochemically (McLachlan and Karn, 1983; Yang et al., 1989). The intersection of the sequences of proteins in each family defines the evolutionarily conserved motor domain. Furthermore, the divergence of tail sequences is consistent with their linkage to a wide variety of different cargoes.

The diversity of structure and function also highlights many questions about the motor mechanism. For example, whereas the original members of the myosin and kinesin families have two heads, physical studies show that there are members of both families that have only one head domain. Furthermore, dyneins can have one, two, or three heads! Is one head per motor sufficient for motility, and might it be that some motors have two heads because, for example, they need two tails? Alternatively, is a multiheaded structure essential for motility, and might single-headed proteins need to oligomerize to form a functional motor? These questions are addressed in the following chapters.

Another interesting question raised by this comparative approach is: What determines the speed of motors? Myosin-based motility ranges from 100 nm/s to 60,000 nm/s, and kinesin-based motility ranges from 20 nm/s to 2000 nm/s (see Table 13.1). What is the structural basis for the different gearings of myosins and kinesins?

Perhaps the most fundamental question is: What determines the directionality of a motor? (Table 13.1 lists the directions of various motors.) This question was emphasized by the discovery that the kinesin-related protein Ncd, which has ~45% sequence identity in the motor domain to conventional kinesin (Endow et al., 1990), moves in the opposite direction (McDonald et al., 1990; Walker et al., 1990). Myosin VI moves in the opposite direction to muscle myosin (Wells et al., 1999). Axonemal and cytoplasmic dyneins are minus-end directed (Sale and Satir, 1977; Paschal et al., 1987): Will a plus-end-directed dynein be discovered?

High-Resolution Structures

The atomic structures of the myosin and kinesin heads have been solved by X-ray crystallography (Figure 12.5). Of great interest is the remarkable structural similarity among a core 180 amino acids within the myosin (Rayment et al., 1993b) and kinesin (Kull et al., 1996) motor domains. The core comprises a 6-stranded β-sheet with three α-helices on both sides. The structural similarity was surprising because the amino acid sequences of the two proteins have little similarity, except for four small conserved motifs that are involved in binding to the nucleotide. These motifs are also found in G-proteins (Smith and Rayment, 1996; Vale, 1996) where they are termed G-1 (the P-loop or the Walker A motif common to many ATPases; Walker et al., 1982), G-2 (switch I), G-3 (switch II), and G-4 (Bourne et al., 1991). G-1, G-2, and G-3 bind to the phosphates and the magnesium ion, whereas G-4 binds to the purine ring. The conserved nucleotide-binding pocket (the **active site**) suggests that the chemistry of the hydrolysis reaction of motor proteins is similar to that of G-proteins

Figure 12.5 Structural similarities between kinesin and myosin
(A) Structure of the kinesin head (Protein Data Bank ID: 1BG2; Kull et al. 1996). (B) Structure of the myosin S1 fragment (Protein Data Bank ID: 2MYS; Rayment et al., 1993a). The darker region within the myosin motor domain has striking structural similarity to the kinesin motor domain. (After Rayment, 1996; courtesy of Ivan Rayment.)

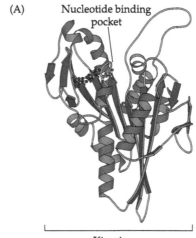

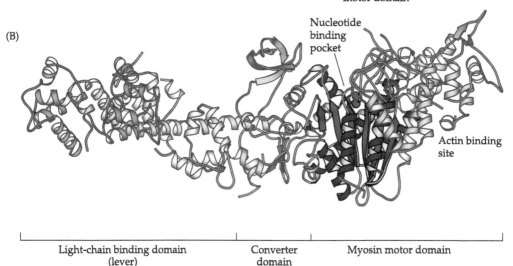

(Smith and Rayment, 1996). The additional structural similarities between myosin and kinesin suggest that nucleotide binding, hydrolysis, and release may trigger similar motions within the motor domains. Thus in addition to having a similar overall domain organization and shape, kinesin and myosin may also have similar hydrolysis mechanisms.

Interestingly, the dynein motor is structurally distinct from kinesin and myosin. Sequence analysis shows that the dynein motor domain is a member of the AAA+ class of ATPases (Neuwald et al., 1999). Like other members of the AAA+ (**A**TPases **a**ssociated with a variety of cellular **a**ctivities) protein superfamily, dynein has 6 AAA modules. Each module probably corresponds to one of the globular subdomains that contributes to the ring-like structure of the dynein head, which is seen by electron microscopy (Samso et al., 1998).

Are the structural similarities between myosin and kinesin due to convergent or divergent evolution? If the structures of the two motor domains had resulted from convergence, we would not expect the order of the secondary structural elements to be important (only their spatial location in the structure), and the direction of the amino acid chains ought not be crucial because both α-helices and β-sheets can pack in parallel or antiparallel geometries. But, in fact, there is only one transposition in the order of the secondary structural elements ($P_1 = 12/11! \cong 3 \cdot 10^{-7}$ if the order is completely irrelevant), and all 12 elements run in the same direction ($P_2 = 2^{-12} \cong 3 \cdot 10^{-4}$ if there is no directional preference). Although this analysis ignores structural constraints that will tend to favor particular sequence orders and directions, it nevertheless suggests that the two structures did not converge ($P \sim P_1 P_2 \sim 10^{-10}$). This leads to the conclusion that the motors diverged from a common, though distant, ancestor. Further structural analysis indicates that this ancestor, the protomotor, is itself probably related to an ancestral G-protein (Kull et al., 1998). Thus there was probably a common ancestral protein that was able to undergo conformational switching depending on whether the bound nucleotide was a triphosphate or diphosphate. The G-proteins refined this switch to modulate interactions with receptor and effector proteins in order to mediate cell signaling, while the kinesin/myosin motor proteins refined this switch to modulate their interactions with their associated filaments in order to mediate cell motility.

The most important revelation of the myosin structure is that the head contains a large subdomain that might function as a molecular lever (Rayment et al., 1993a,b; see Figure 12.5). The domain consists of an 8-nm-long α-helix around which is wrapped the two calmodulin-like light chains. This domain, called the light-chain binding domain, is ideally located to be the lever arm: it extends from near the nucleotide-binding pocket to the distal end of the head where S1 joins S2, and so is correctly positioned to amplify small angstrom-sized movements in the nucleotide-binding pocket into large nanometer-sized motions of the myosin tail.

There is strong crystallographic evidence that the light-chain binding domain does indeed rotate during the ATP hydrolysis cycle. The structure of the "converter" domain that forms the junction between the motor domain and the light-chain binding domain depends on the nucleotide bound at the active site (Fisher et al., 1995). The connection is made by the relay helix (also called the switch II helix), which may act like the connecting rod of an internal combustion engine to couple the opening of the nucleotide-binding pocket (the piston in the cylinder) to rotation of the converter domain (Dominguez et al., 1998; Houdusse et al., 1999; Vale and Milligan, 2000). Though the initial structural studies were performed on myosins whose light-chain binding domains were deleted to facilitate crystallization, modeling studies show that the changes in the converter domain are expected to lead to a rotation of the light-chain binding domain by ~70 degrees (Holmes, 1997; Figure 12.6). This expectation was confirmed by solving the structure of a smooth muscle myosin head that included one of the light chain domains (Dominguez et al., 1998). Comparison of this structure (with ADP·BeF$_x$ in the active site) to the original skeletal myosin structure (with no

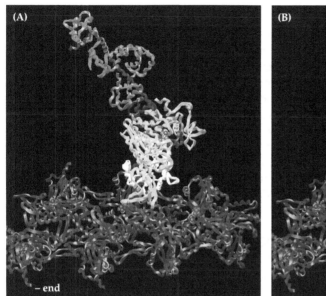
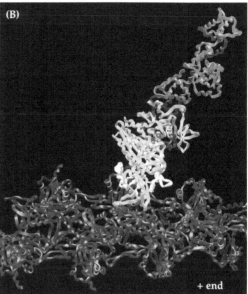

Figure 12.6 Beginning and end of myosin's working stroke
(A) Model of the pre-powerstroke state of *Dictyostelium* myosin docked to actin. The protein has ADP and vanadate in the active site, and this is thought to mimic the structure of the ADP·P_i state that first binds to the filament. (B) Model of the post-powerstroke state, also of *Dictyostelium* myosin but with ADP and BeF_x in the active site. This is thought to resemble the ADP and nucleotide-free states. (See Geeves and Holmes, 1999; courtesy of Ken Holmes.)

nucleotide) revealed a 70 degree rotation of the converter domain with a corresponding rigid-body movement of the light-chain binding domain. Given the orientation of the head when it binds to the actin filament (Rayment et al., 1993a), the rotation associated with the dissociation of nucleotide swings the distal tip of the light-chain binding domain some 10 nm towards the plus end of the actin filament. This is the correct direction to cause the contraction of muscle. These structures give strong support to the hypothesis that muscle contraction is driven by a swinging lever arm. Further support comes from single-molecule experiments showing that the size of the working distance depends on the length of the lever arm (Chapter 15).

The concept of a converter domain that acts as a sort of gearbox proves useful to understanding the reversed directionality of myosin VI. This protein has a large insert in the converter region, and cryoelectron microscopy shows that the rotation of the light chain domain has the opposite sign (Wells et al., 1999). One possibility is that the insert has altered the pivot point for rotation within the converter domain, effectively putting the motor into reverse. If this is the case, a more appropriate analogy is that the converter domain is the "throw" that couples the connecting rod to the crankshaft: Adjustment of the geometry of the throw can reverse the direction of rotation of the crankshaft and make the engine run backwards (see http://www.britannica.com for a description of how an internal combustion engine works).

Kinesin lacks a light-chain binding domain (see Figure 12.5). How could it make movements that are as large, or even larger, than those of myosin? The problem is very serious indeed when one considers that the size of kinesin's motor domain is only $7 \times 4.5 \times 4.5$ nm, yet kinesin is thought to undergo a conformational change of 8 nm as it steps along the microtubule (Chapter 15)! An attractive possibility is that kinesin's second head (Figure 12.7) acts as a lever arm that amplifies smaller motions of the first head.

Even though the dynein superfamily is only very distantly related to the kinesin and myosin superfamilies, dynein nevertheless does have a lever domain, in this case a 15-nm-long microtubule-binding stalk that protrudes from each head (Gee et al., 1997; and see Figures 12.1B and 12.2C).

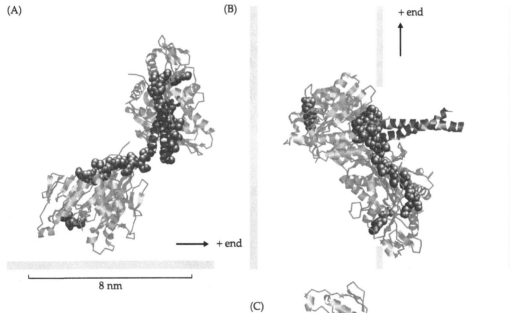

Figure 12.7 Three views of dimeric kinesin
The structure of the conventional kinesin dimer (Protein Data Bank ID: 3KIN; Kozielski et al., 1997) oriented with respect to the microtubule according to the docking of Hirose et al. (1999). The microtubule–motor complex, viewed from the side in (A), from the top in (B), and from behind in (C), looking toward the plus end of the microtubule. The gray lines denote the edges of the protofilaments. The conserved, core motor domains (residues 8–324) and the amino termini (residues 1–7) are shaded light gray. The dimerization domains (residues 339–372, ribbon), the necks (residues 325–338, spacefill), and the nucleotides (ADP, spacefill) are shaded dark gray. (See Schief and Howard, 2001; courtesy of Bill Schief.)

Docking of Motors to Their Filaments

Cryoelectron microscopy provides direct support for the swinging lever arm hypothesis for myosin. Motor proteins bind **stereospecifically** to their associated filaments, meaning that they bind at a unique location on the filament and with a specific orientation. The structure of smooth muscle myosin S1 bound to actin filaments depends on whether ADP is present or not; the differences can be accounted for by postulating that the light-chain binding domain undergoes a rigid-body rotation through ~23 degrees upon ADP dissociation (Figure 12.8; Whittaker et al., 1995). Such a rotation would swing the distal end of the motor domain through some 3.5 nm toward the plus end of the actin filament. Does this motion of smooth muscle myosin correspond to the working stroke of skeletal muscle myosin? Unfortunately, the answer is no. The problem is that such a conformational difference is not seen for skeletal muscle myosin, where the nucleotide-free conformation is indistinguishable from the ADP conformation. Thus the conformational change of skeletal muscle myosin must occur during another part of the hydrolysis cycle, most likely phosphate release (Chapter 14). The ADP-induced swing of smooth muscle myosin may be an adaptation to confer strain dependence on the ADP release step, in order to slow down the ATPase cycle of smooth muscle when it is prevented from shortening (Whittaker et al., 1995; Cremo and Geeves, 1998).

Cryoelectron microscopy of kinesin-decorated tubulin sheets and microtubules hints at how kinesin might move (Figure 12.9). Comparison of micro-

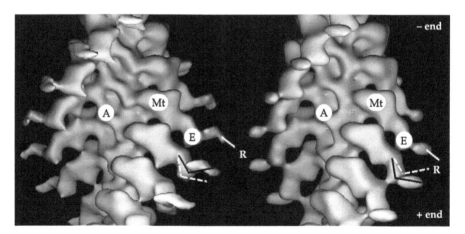

Figure 12.8 Rotation of myosin's light-chain binding domain
Cryoelectron micrographs of actin decorated with the S1 fragment of chicken smooth muscle myosin. The left-hand image is in the presence of ADP and the right-hand one is in the absence of nucleotide. The assignments of the actin monomer (A), the motor subdomain (Mt), and the essential (E) and regulatory (R) light chains of the light-chain binding domain are made by fitting the atomic structures to the helically averaged electron microscopic images. The dissociation of ADP is associated with a movement of the distal tip of the light-chain binding domain through 3.5 nm toward the plus end of the actin filament (bottom of the figure). The black and dashed lines denote the positions of the essential and regulatory light chains in the two different nucleotide conditions. (From Whittaker et al., 1995; courtesy of Ron Milligan.)

tubules decorated with kinesin dimers in the absence of nucleotide and in the presence of the nonhydrolyzable ATP analogue AMP-PNP shows that the binding of AMP-PNP is associated with a large conformational change that moves the unbound head from the left of the bound head to the right of the bound head (Arnal and Wade, 1998; Hirose et al., 1999). This conformational change is consistent with biochemical experiments showing that ATP binding to the attached head greatly accelerates the binding of the free head to the microtubule (Hackney, 1994; Ma and Taylor, 1997; Gilbert et al., 1998). The conformational change is also consistent with single-molecule measurements of the load dependence of motility (Schief and Howard, 2000). Thus the second head of kinesin appears to play an analogous role to the light-chain binding domain of myosin: A small movement in the nucleotide-binding pocket of the attached head is amplified into a large movement that brings the detached head toward the next binding site on the microtubule (Hancock and Howard, 1998, 1999).

For kinesin-related proteins, the "neck" region that forms the junction between the head and the dimerization domain (see Figure 12.7) probably plays the same role as the converter domain of myosin (Howard, 1996; Rice et al., 1999). Dramatic support for this idea comes from the study of kinesin–Ncd chimeras. When

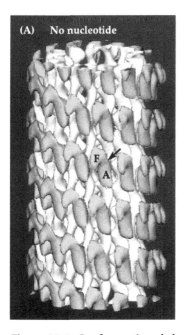
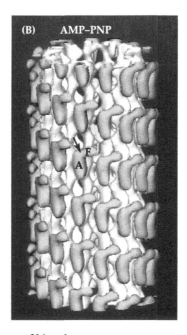

Figure 12.9 Conformational changes of kinesin
Cryoelectron micrographs of dimeric kinesin bound to a microtubule. Kinesin is shaded in dark gray, with the attached head labeled A and the free head labeled F. The microtubule is shaded in light gray. In the absence (or near absence) of nucleotides (A), the free head is to the left of the attached head. Following binding of AMP-PNP (B), there is a large conformational change such that the free head moves to the right of the attached head. The arrows point to the connection between the two heads. (From Arnal and Wade, 1998; courtesy of Dick Wade.)

the conserved motor domain of Ncd (i.e., the region of Ncd that is highly conserved among all kinesin-related proteins) is put into the kinesin "body," the chimera is plus-end directed like kinesin (Case et al., 1997; Henningsen and Schliwa, 1997), whereas when the conserved motor domain of kinesin is put into the Ncd body, the chimera is minus-end directed like Ncd (Endow and Waligora, 1998). Truncation experiments on kinesin (Stewart et al., 1993) place the direction-determining region between the conserved motor domain and the start of the dimerization domain shown in Figure 12.7. The direction-determining sequences are located outside the conserved motor domain, but not by much: When a kinesin motor domain that includes just two more amino acids beyond the conserved core is put into the Ncd body, the chimera is plus-end directed (Endow and Waligora, 1998) like kinesin; and scrambling of 12 amino acids in the coiled coil adjacent to the motor domain of an otherwise wild-type Ncd reverses the directionality (Sablin et al., 1998). The chimeric and truncated proteins move very slowly, indicating that the rigidity of the connection between the motor and the lever (the dimerization domain or the second head) is probably critical for mechanical amplification.

Summary

The structures of motor proteins give tantalizing hints at how the motor reaction might proceed. Kinesin and myosin use a conserved core motor domain that is structurally similar to a core region of G-proteins to generate a nucleotide-dependent conformational change. To transduce this conformational change into a much larger physical motion of a protein requires mechanical amplification. As the resolution of the crossbridges has improved, anatomical candidates for the mechanical lever have been identified. For myosin, the lever is most likely the light-chain binding domain, whereas for kinesin, the lever may be the second head or the dimerization domain. In the case of dynein, the lever may be the 15-nm-long stalk that protrudes from the bulk of the head (Goodenough and Heuser 1984; Gee et al. 1997). The junction between the motor domain and the lever serves as a sort of gearbox to regulate the direction of the rotation. For myosin, the gearbox is the converter domain, whereas for kinesin, it is the neck.

The structural studies provide only a framework for understanding how a motor protein works. An important caveat of the studies is that the atomic structures were solved in the absence of filaments, yet the working stroke takes place when the motor is attached to the filament. The binding to the filament is likely to have a significant effect on the structure of the motor, so the working stroke needs to be measured in an intact system. A second caveat is that structures are static pictures that give no kinetic or energetic information: Photographing an internal combustion engine at top dead center and at the bottom of the down stroke does not explain how it works. In the following chapters I address the workings of motor proteins—their motility, biochemistry, and mechanics—to ask how the conformational changes driven by ATP hydrolysis generate force and power cell motility.

CHAPTER

13

Speeds of Motors

Despite the overall structural similarities between motor proteins, especially within the individual motor families, there are striking functional differences. Different motors go in different directions and at different speeds; some go in huge arrays and some go alone (see Table 13.1). In this chapter, I argue that the speed of a motor as well as the number of motors necessary for movement can be understood using the concept of the duty ratio, the fraction of the time that a motor domain spends attached to its filament. Motors with high duty ratios can maintain continuous attachment to their filaments and so can function on their own. Motors with low duty ratios must oligomerize into large assemblies in order to produce continuous motility. Differences in duty ratio account for much of the variation in speeds of different motors.

The Speeds of Motors in Vivo

Motor proteins move at a variety of speeds inside cells. For example, during the unloaded contraction of rat extensor muscle, one of the fastest vertebrate skeletal muscles, the tissue shortens in length by 20% in only 10 ms (37°C; Close, 1964). This corresponds to a movement of the myosin crossbridges along the actin filaments at a speed of 25,000 nm/s. At the other extreme, chromosomes move along microtubules during anaphase of mitosis at a speed of only 5 to 10 nm/s (Nicklas, 1983). Table 13.1 shows the in vivo speeds of several motor proteins from the myosin, dynein, and kinesin families. The range of speeds is remarkable. Even within a motor family the range is large. The quickest myosins move over one hundred times faster than the slowest myosins, those responsible for the transport of organelles and the contraction of smooth muscle. Similarly, the quickest kinesins, those responsible for organelle trans-

Table 13.1 Motor speeds in vivo and in vitro

Motor	Speed[a] in vivo (nm/s)	Speed[b] in vitro (nm/s)	In vitro ATPase[c] (s^{-1})	Function
Myosins				
1. Myosin IB	ND[d]	200	6	Amoeboid motility, hair cell adaptation
2. Myosin II	6000	8000	20	Fast skeletal muscle
3. Myosin II	200	250	1.2	Smooth muscle contraction
4. Myosin V	200	350	5	Vesicle transport
5. Myosin VI	ND	−58	0.8	Vesicle transport?
6. Myosin XI	60,000	60,000	ND	Cytoplasmic streaming in algae
Dyneins				
7. Axonemal	−7000	−4500	10	Sperm and cilial motility
8. Cytoplasmic	−1100	−1250	2	Retrograde axonal transport, mitosis, transport in flagella
Kinesins				
9. Conventional	1800	840	44	Anterograde axonal transport
10. Nkin	800	1800	78	Secretory vesicle transport
11. Unc104/KIF	690	1200	110	Transport of synaptic vesicle precursors and mitochondria
12. Fla10/KinII	2000	400	ND	Transport in flagella, axons, melanocytes
13. BimC/Eg5	18	60	2	Mitosis and meiosis
14. Ncd	ND	−90	1	Meiosis and mitosis

[a]*In vivo speed* in cells or extracts applies to motion of the motor relative to the filament without external load. A positive speed denotes motion towards the plus (rapidly polymerizing) end of the filament. A negative speed denotes motion toward the minus end.
[b]*In vitro speed* of purified motors at high ATP concentration.
[c]*ATPase* is the maximum filament-activated rate of hydrolysis per head per second measured in solution at high ATP and filament concentrations.
[d]ND = not determined.

Notes

Speed: Zot et al. (1992). ATPase: Ostap and Pollard (1996).

In vivo speed: glycerinated rabbit psoas fibers at 25°C; Pate et al. (1994); Cooke et al. (1988). See Appendix 13.1. In vitro speed and ATPase: rabbit psoas HMM at 30°C; Toyoshima et al. (1987).

In vivo speed: chicken gizzard at 20°C; Siemankowski et al., 1985. The speed was calculated using a "sarcomere" length of 2.5 μm (see Appendix 13.1). In vitro speed and ATPase: phosphorylated turkey gizzard myosin at 22°C and 37°C, respectively; Warshaw et al. (1990).

In vivo speed: yeast Myo2 at 37°C; Govindan et al. (1995). With functional Myo2, a mother cell is able to transport 1 vesicle per 17 s, over ~4 μm to the tip of the bud. In vitro speed and ATPase: native chicken brain myosin V at 24°C and 37°C; Cheney (1993).

Recombinant truncated pig protein at 25°C; Wells et al. (1999).

Nitella myosin; Rivolta et al. (1995). *Chara* myosin at 23°C; Higashi-Fujime et al. (1995); Morimatsu et al. (2000).

In vivo speed: reactivated sea urchin sperm at 25°C using parameters from Brennen and Winet (1977) (see Appendix 13.2). In vitro speed and ATPase: native sea urchin outer-arm dynein; 25°C; Yokota and Mabuchi (1994). Outer-arm dynein determines the speed.

In vivo speed: retrograde transport in squid axoplasm at room temperature; Brady et al. (1990). In vitro speed: native bovine brain cytoplasmic dynein at 30°C; Paschal et al. (1987). ATPase: native bovine brain cytoplasmic dynein at 37°C; Shpetner et al. (1988).

In vivo speed: anterograde transport in squid axoplasm at room temperature; Brady et al. (1990). In vitro speed and ATPase: recombinant *Drosophila* full-length conventional kinesin tetramer; single-motor assays at 25°C; Coy et al. (1999).

In vivo speed: This corresponds to an elongation rate of *Neurospora* hyphae of 68 mm/day at 25°C; Seiler et al. (1997, 2000). In vivo speed and ATPase: recombinant *Neurospora* dimer at 24°C; Crevel et al. (1999).

In vivo speed: Mouse KIF1A (Okada et al. [1995]; Yonekawa et al. [1998]) is thought to move synaptotagmin-containing vesicles; Nakata et al. (1998). In vitro speed: recombinant full-length KIF1A; Okada et al. (1995). ATPase: recombinant truncated KIF1A; Okada and Hirokawa (1999).

In vitro speed: base-to-tip movement of "rafts" in *Chlamydomonas* flagella mediated by Fla10 at room temperature; Kozminski et al. (1993, 1995). In vitro: sea urchin egg Kin II; Cole et al. (1993).

In vivo speed: yeast spindle elongation during mitosis at 20–24°C; Yeh et al. (1995). This is attributed to the action of the related Kip1 and Cin8 kinesins; Saunders and Hoyt (1992). In vitro speed and ATPase: recombinant *Xenopus* Eg5 dimer at 20°C; Lockhart and Cross (1996).

Recombinant *Drosophila* Ncd dimer (GST-MC5) at room temperature; Chandra et al. (1993).

Key:

Motors operate in large arrays (10^4 to 10^9)
Motors operate in small arrays (10 to 1000)
Motors operate alone or in small numbers (<10)
Unknown

port, move one hundred times faster than mitotic motors. The speed of a specific motor may also vary depending on physiological conditions. For example, imposing a large enough load on a muscle will stall the myosins. In this chapter we consider the unloaded speeds of motors, deferring to Chapter 15 consideration of how load affects speed.

Rowers and Porters

Some motors, like muscle myosin and axonemal dynein, operate in huge arrays. For example, a large muscle fiber may have as many as a billion myosin molecules. Even within a sarcomere, the anatomical unit of muscle contraction (see Figure 1.1C), there are thousands of myosin-containing thick filaments in parallel. Because each thick filament contains ~600 crossbridges, it can be likened to a galley ship with the crossbridges forming oars that row each side of the thick filament into the array of actin filaments. On the other hand, conventional kinesin operates alone or in small numbers: Electron micrographs show at most a few crossbridges between vesicles and microtubules (see Figure 12.1; Miller and Lasek, 1985; Ashkin et al., 1990). For this reason, kinesin has been likened to a porter (Leibler and Huse, 1993).

Interestingly, there are members of the myosin and dynein families that function alone or in small numbers to move membrane-bounded vesicles and organelles. Myosin V and cytoplasmic dynein are examples of such motors. And conversely, there are kinesin-related proteins that function in macromolecular assemblies such as the mitotic spindle. The oligomerization state has been indicated by the shading in Table 13.1.

In Vitro Motility Assays

The study of motor proteins was revolutionized by the development of **in vitro motility assays** in which the motility of purified motor proteins along purified cytoskeletal filaments is reconstituted in cell-free conditions. An important milestone in this development was the visualization of fluorescent beads coated with purified myosin moving along actin cables in the cytoplasm of the alga *Nitella* (Sheetz and Spudich, 1983). This was quickly followed by the first completely reconstituted assay in which motor-coated beads were shown to move along oriented filaments made from purified actin that had been bound to the surface of a microscope slide (Spudich et al., 1985). Though "threads" of actin and myosin had been known to contract in the presence of ATP (Szent-Györgyi, 1941), this contraction was very slow. The significance of the new findings was that it proved that myosin (together with actin) was sufficient to produce movement at rates consistent with the speeds of muscle contraction and cell motility. At about the same time, technical developments in light microscopy allowed the visualization of individual microtubules by differential interference contrast microscopy (Allen et al., 1981, 1985) and individual actin filaments by fluorescence microscopy (Yanagida et al., 1984), greatly facilitating the in vitro assays and allowing the speed and direction of many motors to be measured. Further refinement led to visualization of movement by single-motor molecules (Howard et al., 1989). The combination of these assays with increasingly sophisticated optical and mechanical techniques has allowed measurement of the conformational changes of individual motor molecules as they generate force (Chapter 15). These techniques have now been applied to several other molecular motors, such as RNA polymerase (Yin et al., 1995; Wang et al., 1998), ATP synthase (Noji et al., 1997; Yasuda et al., 1998), and DNA polymerase (Wuite et al., 2000).

There are two geometries used in in vitro motility assays: the **bead assay** and the **gliding assay**. In the bead assay, filaments are fixed to a substrate, such as a microscope slide, and motors are attached to small plastic or glass beads with typical diameters of 1 µm. The motion of the beads along the filaments in the presence of ATP can then be visualized using a light microscope (Figure 13.1A). In the gliding assay, the motors themselves are fixed to the substrate, and the filaments are observed to diffuse down from solution, attach to, and glide along the motor-coated surface (Figure 13.1B). Visualization of the beads, actin filaments, and microtubules is readily accomplished using fluorescence microscopy; beads and microtubules can also be seen using other light-microscopic techniques such as differential-interference contrast and darkfield microscopy (Spencer, 1982). With improved fluorescence sensitivity, it is even possible to image individual fluorescently labeled motors (Funatsu et al., 1995) and to watch them move along filaments (Vale et al., 1996; Pierce et al., 1997). The motions can be recorded on videotape and the speed measured by tracking the centroid of the bead or the leading edge of the filament from frame-to-frame (see Scholey, 1993 for detailed methods). There is good overall agreement between the speed of a motor protein in vitro and the speed of the cellular motion that is attributed to the motor (see Table 13.1).

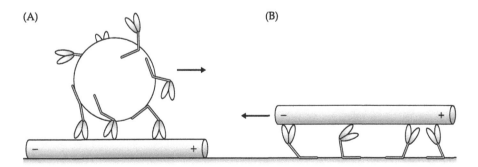

Figure 13.1 In vitro motility assays
(A) Bead assay. (B) Gliding assay. A motor that moves the bead toward the plus end of the filament will cause a filament to glide with its minus end leading.

The direction of movement of a motor along its associated polar filament can be ascertained readily in the in vitro motility assay. For example, the minus end of a microtubule can be fluorescently labeled to distinguish it from the plus end. In gliding assays using kinesin, such "polarity-marked" microtubules always move with their minus ends leading (Figure 13.2), showing that the kinesin motors try to move toward the plus end but, because they are fixed to the substrate, the microtubule moves the other way (see Figure 13.1B). Using these assays, it was discovered that the kinesin-related proteins Ncd (Walker et al., 1990) and Kar3 (Endow et al., 1994) were minus-end-directed motors. These discoveries were very surprising at the time because of the high degree of sequence similarity between these motors and kinesin. However, recently a myosin VI has been shown to have reversed directionality, and, as explained in the last chapter, these discoveries have led to the notion of a converter domain adjacent to the motor domain that can act like a gearbox to switch the direction of movement.

In vitro motility assays have also provided biochemical information about motor proteins. For all motor proteins studied, removing the nucleotides from solution results in a very strong **rigor** attachment between the motor and filament. Consistent with this, movement in the in vitro motility assay ceases. In the case of kinesin (but not myosin), AMP-PNP, a nonhydrolyzable analogue of ATP, also promotes rigor-like binding in motility assays (Lasek and Brady, 1985). This finding led to the discovery of kinesin (Brady, 1985; Vale et al., 1985). In vitro assays can also be used to estimate rate constants of individual molecules, providing the information that would traditionally be obtained using stopped-flow methods of enzyme kinetics. For example, the rate constants

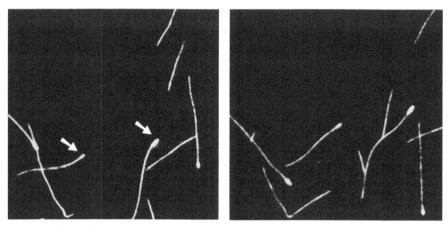

Figure 13.2 Polarity assay
Gliding of polarity-marked, fluorescent microtubules over a dense lawn of kinesin. The two images are taken 15 seconds apart. The minus ends of four microtubules are brightly labeled (two with arrows), and in all cases the minus end leads, indicating that kinesin attached to the surface is actually moving toward the microtubules' plus ends. (From Howard and Hyman, 1993.)

for the unbinding of kinesin from microtubules can be estimated from measurement of the average time that a bead coated with just one kinesin molecule remains attached to microtubules (Hancock and Howard, 1999). These measurements show that the unbinding of kinesin from a microtubule is accelerated by hydrolysis and phosphate release.

The mechanical loads in these assays are very small. The viscous drag on a 20-μm-long microtubule moving at 1 μm/s through a solution with the viscosity of water (~1 mPa·s) is only ~0.14 pN (Hunt et al., 1994; Chapter 6). Yet a single kinesin molecule can move such a long microtubule just as quickly as it can move a 1-μm-long microtubule (Hunt et al., 1994), indicating that even a single kinesin molecule is capable of producing forces much larger than 0.1 pN. The drag in the bead assay is even smaller because the beads are typically only ~1 μm in diameter. Furthermore, in typical motility assays there are often a large number of motors moving the one filament or bead. Thus in both the gliding and bead assays, the loads are usually minute. In Chapter 15 we will consider the influence of large loads on the motor mechanism.

Processive and Nonprocessive Motors

In vitro motility assays have shown that conventional kinesin is a **processive** motor: A single molecule of kinesin can move continuously along the surface of a microtubule for up to several microns (Howard et al., 1989; Block et al., 1990). This distance corresponds to hundreds of 8 nm steps (Chapter 15). The evidence for processivity is that motion can be observed at a very low density of kinesin on a surface (<1 molecules/μm^2), that gliding microtubules swivel about a single point on the surface (at which the motor is presumed to be located), and that the rate at which microtubules bind and begin to move across a kinesin-coated surface is linearly proportional to the density of motors on the surface. Subsequent force measurements (Hunt et al., 1994; Svoboda and Block, 1994; Meyhöfer and Howard, 1995) discussed in Chapter 15 establish this processivity almost beyond doubt by showing that there is an indivisible force-generating unit that produces ~6 pN of force. The only remaining possibility, that this motor unit is a small, constant-sized aggregate of kinesins, is unlikely because the formation of such an aggregate would be too slow at the concentration of kinesin used to coat the surfaces (Howard et al., 1989). Biochemical experiments also confirm that kinesin is processive: kinesin hydrolyzes on average about 125 molecules of ATP following its initial binding to the microtubule (Hackney, 1995), consistent with the motor taking 125 8 nm steps (1 step/ATP) before dissociating. Some kinesin-related proteins are also processive (Nkin, Crevel et al., 1999; KIF1A, Okada and Hirokawa, 1999); others are not (Ncd, deCastro et al., 1999).

In contrast to conventional kinesin, muscle myosin II is not a processive motor. A threshold density of myosin on the surface is required for continuous motility of actin filaments. Below the threshold, the filaments do not associate with the surface for long enough times for movement to be detected by video microscopy. The threshold density is ~4000 molecules/μm^2 to move actin filaments longer than ~0.040 µm (Harada et al., 1990) and ~600 HMM molecules/μm^2 to move actin filaments longer than 1.1 µm (Toyoshima et al., 1990). If each myosin head can reach 30 nm to contact an actin filament, then there are respectively 48 and 84 myosin molecules on average able to interact with these filaments at these densities. High-resolution displacement measurements also show that many myosin molecules are required for continuous motility: In the presence of ATP, actin filaments make only transient attachments to surfaces sparsely coated with myosin and move distances less than 20 nm before dissociating (Finer et al., 1994; Molloy et al., 1995; Chapter 15). Thus skeletal muscle myosin II is not processive. But myosin V, a vesicle transporter like conventional kinesin, is processive (Mehta et al., 1999; Rief et al., 2000; Sakamoto et al., 2000; Walker et al., 2000).

Like the other motors, some dyneins are processive and some are not. Densities of outer-arm dynein of ~1000/μm^2 are required for continuous gliding of microtubules (Vale and Toyoshima, 1988; Sale et al., 1993; Hamasaki et al., 1995), suggesting that outer-arm dynein also requires large assemblies for continuous motility. This is supported by high-resolution mechanical studies showing that outer-arm dynein is not processive at the high ATP concentrations found in cells (Hirakawa et al., 2000); interestingly, though, it is processive at low ATP concentrations. By contrast, high-resolution assays provide evidence that inner-arm dynein is processive even at high ATP concentrations (Shingyoji et al., 1998; Sakakibara et al., 1999). Cytoplasmic dynein is processive (King and Schroer, 2000); and its processivity is increased (i.e., it moves further before dissociating) by the dynactin complex, which is the dynein receptor thought to anchor cytoplasmic dynein to its membranous cargo.

Processivity, or the lack thereof, is reflected in the dependence of the speed of movement on the number of participating motors. The speed of kinesin is independent of the density: One motor can move a microtubule as quickly as ten or a hundred motors (Figure 13.3A). The speed is also independent of filament length (Howard et al., 1989). Because the load in these in vitro assays is small, this is analogous to a group of people carrying a pole: Provided that the pole is light enough, one person can carry it as quickly as a dozen people, irrespective of its length. This analogy illustrates another point: If many people are carrying a pole, a certain amount of coordination is required; otherwise, the speed may *decrease* as the number of helpers increases. Thus measurements of speed and motor number constrain models for the interaction of motors. When the density of myosin or dynein exceeds a certain threshold, the speed is also independent of density and filament length, as found for kinesin. However, in the case of myosin, it is possible to observe actin gliding at densities below threshold. This is done by adding methyl cellulose to the buffer solution. This highly elongated polymer inhibits diffusion of actin filaments perpendicular (but not parallel) to their long axes; it thereby promotes motility at low motor densities by slowing the diffusion of the filaments away from a surface during brief moments when all the motors are unbound. In the presence of methyl cellulose, the speed of actin filaments of fixed length increases with the myosin density (Figure 13.3B; Uyeda

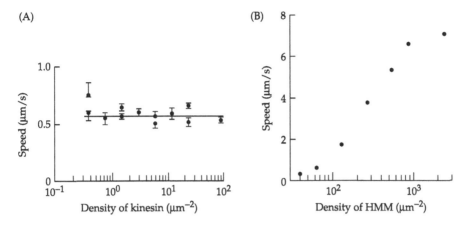

Figure 13.3 Dependence of speed on motor density
(A) The speed of microtubules as a function of the density of kinesin. At a nominal density of 6 molecules/μm², it is estimated that on average one motor molecule is interacting with each microtubule. Average microtubule length is 2.5 ± 1.4 μm. (B) The speed of 2-μm-long actin filaments as a function of the density of myosin moving in buffer solution containing methyl cellulose. (A after Howard et al., 1989, 1993; B after Uyeda et al., 1990.)

et al., 1990). The speed reaches half maximum when the number of heads interacting with the actin filament equals about 50 (assuming a reach of 30 nm), again consistent with myosin not being processive.

The Hydrolysis Cycle and the Duty Ratio

The functional differences between different motors—their speeds and processivity—can be understood using the concept of the duty ratio, the fraction of the time that each motor domain spends attached to its filament. The concept first arose (initially called the duty cycle) to explain the higher force generated by smooth muscle compared to skeletal muscle (Dillon and Murphy, 1982). The development of this concept requires the notion that motor proteins have mechanical and chemical cycles.

An important, early insight into how motor proteins work arose from a simple observation: Motor proteins move along their filaments through distances that are large compared to molecular dimensions (e.g., Huxley, 1980). For example, the thick filament can slide up to 0.7 μm along the thin filament as a muscle contracts (see Figure 1.1), and individual microtubules glide for many microns over a kinesin-coated surface (e.g., see Figure 13.2). Because these distances are much larger than the crossbridges, the motor reaction must be a *cyclic* one in which the motor repetitively binds to and unbinds from the filament. During each **crossbridge cycle**, we imagine that a motor domain spends an average time attached to the filament, τ_{on}, during which it makes its **working**

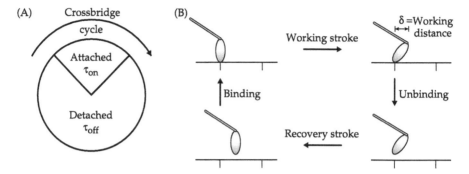

Figure 13.4 The crossbridge cycle
(A) During one cycle of ATP hydrolysis, each head, or crossbridge, spends time τ_{on} attached to the filament and time τ_{off} detached from the filament. (B) It is hypothesized that a crossbridge makes a working stroke during the attached phase and makes a recovery stroke during the detached phase. By recovering to its initial conformation while detached, the motor avoids stepping backward, and so progresses a distance equal to the working stroke during each cycle.

stroke and an average time detached from the filament, τ_{off}, during which it makes its **recovery stroke**. In this way it returns to its initial conformation for the beginning of a new cycle (Figure 13.4). By recovering during the detached phase, the motor avoids taking a step backwards, and thus progresses through a distance along the filament equal to the **working distance**.

We define the **duty ratio**, r, as the fraction of the time that each head spends in its attached phase

$$r = \frac{\tau_{on}}{\tau_{on} + \tau_{off}} = \frac{\tau_{on}}{\tau_{total}} \quad (13.1)$$

The duty ratio of a crossbridge can in principle be large (~1) if it spends most of its time attached, or small (~0), if it spends most of its time detached.

Differences in duty ratio account for the finding that some motors are processive and others are not. The minimum number of heads required for continuous movement, N_{min}, is related to the duty ratio by

$$r \cong \frac{1}{N_{min}} \quad (13.2)$$

as this guarantees that there will usually be at least one head bound to the filament (Harada et al., 1990; Uyeda et al., 1990; Leibler and Huse, 1993). Because a two-headed molecule of conventional kinesin is able to maintain continuous attachment to the microtubule, its duty ratio must be at least 0.5 for each head; otherwise, there will be times when neither head is attached and the motor will diffuse away from the filament. Likewise, two-headed myosin V and cytoplasmic dynein must have duty ratios ≥0.5 because they, too, are processive. On the other hand, because skeletal muscle myosin and outer-arm dynein must be in large assemblies with at least 50 to 100 crossbridges to move, their duty ratios must be small, ~0.01 to 0.02, the reciprocal of the minimum number of heads needed for continuous motility.

If there is one-to-one coupling between mechanical cycles (binding, working stroke, unbinding, and recovery) and chemical cycles (the ATP hydrolysis reaction), we expect that the speed of a motor, v, is equal to

$$v = k_{\text{ATPase}}\Delta \tag{13.3}$$

where Δ is the distance traveled by each head relative to the filament per mechanical cycle, and k_{ATPase} is the rate at which each head hydrolyzes a molecule of ATP. The speed is easy to measure in a motility assay. The ATPase rate is more difficult to measure because in a motility assay not all the motors are moving (e.g., only a small fraction of the motors on a surface are interacting with filaments). For this reason, the ATPase rate is usually measured with motors in solution (Figure 13.5A). For all motors studied so far, addition of filaments to the solution increases the ATPase rates (e.g., see Figure 13.5B), showing that the chemical cycle is coupled to the mechanical cycle. Conversely, the speed of movement increases with the ATP concentration, showing that the mechanics is coupled back to the chemistry. The maximum ATPase rate measured in solution (k_{cat}) is often assumed to correspond to the ATPase during motility. The k_{cat} values of several motor proteins are shown in Table 13.1.

It might have been expected that the different speeds of different motors could be explained by differences in ATPase rates. But Table 13.1 shows that this is not true in general. For example, skeletal muscle myosin moves 40 times faster than myosin ID, yet consumes ATP only three times faster. The difference is even more dramatic when comparing skeletal muscle myosin with kinesin; myosin moves ten times faster but has only half the ATPase rate. However, within subfamilies, such as the myosin II subfamily, there is a closer correlation between the speed and the ATPase rate: For example, Bárány (1967)

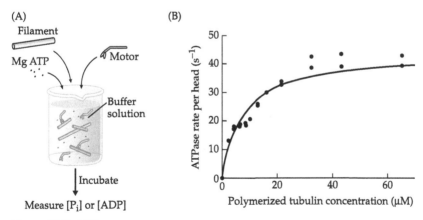

Figure 13.5 ATPase assays
(A) The ATPase rate of a motor is measured by adding ATP to a solution containing a known number of active motor molecules and using colorimetric methods or radioactivity assays to measure the amount of ADP or phosphate formed over a known period of time, typically minutes. (B) The dependence of the ATPase rate of kinesin (expressed as ATP per two-headed molecule) on the microtubule concentration. Extrapolation to infinite microtubule concentration gives a maximum rate of 94 s^{-1} per molecule. (After Coy et al., 1999b.)

showed that the contraction speed of 16 types of muscles varies 100-fold, yet, with one exception, the ratio of the speed to the ATPase rate varies by less than a factor of two.

Another unexpected finding is that some motors move a very large distance during the time they take to hydrolyze a molecule of ATP. For example, dividing the speed by the ATPase rate gives a distance of 400 nm/ATP for skeletal muscle myosin and 450 nm/ATP for outer-arm dynein! These findings are confirmed by more well-controlled experiments in which the ATPase rate and the speed are measured in similar buffers and geometries: Muscle myosin HMM molecules in vitro move 190 nm/ATP (Toyoshima et al., 1987), and the crossbridges in myofibrils move 140 nm/ATP (Ma and Taylor, 1994). These distances are perplexing because they are an order of magnitude larger than the dimensions of the crossbridges! These findings generated much controversy in the myosin field (Harada et al., 1990; Uyeda et al., 1990); they led to the step-size paradox and the suggestion that myosin might take multiple mechanical steps for each ATP that it hydrolyzed (Yanagida et al., 1985; Kitamura et al., 1999).

The duty ratio concept provides a simple explanation for the step-size paradox. Consider a filament moving at constant speed, v, over an array of fixed motor proteins as occurs during filament sliding in muscle or in an in vitro gliding assay. We suppose that there are enough heads interacting with the filament to ensure continuous motility—that is, at least one kinesin molecule or at least 50–100 myosin heads. If each head is attached for time τ_{on} and moves through the working distance δ, then

$$v = \delta/\tau_{on} \tag{13.4}$$

(Harada et al., 1990; Uyeda et al., 1990; Pate et al., 1993). On the other hand, because the cycle is driven by ATP hydrolysis, we expect that the total cycle time, $\tau_{total} = 1/k_{ATPase}$, where k_{ATPase} is the ATPase rate. Substituting these expressions for τ_{on} and τ_{total} into Equations 13.1 and 13.4 we obtain another expression for the duty ratio:

$$r = \frac{\tau_{on}}{\tau_{total}} = \frac{\delta \cdot k_{ATPase}}{v} = \frac{\delta}{\Delta} \tag{13.5}$$

Hence the resolution of the paradox is that myosin has a low duty ratio: With $r = 0.02$ and an ATPase rate of 20/s per head, the attached time is only 1ms and a speed of 5000 nm/s can be reached with a working distance of only 5 nm (Equation 13.4), well within the dimension of the crossbridge. The crucial point is that each of the hundred or so crossbridges moving the filament contributes only 5 of the 250 nm moved while it hydrolyzes one molecule of ATP; while this motor is detached, the other myosins sweep the filament along the rest of the way. Because each half of a thick filament contains 300 crossbridges (Bagshaw, 1993), a duty ratio of only 2% in a rapidly contracting muscle still means that at all times there will be 6 or so crossbridges maintaining contact between the thin and thick filaments.

The low duty ratio of myosin II contrasts with that of processive motors such as conventional kinesin, myosin V, and cytoplasmic dynein. Because kinesin has a high duty ratio it needs a high ATPase rate to attain even moderate speeds: A working distance of 8 nm, a speed of 800 nm/s, and an ATPase rate of 50/s implies that $r = 0.5$, consistent with the duty ratio required to account for kinesin's processivity. Likewise, using values from Table 13.1, the duty ratios for Nkin and myosin V are ~0.5. However, values in Table 13.1 suggest that the duty ratio of cytoplasmic dynein is <<1, inconsistent with its processivity (King and Schroer, 2000); a likely explanation is that the ATPase activity measured in solution underestimates the ATPase activity during motility.

Because both the distance moved per ATP, Δ, and the duty ratio have been measured, it is possible to use Equation 13.5 to estimate the working distance, δ. For myosin, $\Delta = 200$ to 400 nm, which gives a working stroke distance in the range of 2 nm to 8 nm if the duty ratio is 0.01 to 0.02. This is in good agreement with direct measurement of the working distance using single-molecule techniques (Molloy et al., 1995a; see also Chapter 15). *This working distance is no larger than the molecular dimension of the heads and is therefore consistent with myosin making just one mechanical cycle per chemical cycle of ATP hydrolysis.* This important conclusion is confirmed by experiments using single-molecule fluorescence to detect ATP hydrolysis and optical tweezers to detect steps: There is a one-to-one coupling of the mechanical and chemical cycles of myosin (Ishijima et al., 1998), though some uncorrelated events have been interpreted differently.* In the case of kinesin, careful comparison of the distance moved per ATP (see Table 13.1, Figure 13.4; Coy et al., 1999b; Iwatani et al., 1999) with the step size of 8 nm (see Chapter 15) shows that the coupling is also tight. *There is exactly one step for each molecule of ATP that is hydrolyzed.* Thus there is now good evidence that under the low loads characteristic of these motility assays, both kinesin and myosin are tightly coupled motors.

The coupling has not been directly measured at high loads. However, high loads decrease both the speed of contraction and the ATPase activity of muscle (Fenn, 1924; Kushmerick and Davies, 1969), as expected for tight coupling; though the maintained ATPase activity of **isometrically contracting** muscle—fully activated muscle held so it cannot shorten—shows that the coupling is not one-to-one at the very highest loads. High loads decrease the speed of kinesin (Hunt et al., 1994; Svoboda and Block, 1994b; Meyhöfer and Howard, 1995) by decreasing the rate of stepping and (and therefore the rate of completion of mechanical cycles). Statistical analysis of the timing of the steps at high forces suggests that kinesin remains tightly coupled under load (Visscher et al., 1999), though a definitive statement must wait for an ATPase measurement at high load.

---

*The Yanagida lab argues that the working distance of myosin (the distance moved while it is attached to the actin filament) is large compared with the size of the head (71 nm in Saito et al., 1994; 11–30 nm in Kitamura et al., 1999). However, in the former paper, the number of heads in the synthetic filaments may have been underestimated. Thus their value for the duty ratio, 0.16, may be an overestimate, which would lead to an overestimate of the working distance. The results in the latter paper, which cannot be reconciled with a 5 nm working stroke, are not consistent with results obtained from other groups (Molloy et al., 1995a; Mehta et al., 1997, Ruff et al., 2000).

The analysis of molecular motion in terms of duty ratios and working strokes is a useful way of synthesizing a large body of experimental data on the speeds of different motors. However, this analysis is necessarily a simplification of the motor reaction. I have not attempted to account for the force and energetics of motors, nor for the likely dependence on strain of the binding and unbinding of motor filaments, nor the more detailed interactions between motors moving the one filament. I defer a more complete analysis to Chapter 16.

Analogies to Internal Combustion Engines and Animal Locomotion

The rotating crossbridge is analogous to a two-stroke internal combustion engine. In a two-stroke engine, the fuel mixture is injected into the cylinder and compressed during the compression stroke, creating a high chemical energy state at top dead center. Following ignition, the chemical energy is converted to heat, which creates the high pressure that drives the expansion of the gases during the power stroke (for information on how an internal combustion engine works, go to http://www.britannica.com). By analogy, we can imagine that myosin binding to actin creates the spark that triggers the working stroke. Where the analogy breaks down is that in a two-stroke engine, the duration of the working and recovery strokes are the same because of the coupling of the pistons to the crankshaft; the duty ratio is 0.5. However, motor proteins are not so constrained.

It is instructive to continue the analogy to an internal combustion engine. If kinesin's heads really do alternate between attached and detached phase with a duty ratio of 0.5—a model that we will examine in later chapters—then its motion is analogous to that of a two-cylinder, two-stroke engine in which one piston (analogous to one head) is always in its power stroke phase of the cycle. The analogy for myosin is to an engine with more strokes. A four-stroke engine has an intake, a compression, an expansion, and an exhaust stroke, and therefore spends only 25% of its time in its working stroke (the expansion stroke). Thus, in a loose sense, we can think of unloaded myosin, with a duty ratio of 1%, as being a 100-stroke engine; smooth motion requires 100 cylinders.

Another useful analogy is to walking and running. The gait of a biped is defined as a walk if "the feet move alternatively with one foot not clear of the ground before the other touches" (Merriam Webster's Collegiate Dictionary, 10th Edition). While walking, each foot must spend ≥50% of the time on the ground. On the other hand, during a run, there are times when neither foot is in contact with the ground. Thus a two-headed motor protein walks if it is always in contact with the filament, and it runs if it spends time completely detached from the filament. Using these definitions, kinesin walks along the microtubule (Cross, 1995) and myosin runs along the actin filament. A walking motor moves one working distance per ATP hydrolyzed. A running motor, by contrast, can move a much greater distance because it is carried by its fellow motors while it is detached. In this way the total distance moved can be much larger than the molecular distance. Where the analogy breaks down this time is that whereas the runner relies on momentum to carry her through the air, the motor, which has no inertia (Chapters 2 and 3), must rely on its neighbors to carry it forward while it is detached from its filament.

Summary: Adaptation to Function

The high duty ratios of conventional kinesin and myosin V appear to be adaptations for processivity; the low duty ratios of myosin II and outer-arm dynein appear to be adaptations for high speed. With a working distance of 8 nm and an ATPase rate of ~50 s^{-1}, kinesin attains a top speed of only 800 nm/s. By contrast, with a low duty ratio of 0.01, myosin and dynein can attain speeds of 10 μm/s with smaller working distances and ATPase rates. The consequent high speeds of muscle contraction and sperm motility can clearly confer selective advantage to the organism and to the germ cell. Another advantage of a low duty ratio is that a large distance can be moved during a single hydrolysis cycle: An unloaded muscle can shorten from an initial sarcomere length of 2.5 μm down to an almost fully contracted length of 1.9 μm within about the time it takes to complete the hydrolysis of a single molecule of ATP, yet the thin and thick filaments have slid 300 nm, perhaps 30 times the individual working distance.

CHAPTER

14

ATP Hydrolysis

In the last chapter we saw that motor proteins undergo cyclic reactions in which the moving motor alternately attaches to and detaches from its filament. Directed movement along the filament is thought to result from the motor undergoing a conformational change while it is attached. In this chapter, we look in detail at how changes in the nucleotide state modify the affinity of a motor for its filament, and how these changes drive the conformational changes responsible for the working stroke.

ATP

Adenosine triphosphate, ATP (Figure 14.1), was discovered in muscle extracts in 1929 (Lohmann, 1929) and was shown to be hydrolyzed by actomyosin in 1939 (Engelhardt and Ljubimowa, 1939). Its central role as the energy currency of cells was realized in 1941 (Lipmann, 1941). The hydrolysis of the gamma phosphate can be summarized by the following reaction

$$\text{ATP} \rightleftharpoons \text{ADP} + P_i \qquad K_{eq} = \frac{[\text{ADP}]_{eq} \cdot [P_i]_{eq}}{[\text{ATP}]_{eq}} = 4.9 \times 10^5 \text{ M} \qquad (14.1)$$

K_{eq} is the equilibrium constant which, in this case, has the same unit as concentration (M). [ATP] refers to the sum of all the species with different charges and ligands: [ATP] = [MgATP^{2-}] + [MgATP$^-$] + [MgATP] + [ATP^{4-}] + [ATP^{3-}] + [ATP^{2-}] + [ATP$^-$] + etc. ADP and P_i (the phosphate ion) also exist as several species, and again [ADP] and [P_i] refer to the sum of all these species. Because of all these species, the equilibrium constant depends on several factors, such as the free magnesium concentration, the pH, and ionic strength. The reac-

Figure 14.1 The structure of ATP

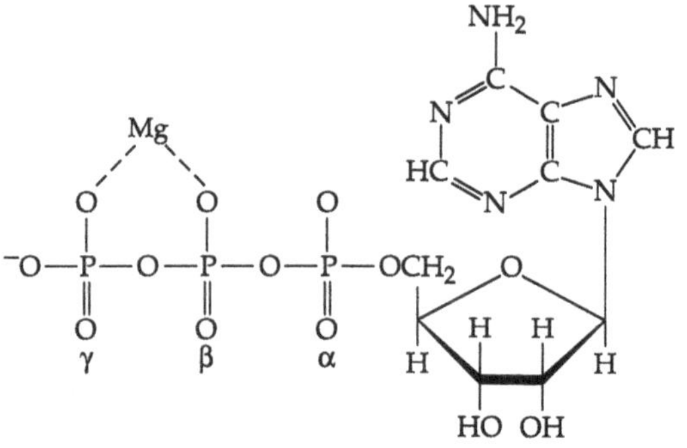

tion also involves a hydroxyl ion (see Figure 14.2), but this is omitted from the equilibrium equation (Equation 14.1) because the reaction takes place at a fixed pH (which fixes the hydroxyl ion concentration). Under conditions similar to the cytoplasm of vertebrate cells—$[Mg^{2+}]_{free}$ = 1 mM, pH = 7, ionic strength = 250 mM, and 25°C—the equilibrium constant is 4.9×10^5 M (Alberty and Goldberg, 1992). This means that the products ADP and P_i are strongly favored: If 1 mM ATP is added to a cytoplasm-like buffer solution and allowed to come to equilibrium, the ATP will be almost completely hydrolyzed so that there is only ~1 pM left in solution, whereas the equilibrium concentrations of ADP and P_i will both be ~1 mM. The reaction is said to proceed strongly to the right. However, equilibration is extremely slow: It will take over a week at 0°C (Dawson, 1986). It is this stability that makes the gamma phosphate bond such an ideal high-energy intermediate.

The free energy of the hydrolysis reaction depends on both the standard free energy as well as the concentration of ATP, ADP, and P_i:

$$\Delta G = \Delta G_0 - kT \ln \frac{[ATP]_c}{[ADP]_c [P_i]_c} \qquad \Delta G_0 = -54 \times 10^{-21} \text{ J} \qquad (14.2)$$

where the subscript c refers to the cellular or buffer concentration. If the reaction is at equilibrium, the free energy change equals zero because the two terms on the right cancel. On the other hand, if the reagents and products are in their standard chemical state, which by convention is 1 M, then the log term vanishes and the free energy equals the standard free energy of -54×10^{-21} J = $-kT\ln(4.9 \times 10^5)$. This is why ΔG_0 is called the standard free energy. Estimates of the absolute value of the standard free energy vary somewhat in the literature: for example, 51×10^{-21} J (Stryer, 1995), 49×10^{-21} J (Woledge et al., 1985), and 47×10^{-21} J (Cardon and Boyer, 1978), all at pH 7. We will use the Alberty and Goldberg value of 54×10^{-21} J.

A short digression on units: In keeping with the molecular spirit of this book, I refer to energy per molecule. This differs from most chemistry and biochemistry textbooks, which refer to energy per mole of molecules (which is 6.022×10^{23} times larger). A second difference is that I use the SI unit for energy, the Joule, whereas most American biochemistry textbooks and journals use the non-SI unit, the calorie. The advantage of using Joules per molecule as the unit of energy becomes immediately clear when one writes 10^{-21} J = 1 pN·nm. An energy of 1 kT (~4×10^{-21} J) corresponds to the work done by a force of 1 pN acting through 4 nm. By contrast, it is not transparent that a kcal/mol (~1.7 RT) per angstrom (another unit preferred by biochemists) even refers to a force. And when one calculates the force, it turns out to be ~4×10^{13} N/mol, a force equal to the weight of 4 km^3 of water! Using SI units per molecule makes it easy to relate the energetics to molecular mechanical parameters such as force (pN), displacement (nm), and stiffness (pN/nm).

In cells, the concentration of ATP is ~1 mM, that of ADP is ~10 μM, and that of phosphate is ~1 mM. Therefore, the cellular free energy is significantly greater than the standard free energy, and equals -101×10^{-21} J, corresponding to about $-25\ kT$. It is important to realize that the reason why the free energy is so large is that metabolic processes within the cell maintain the ATP, ADP, and P_i concentrations very far from equilibrium. The absolute value for the cellular free energy is a little soft because the $[Mg^{2+}]_{free}$ and pH, as well as the ATP, ADP, and phosphate concentrations, can vary from cell to cell, or from time to time in the same cell. This leads to variations in free energy of a few kT. For most purposes we will assume that the free energy lies between -20 and $-25\ kT$.

The hydrolysis of ATP occurs via an "in-line" mechanism in which the gamma phosphate is attacked by a hydroxyl ion and a pentaphosphate intermediate is formed (Figure 14.2). This mechanism is consistent with isotope studies showing that the oxygen from ^{18}O-labeled water is incorporated into the leaving phosphate rather than into the ADP. An in-line mechanism is also supported by X-ray structures of myosin showing a well-ordered water molecule in the attack position roughly opposite the oxygen that bridges the β and γ phosphates (Fisher et al., 1995). The pentaphosphate intermediate is likely to form a trigonal-bipyramidal structure. This is supported by the finding that vanadate binds tightly along with ADP in the active sites of myosin, dynein, and kinesin; if the motor catalyzes the hydrolysis reaction by preferentially binding to and stabilizing the transition state, then it is expected that vanadate acts as a "transition state inhibitor" because it adopts a trigonal-bipyramidal structure in crystals (Pauling, 1970).

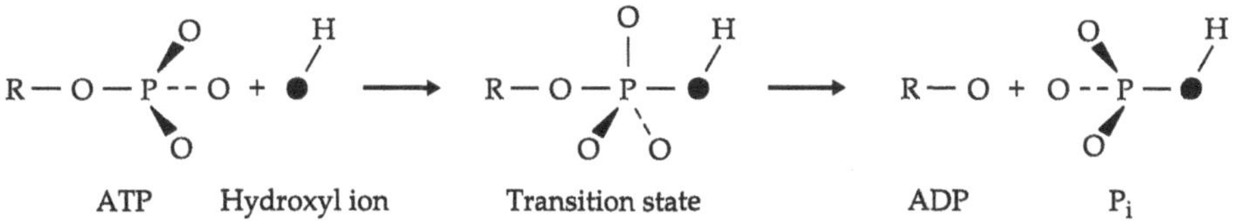

Figure 14.2 In-line hydrolysis mechanism
The middle panel corresponds to the trigonal-bipyramidal transition state. The filled oxygen denotes a heavy isotope used to show that water-derived oxygen is incorporated into the leaving (gamma) phosphate group.

Coupling Chemical Changes to Conformational Changes

The central question of this chapter is: How do nucleotides regulate the association of the motor with the filament? Motor proteins have four **chemical states**—nucleotide-free (M), ATP-bound (M·T), products-bound (M·D·P), and ADP-bound (M·D)—that are occupied sequentially during the hydrolysis reaction:

$$M \rightleftharpoons M \cdot T \rightleftharpoons M \cdot D \cdot P \rightleftharpoons M \cdot D \quad (14.3)$$

(the phosphate-bound state is not occupied because the phosphate leaves before the ADP; see discussion in Woledge et al., 1985). Motor proteins also have at least two **mechanical states**—attached to the filament and detached. Because these states are not mutually exclusive, the number of possible states for a single crossbridge is 8 (= 4 × 2). In fact, the total number is even higher because we expect that each chemical or mechanical state may comprise more than one structural state (defined in Chapter 5), specifically the pre-working-stroke and the post-working-stroke states discussed in the last two chapters.

By definition, the elucidation of the chemical mechanism of a motor reaction entails measurement of all the rate constants between these states (= 24 for the case of 8 states because there are 3 transitions out of each state: one to the next nucleotide state, one to the previous nucleotide state, and one for detachment or attachment). Because there are many states, there are many possible pathways for ATP hydrolysis, meaning there are many sequences of states that define a complete hydrolysis cycle. (For example, a complete hydrolysis cycle could take place while the motor is always attached or while it is always detached.) All paths are possible. But some paths are more likely than others, and we define the **hydrolysis cycle** as the most likely path (or paths).

The situation is potentially a lot more complex for two-headed motors because the number of states increases to 64 (= 8^2) and the number of rate constants increases even more, making it impractical to measure them all. Fortunately, the two heads of myosin II operate independently (Taylor, 1979), so the mechanism need be determined for just one head. However the two heads of conventional kinesin are coordinated and do not operate independently (see below); the situation is therefore considerably more complex, and, as a result, the full mechanism is not known, though the most likely path (i.e., hydrolysis cycle) is known. In the rest of this chapter, we will look at the cycles of myosin II and conventional kinesin in detail.

Hydrolysis of ATP by Skeletal Muscle Myosin

The key to understanding the hydrolysis cycle of myosin is that the release of phosphate is catalyzed by the binding of myosin to the actin filament, whereas the release of myosin from actin filament is catalyzed by the binding of ATP. In this way, a cycle of ATP hydrolysis is tightly coupled to a cycle of attachment to and detachment from the filament.

Without Actin

In the absence of actin, myosin has a low ATPase activity, equal to about 0.1 s^{-1}. After binding ATP, myosin hydrolyzes it to ADP and P_i, and then sequentially releases the products, first P_i then ADP. At the high ATP concentration found in cells, ATP binding is fast, and the rate-limiting step is the release of phosphate, so that the main species is M·D·P. The rate constants in the kinetic pathway for myosin S1 are shown in Table 14.1.

These biochemical data provide important information about myosin's nucleotide binding pocket. The gamma phosphate fits snugly into the protein where it binds very strongly. This follows because myosin has a much higher affinity for ATP (~10 pM) than for ADP (~4 μM)—this can only be due to the favorable interactions of the gamma phosphate with the protein. Another way of viewing the higher affinity of ATP over ADP is that the free-energy drop associated with hydrolysis and phosphate release from myosin-ATP (8 kT, top row of Table 14.1) is much smaller than that from ATP alone (25 kT, bottom row of Table 14.1): the enzyme holds the phosphate very tightly.

The chemical mechanism for the myosin S1 ATPase is known because the forward and backward rate constants of each step have been measured. The binding of ATP and ADP is measured using a stopped-flow apparatus (Bagshaw, 1993): Solutions containing myosin and ATP or ADP are rapidly

Table 14.1 Hydrolysis cycle of rabbit skeletal muscle myosin S1 without actin

(–19 kT)	100s^{-1}	(–21 kT)	0.1 s^{-1}	(–27 kT)
M·T	$\rightleftharpoons$	M·D·P	$\rightleftharpoons$	M·D + P
	10s^{-1}		1$M^{-1}s^{-1}$	
2μ$M^{-1}s^{-1}$ ↕ ~$10^{-4}s^{-1}$				0.5μ$M^{-1}s^{-1}$ ↕ 2s^{-1}
M + T		K_0		**M + D + P**
(0)		$\rightleftharpoons$		(–25 kT)

Notes: M = myosin S1, T = ATP, D = ADP, P = phosphate. The terms in parentheses are approximate energies assuming that [ATP] = 4 mM, [P_i] = 2 mM, and [ADP] = 20 μM as found in muscle. Overall ATPase: k_{cat} = 0.1 s^{-1}, K_M(ATP) = 50 nM at 20°C (Bagshaw, 1993). The rate constants are based on the following experimental data. Differences between different labs can, for the most part, be attributed to differences in ionic conditions (e.g., pH, pMg, ionic strength). Where possible, I have chosen values more appropriate to cellular conditions. The constants (K) are expressed as dissociation constants because this gives them convenient units (molar, M). k_+ is the forward rate constant, k_- is the reverse rate constant.

$K_0 = 5 \times 10^5$ M (Equation 14.1).

K_1: M + T $\rightleftharpoons$ MT. K_1 = 60 pM (pH 7, Cardon and Boyer, 1978), = 3 pM (pH 8, Goody et al., 1977); k_+ (Geeves, 1991); k_- (Cardon and Boyer, 1978).

K_2: M·T $\rightleftharpoons$ M·D·P (Bagshaw, 1993).

K_3: M·D·P $\rightleftharpoons$ M·D + P. $K_2K_3 \cong 1$ M (Cardon and Boyer, 1978, Fig. 5); k_+ from steady-state ATPase (Bagshaw, 1993); k_- = 4 $M^{-1}s^{-1}$ (Sleep et al., 1978), = 0.23 $M^{-1}s^{-1}$ (Webb et al., 1978).

K_4: M·D $\rightleftharpoons$ M+D. K_4 (Cardon and Boyer, 1978); k_+ = 1.4 s^{-1} (Bagshaw et al., 1974); k_- = 1 μ$M^{-1}s^{-1}$ (Bagshaw et al., 1974).

Note: $K_2K_4K_3/K_1 = 2 \times 10^5$ M $\cong 5 \times 10^5$ M = K_0, as expected.

mixed (time constant ~1 ms), and the binding of nucleotide is monitored by the accompanying fluorescence change. ATP hydrolysis is measured in a quenched-flow apparatus: Solutions containing myosin and ATP are rapidly mixed, then quenched with acid at various times after mixing. The acid denatures the protein (thereby stopping the hydrolysis reaction), and the amount of nucleotide in its ATP and ADP forms is assayed. If hydrolysis were irreversible, one would expect that when myosin is in great excess over ATP, essentially all the ATP would be hydrolyzed within a second because the hydrolysis rate is ~100 s^{-1}. However, it is found that about 10% of the nucleotide remains in the ATP form (Bagshaw and Trentham, 1973). This indicates that the hydrolysis reaction is reversible. Reversibility is confirmed by isotope studies showing that the gamma phosphate can be labeled with more than one ^{18}O when the reaction is carried out in ^{18}O-labeled water (Bagshaw et al., 1975). Given Figure 14.2, the only way that this can happen is (1) an ATP is hydrolyzed with incorporation of ^{18}O into the phosphate ion; (2) the phosphate ion rotates while it is still held by the protein; (3) the reaction reverses (remaking ATP), and an unlabeled oxygen is lost (due to the rotation); and (4) the ATP is hydrolyzed again with incorporation of another ^{18}O into the phosphate ion. The rate constant for ATP release can be measured from the rate at which ^{18}O-labeled ATP is liberated back into solution (Cardon and Boyer, 1978): It is very low, $\cong 10^{-4}$ s^{-1}, indicating that myosin has a very high affinity for ATP. Because the forward and reverse rate constants of each step have been measured, the free-energy drop at each step can be estimated; the sum of the free energy decrements is found to be equal to the free-energy decrease of the uncatalyzed reaction, providing an important test of internal consistency for the scheme shown in Table 14.1.

With Actin

The reaction of myosin S1 with actin and ATP is not known in detail because not all the rate constants have been measured. However, by combining biochemical data measured in solution with mechanical data from muscle it is possible to infer the likely kinetic mechanism shown in Table 14.2. The main features of the hydrolysis mechanism follow from the fact that myosin binds strongly to actin and to the gamma phosphate, but not to both at the same time.

Actin accelerates myosin's ATPase rate 200-fold, to ~25 s^{-1} per head (Bagshaw, 1993). Because the release of phosphate is the rate-limiting step in the absence of actin, its release must also be accelerated at least 200-fold. Thus actin binding catalyzes the release of phosphate. On the other hand, in the absence of ATP, myosin binds very strongly to actin and dissociates slowly (~1 s^{-1}). The binding of ATP catalyzes the dissociation (Szent-Györgyi, 1941; Eisenberg and Moos, 1968), accelerating the rate more than 1000-fold (Lymn and Taylor, 1971). Thus we see the essence of how the mechanical and chemical cycles are coupled: A mechanical step (the binding to the actin filament) is required to catalyze a chemical step (the release of phosphate), then a chemical step (the binding of ATP) is required to catalyze a mechanical step (the dissociation from the filament).

Table 14.2 Actin-myosin hydrolysis cycle (rabbit skeletal muscle)

(+8 kT)		(0)		(−12 kT)		(−15 kT)		(−17 kT)
			$\geq 10^4 s^{-1}$		$10^3 s^{-1}$		$20{,}000 s^{-1}$	
A·M·T		A·M·D·P	→	A·M·D	⇌	A·M	⇌	A·M·T
					$100 s^{-1}$		$2000 s^{-1}$	
↕	$30 s^{-1}$ ↕ $300 s^{-1}$		$0.2 s^{-1}$ ↓		↓$1 s^{-1}$		$0.4 s^{-1}$ ↕ $2000 s^{-1}$	
	$100 s^{-1}$							
M·T	⇌	M·D·P	→	M·D		M	←	M·T ⇌
	$10 s^{-1}$		$0.1 s^{-1}$				$\sim 10^{-4} s^{-1}$	
(0)		(−2 kT)		(−8 kT)		(−6 kT)		(−25 kT)

Notes: (See Figure 14.3 for a structural interpretation of this reaction.) M = myosin, T = ATP, D = ADP, P = phosphate, A = actin. The most commonly followed path is shaded. In parentheses are the approximate free energies of the states assuming [ATP] = 4 mM, [P_i] = 2 mM, [ADP] = 20 μM. The actin concentration is taken to equal 1 mM, its concentration in skeletal muscle (this is fairly arbitrary because the effective concentration in a muscle depends on the accessibility of the actin site to the myosin head). Overall ATPase: k_{cat} = 25 s^{-1}, K_M(ATP) = 10 μM, K_M(actin) ≅ 100 μM (low ionic strength) at 20°C (Bagshaw, 1993). The rate constants are based on the following experimental data taken from isolated S1 and/or myofibrils under roughly physiological ionic strength. The values vary by up to a factor of 10 between laboratories. The constants (K) are dissociation constants and k_+ (k_-) is the rate constant in the clockwise (counterclockwise) direction.

	$K_{2'}$		$K_{3'}$		$K_{4'}$		$K_{1'}$		
A·M·T	⇌	A·M·D·P	⇌	A·M·D	⇌	A·M	⇌	A·M·T	attached (strong)
K_T ↕	(2)	K_{DP} ↕	(3)	K_D ↕	(4)	K_ϕ ↕	(1)	K_T ↕	
M·T	⇌	M·D·P	⇌	M·D	⇌	M	⇌	M·T	detached (weak)
	K_2		K_3		K_4		K_1		

Subcycle (1)
K_1 = 50 pM (from Table 14.1)
K_ϕ = 500 nM; $k_{+\phi}$ = 2 μ$M^{-1}s^{-1}$; $k_{-\phi}$ = 1 s^{-1} (Highsmith, 1976; Margossian and Lowey, 1978; Geeves, 1991)
$K_{1'}$ = 500 μM; $k_{-1'}$ = 2000 s^{-1}; $k_{+1'}$ = 4 μ$M^{-1}s^{-1}$
K_T = 5 M; k_{+T} = 2000 s^{-1}; k_{-T} = 400 $M^{-1}s^{-1}$
Data from several different experiments are needed to obtain the rate constants for $K_{1'}$ and K_T:
1. $k_{-1'} \geq$ 500 s^{-1} because there is no lag (Millar and Geeves, 1983; Ma and Taylor, 1994).
2. $k_{-1'} \leq$ 4000 s^{-1} because otherwise $k_{+1'} \geq 10^7$ $M^{-1}s^{-1}$, which is the diffusion limit (see Chapter 5).
3. k_{+T} = 800 to 5000 s^{-1} (Ma and Taylor, 1994; Geeves, 1991).
4. $k_{-1'} \cdot k_{-T}/(k_{+T} + k_{-1'}) \cong$ 200 $M^{-1}s^{-1}$ because release of T from MT is very slow in the absence of actin (Cardon and Boyer, 1978). Using the value for $k_{-1'}$ = 2000 and k_{+T} = 2000 s^{-1} gives k_{-T} = 400 $M^{-1}s^{-1}$ and therefore K_T = 5 M.
5. $(k_{+T} + k_{-1'})/k_{+1'}$ = 1000 μM ≅ 500 μM (Millar and Geeves, 1983; Ma and Taylor, 1994). 6. $K_{1'} = K_T K_1/K_\phi$.

Subcycle (4)
K_4 = 4 μM (from Table 14.1)
K_D = 5 μM; k_{+D} = 4 × 10^4 $M^{-1}s^{-1}$; k_{-D} = 0.2 s^{-1} (Highsmith, 1976; Geeves, 1991)
$K_{4'}$ = 0.2 mM; $k_{+4'}$ = 1000 s^{-1}; $k_{-4'}$ = 5 × 10^6 $M^{-1}s^{-1}$. $k_{+4'}$ > 400 s^{-1} at 15°C (Greene and Eisenberg, 1980; Geeves, 1991; Siemankowski et al., 1985), and it cannot be too much larger otherwise $k_{-4'}$ will exceed the diffusion limit.
Check: $K_4 K_D / K_\phi K_{4'}$ = 0.2 ≅ 1

Subcycle (3)
K_{DP} = 10 mM; k_{+DP} = 30 s^{-1} [A]/([A] + 100 μM) depending on ionic strength (see text)
$K_{3'}$ = 600 M; $k_{+4'} \geq$ 3000 s^{-1}; $k_{-4'} = k_{+4'}/K_{4'} \geq$ 5 $M^{-1}s^{-1}$ (see text)

Subcycle (2)
K_2, k_{+2}, k_{-2} (from Table 14.1)

Another way of describing the hydrolysis cycle is to say that myosin alternates between **weakly bound** states (low affinity for actin in the ATP and ADP·P_i states) and **strongly bound** states (high affinity for actin in the ADP and nucleotide-free states). These terms are potentially confusing because surely myosin is bound to actin or it is not. However, the terms arose because different assays for measuring the binding of myosin S1 to actin gave different answers. Binding can be measured in several ways. In a pelleting assay, the fraction of myosin bound to actin is determined by centrifuging the solution. A centrifugal acceleration and run time are chosen so that the more massive actin filaments spin down to the bottom of the tube but the free myosin stays in the supernatant; the fraction of myosin in the pellet is approximately equal to the fraction that is bound to actin in the solution (the truth of this statement is not obvious!). Alternatively, the binding can be assayed by light scattering. The scattering of near-UV light by actin filaments is greatly enhanced by the binding of the large myosin heads. A third assay detects the decrease in fluorescence when myosin binds to pyrene-labeled actin. All three methods show that in the absence of nucleotide or in the presence of ADP, S1 binds tightly to actin (Highsmith, 1976; Margossian and Lowey, 1978; Geeves, 1991). In the presence of ATP, all three methods show that the binding is much weaker, but the pyrene fluorescence indicates a much lower affinity than the other assays. The state with increased pyrene fluorescence is called strongly bound. The state that has no enhanced fluorescence but still appears to be a bound state by centrifugation and light scattering is called weakly bound. The weakly bound state has also been detected in muscle fibers (Brenner et al., 1982); its lifetime is only ~10 μs, sufficiently short that it does not slow contraction (Schoenberg and Eisenberg, 1985). The short lifetime indicates that the weakly bound state is in rapid equilibrium with the detached state (Geeves, 1991). Because the weakly bound state is too short-lived to maintain tension in muscle, and because it catalyzes neither the release of ATP from myosin (as measured by ^{18}O incorporation into ATP; Cardon and Boyer, 1978) nor the release of phosphate from M·D·P (see below), I have grouped the weakly bound state with the detached state in Table 14.2. It is not clear what role the weakly bound state plays. It could just be a nonspecific interaction between the two proteins. An alternative idea is that it is on the pathway to strong binding (Geeves et al., 1984), though this is probably not the case, as we will see. From a practical point of view, it does not matter what the weakly bound state is because models with just two association states (attached and detached) can explain most of the structural and mechanical results anyway.

The release of phosphate after M·D·P binds to actin is thought to produce a highly strained A·M·D* state that generates force. The relaxation of this strain is the driving force for the working stroke and the sliding of the filaments (Figure 14.3). In other words, the working stroke corresponds to the transition from the unstrained A·M·D·P state to the unstrained A·M·D state. The evidence for this is as follows.

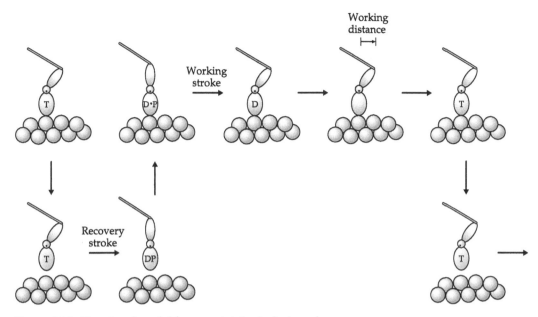

Figure 14.3 Structural model for myosin's hydrolysis cycle
This figure shows an interpretation of the biochemical data in Table14.2 in terms of the structural data of Chapter 12 and the motility data of Chapter 13. The working stroke occurs in the ADP state, and the recovery stroke occurs in the ADP·P$_i$ state.

1. Skeletal muscle myosin appears to undergo little conformational change upon dissociation of ADP (Chapter 12), suggesting that the working stroke occurs prior to ADP dissociation.

2. In solution there is only a small free energy drop in the A·M·D → A·M → A·M·T steps (4.5 kT, Table 14.2). But the efficiency of muscle is about 60% (see Figure 16.5B; Woledge et al., 1985), corresponding to an energy of 15 kT. A simple idea is to suppose that A·M·D·P has the same high free energy as the strained A·M·D* state, perhaps about 12 kT greater than the unstrained A·M·D state. Another way of saying this is that the flux through A·M·D·P → A·M·D should be greater than that through M·D·P → M·D even in the presence of a large opposing force. Therefore, in the absence of load, the former transition should be very fast (as it is expected to be slowed by the load).

3. In isometrically contracting muscle (i.e., muscle prevented from shortening), the rate of incorporation of radioactively labeled P_i from solution into ATP is 500 times greater than that for myosin and actin in solution (Bowater and Sleep, 1988). This is consistent with the idea that pulling back on the A·M·D state stabilizes the strained A·M·D* state to which phosphate can bind more readily. The dissociation constant for phosphate binding to myosin in isometrically contracting muscle is ~3 mM.

4. Adding phosphate to isometrically contracting muscle decreases the force, whereas it has no effect on the contraction speed of unloaded muscle (Cooke and Pate, 1985; Hibberd et al., 1985; Homsher and Millar, 1990; Fortune et al., 1991; Dantzig et al., 1992). This suggests that formation of A·M·D·P reverses the strain-producing conformational change. The binding of phosphate to these strained crossbridges has a dissociation constant of 1 to 10 mM (Homsher and Millar, 1990; Fortune et al., 1991; Dantzig et al., 1992), consistent with the dissociation constant inferred from the isotope experiments. In the absence of tension, the dissociation constant of phosphate from A·M·D·P could be as high as 500 M ($\cong 3$ mM $\times \exp(12)$).

Taken together, these arguments suggest that the binding of M·D·P to actin creates a large force that pushes out the phosphate and leaves the crossbridge in a highly strained A·M·D* state.

What is the rate constant for the binding of M·D·P to actin? Because all the other steps in the pathway are faster than the maximum actin-activated ATPase rate of ~25 s^{-1} measured in solution (Ma and Taylor, 1994), myofibrils (Ma and Taylor, 1994), and permeabilized muscle fibers (He et al., 1997), the transition M·D·P $\rightarrow$ A·M·D·P $\rightarrow$ A·M·D must be rate limiting, with an overall rate of ~30 s^{-1} (to give an overall rate of ATPase 25 s^{-1}). The most likely scheme is that the first step is rate limiting ($k_1 \cong 30$ s^{-1}) and the second step is very fast ($k_2 \geq 3000$ s^{-1} >> 300 s^{-1} $\cong k_{-1}$). The alternate scheme is that the first step is in rapid equilibrium with [A·M·D·P]/[M·D·P] = $k_1/k_{-1} \cong 0.1$ and the second step quite slow ($k_2 = 300$ s^{-1}) to give an overall rate of 30 s^{-1}. The second scheme corresponds to a weakly bound to strongly bound transition. However, if this were true, then shifting the equilibrium toward A·M·D·P is expected to increase the overall ATPase rate. But this is not observed. First, increasing the actin concentration beyond ~10 µM does not increase the solution ATPase rate, even though it does increase the weak binding of M·D·P to actin (see Figure 14.8; Stein et al., 1979). Second, decreasing the ionic strength (which is expected to favor the weakly bound state) has little effect on the ATPase rate in myofibrils (Ma and Taylor, 1994). Thus a weakly bound state may not be on the pathway to strong binding.

Hydrolysis of ATP by Conventional Kinesin

Just as myosin's hydrolysis reaction is tightly coupled to the binding and unbinding of myosin from the actin filament, kinesin's hydrolysis reaction is tightly coupled to the binding and unbinding of kinesin from the microtubule. However, the details are different. One key difference is that in the presence of ATP, kinesin binds strongly to the microtubule, whereas myosin binds weakly to the actin filament. Another key difference is that in the absence of the associated filament, kinesin releases phosphate quickly, whereas myosin releases phosphate slowly. Tight coupling between the kinesin's hydrolysis reaction and its binding to and unbinding from the microtubule is achieved because the release of ADP is catalyzed by the binding of kinesin to the microtubule, whereas the unbinding of kinesin from the microtubule requires the hydrolysis of the ATP.

Without Microtubules

The key to the ATP hydrolysis reaction of kinesin in the *absence* of microtubules is that the release of the product ADP is very slow and rate limiting (Hackney, 1988). This slow release of ADP, measured by following the release of radioactively labeled ADP from kinesin in solution (Hackney 1988), limits the maximum ATPase rate measured at high ATP concentration to $0.01~s^{-1}$ per kinesin head. The slow release of ADP is a reflection of the high affinity that kinesin has for ADP, ~1 nM. By contrast, kinesin has a very low affinity for ATP (~10µm, as determined by experiments with AMP-PNP; Rosenfeld et al., 1996). Thus kinesin, unlike myosin, does not bind the gamma phosphate tightly.

With Microtubules

Microtubules stimulate the ATPase rate 5000-fold, to ~$50~s^{-1}$ per kinesin head (Hackney, 1995; Coy et al., 1999). Because ADP release is the rate-limiting step in the absence of microtubules, it must be accelerated by at least a factor of 5000. This increase in the rate of ADP release by microtubules is also reflected in the increase in the ATP concentration necessary for the half-maximum ATPase rate: The K_M for ATP increases from 3 nM (without microtubules) to 40 µM (with microtubules). In addition to accelerating ADP release, binding to microtubules also accelerates the hydrolysis step, which is only $10~s^{-1}$ in the absence of microtubules (Ma and Taylor, 1995a,b).

Like myosin, the rate of unbinding of kinesin from microtubules depends on the nucleotide. However, the nucleotide dependence is very different. Single-molecule measurements show that K and K·T detach only very slowly from microtubules, whereas K·D·P and K·D detach much faster (Hancock and Howard, 1999). This suggests a cycle that is shifted 90 degrees in phase relative to myosin's cycle, with hydrolysis catalyzing unbinding from the microtubule and microtubule binding catalyzing ADP release (Figure 14.4). As we will see below, the hydrolysis cycles of the two heads of conventional kinesin are highly coordinated; consequently, the one-headed pathway of the type shown in Figure 14.4 is not applicable to two-headed kinesin. However, the pathway may apply to kinesin-related proteins such as Ncd and Eg5 that are not processive and therefore may not have coordinated heads like kinesin (Lockhart and Cross, 1994, 1996; Lockhart et al., 1995; deCastro et al., 1999).

Coordination of the Heads

It is not possible to describe conventional kinesin using a kinetic scheme like that of myosin because, as mentioned above, the hydrolysis cycles of kinesin's

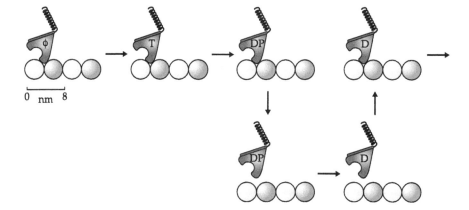

Figure 14.4 Hydrolysis cycle of one-headed kinesin with microtubules
The pathway is based on single-molecule measurements of the rate of detachment of kinesin from microtubules in the various nucleotide states. These rates are given in the following table. All rates have units s^{-1}. The table also contains the corresponding rates of detachment of myosin from actin.

Motor	φ	ATP	ADP·P$_i$	ADP
Kinesin	0.02	0.01[a]	3.8	3.9
Myosin	1	2000	300	0.2

Source: Hancock and Howard, 1999.
[a]Measured using the ATP analogue AMP-PNP.

two heads are highly coordinated. Coordination is necessary to account for kinesin's processivity: Single-molecule force experiments show that during each hydrolysis cycle kinesin can spend no more than 1 μs detached; otherwise, the applied force would pull it backwards (Meyhöfer and Howard, 1995; Chapter 15). Such tight association suggests that the two heads move by a **hand-over-hand** mechanism in which the release of the trailing head is contingent on the binding of the leading head (Figure 14.5).

The hand-over-hand mechanism is supported by the following evidence (Schief and Howard, 2001).

1. When a kinesin dimer with ADP bound to each head is added to microtubules, only 50% of the ADP is released immediately (with a rate of ~100 s^{-1}), while the other half comes off with a similar low rate observed in the absence of microtubules (0.01 s^{-1}) (Hackney, 1994a; Ma and Taylor, 1997; Gilbert et al., 1998). This shows that when one head in the dimer is attached to the microtubule and has no bound nucleotide, the other head cannot attach in a conformation that promotes ADP release. Addition of ATP causes the rapid release of the ADP from the second head (rate ~100 s^{-1}), showing that ATP binding to the attached head initiates a conformational change that accelerates the binding of the second head by about 5000-fold or more.

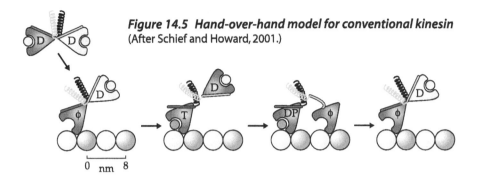

Figure 14.5 Hand-over-hand model for conventional kinesin (After Schief and Howard, 2001.)

Hydrolysis is not necessary for this acceleration because AMP-PNP also accelerates ADP release, though only up to ~20 s^{-1} (Ma and Taylor, 1997; Gilbert et al., 1998). Similar results are obtained for Nkin, a processive kinesin-related protein from the fungus *Neurospora*, which is a distant member of the conventional kinesin subfamily (Crevel et al., 1999).

2. One-headed kinesin detaches slowly from microtubules (rate ~3 s^{-1}) even at high ATP concentration (Hancock and Howard, 1999). But each head in an active dimer detaches 50 times per second as it walks along a microtubule, taking 100 steps per second (Chapters 13, 15). It follows, therefore, that the binding of the second head must accelerate the release of the trailing head more than 10-fold. Presumably the binding of the second head causes intramolecular strain that catalyzes the release of the trailing head.

3. One-headed kinesin has a low ATPase rate, ~3 s^{-1}, similar to its rate of detachment from the microtubule (Hancock and Howard, 1999). This rate is much less than the ATPase rate of the two-headed protein (~50 s^{-1} per head; see Figure 13.5); again, this indicates that there is a coordination of the two heads, with the binding of the second head presumably accelerating a slow step in the other head (which is likely to be its detachment from the microtubule in the ADP·P$_i$ state, as shown in Figure 14.4). This result is somewhat controversial because some highly truncated one-headed kinesin constructs have a very high ATPase activity (e.g., Stewart et al., 1993; Huang and Hackney, 1994); it is likely, however, that ATP hydrolysis of these short constructs, which move slowly or not at all, has become uncoupled from motility (Stewart et al., 1993; Hancock and Howard, 1999).

Taken together, these experiments show that the mechanical and chemical cycles of kinesin are coupled, but in a different way to myosin: A chemical step (the binding of ATP to head 1) catalyzes a mechanical step (the attachment of head 2 to the microtubule), which in turn catalyzes a chemical step (the release of its ADP from head 2), which in turn catalyzes a mechanical step (the detachment of head 1 from the microtubule). The release of phosphate from head 1 then completes the cycle. In other words, the coupling occurs via a coupling between the heads, as shown in Figure 14.5.

During one nucleotide hydrolysis cycle kinesin moves its load through 8 nm. In a two-headed mechanism, many of the transitions are expected to move the load (which is connected to the heads via the dimerization domain). Mechanical experiments described in Chapters 15 and 16 have identified some of the transitions whose rates depend on the load. Of particular importance is the binding of ATP to the bound head: This is associated with movement of the load through 1 to 2 nm.

Functional Differences between Kinesin and Myosin ATPase Cycles

Biochemical experiments add further support to the picture developed in the last chapter that kinesin has a high duty ratio and myosin has a low duty ratio.

Kinesin Is Attached during Its Rate-Limiting Step, but Myosin Is Detached

The ATPase rates of both kinesin and myosin increase as the concentration of the appropriate filaments is increased (Figure 14.6). For kinesin, the microtubule concentration required for half-maximal ATPase is similar to that required for physical binding of the motor to the microtubule (Figure 14.6A). This shows that the rate-limiting step(s) occurs while kinesin is bound to the microtubule, as expected if kinesin walks along the filament and the duty ratio is greater than 0.5. By contrast, the ATPase of myosin S1 saturates at an actin concentration much lower than that needed to saturate the binding (Figure

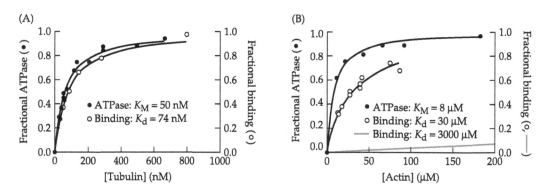

Figure 14.6 Dependence of ATPase on filament concentration
(A) Dependence on microtubule concentration of kinesin's ATPase activity (closed circles) and physical binding to the filament (open circles). *Drosophila* kinesin dimer. (B) Dependence on actin concentration of myosin's ATPase activity (closed circles) and physical binding to the filament measured by a pelleting assay (open circles) and by quenching of fluorescence of pyrene-labeled actin by myosin (gray line). Rabbit skeletal muscle myosin S1 fragment, low ionic strength. (A after Hackney, 1994b; ATPase and physical binding in B from Stein et al., 1979; fluorescence data from Geeves, 1991.)

14.6B). This is consistent with a low duty ratio because the rate-limiting step is occurring while myosin is detached and in its M·T or M·D·P state.

Biochemical Evidence for Kinesin's Processivity

Biochemical experiments confirm that kinesin is processive (Hackney, 1995). At a low microtubule concentration (16 nM tubulin dimer), it takes about 25 s for kinesin to diffuse to and initially bind to a microtubule (measured from the rate of microtubule-stimulated release of radiolabeled ADP). This corresponds to a second-order rate constant of 2.5×10^6 $M^{-1}s^{-1}$, characteristic of a fast, diffusion-limited protein–protein interaction (Northrup and Erickson, 1992). Yet, at the same microtubule concentration, the average ATPase rate is 5 s^{-1}, corresponding to one ATP hydrolyzed per 0.2 s. This indicates that kinesin must hydrolyze about 125 molecules of ATP (25 s/0.2 s) following the initial binding to the microtubule. This is consistent with processive movement over a distance of about 1 µm (~125 × 8 nm), similar to that measured in motility assays (Howard et al., 1989; Block et al., 1990; Coy et al., 1999). By contrast, myosin hydrolyzes only one ATP per encounter with an actin filament (Hackney, 1996).

Biochemical Evidence that Myosin Has a Low Duty Ratio

Biochemical experiments confirm that myosin has a low duty ratio. The ATP concentration needed for half-maximal ATPase of myosin in solution, K_{ATPase}, is ~10 µM (Bagshaw, 1993). By contrast, the ATP concentration needed for half-maximal sliding speed, K_{speed}, is ~200 µM (motility assay, Harada et al., 1990; Figure 14.7; myofibrils, Ma and Taylor, 1994; permeabilized muscle fibers, Pate and Cooke, 1989). To make sense of these data, let k_{+1}, be the second-order rate

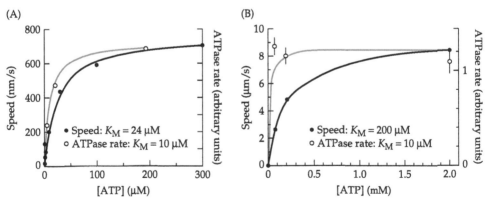

Figure 14.7 The dependence of the speed and the ATPase rate on the ATP concentration
(A) Bovine brain kinesin: Speed is from Howard et al. (1989), ATPase is from Hackney (1988).
(B) Myosin's speed and ATPase measured in a motility assay (Harada et al., 1990).

constant for the binding of ATP to A·M, τ_{on} be the attached time (at very high ATP concentration), and τ_{total} equal the total hydrolysis-cycle time at high ATP concentration. The interpretation of K_{speed} is that this is the ATP concentration at which the motor spends time τ_{on} waiting for ATP to bind (thereby doubling the attached time and halving the speed). Thus $k_{+1'} \times K_{speed} = 1/\tau_{on}$. On the other hand, the interpretation of K_{ATPase} is that this is the ATP concentration at which the motor spends time τ_{total} waiting for ATP to bind (thereby doubling the hydrolysis cycle time and halving the ATPase rate). Thus $k_{+1'} \times K_{ATPase} = 1/\tau_{total}$. Dividing the latter expression by the former gives the duty ratio:

$$\frac{K_{ATPase}}{K_{speed}} = \frac{\tau_{on}}{\tau_{total}} = r \qquad (14.4)$$

Thus we obtain a duty ratio for myosin of ~0.05 (=10 μM/200 μM), in good agreement with that obtained from the minimum number of motors necessary for continuous motility (Equation 13.2) and from the ratio of the speed to the ATPase (Equation 13.5). For kinesin, $K_{speed} \cong K_{ATPase}$ (Hackney, 1988, 1994b; Howard et al., 1989; Hua et al., 1997; see Figure 14.7), consistent with a high duty ratio of 0.5 to 1, depending on the nature of the coordination between the heads. For Ncd, K_{speed} = 235 μM and K_{ATPase} = 23 μM, showing that Ncd has a low duty ratio and is not processive (deCastro et al., 1999).

Summary

Nucleotide chemistry plays three key roles in chemomechanical transduction by motor proteins. First, it regulates the attachment and detachment of the motor from the filament. Second, it drives the working stroke while the head is attached, and the recovery stroke while the filament is detached, so that movement is directed. And third, because the chemical steps are contingent on the completion of mechanical steps, the coupling is tight, ensuring that as few ATPs as necessary are hydrolyzed during movement.

Despite the similarity in structure between kinesin and myosin, the mechanisms of conventional kinesin and skeletal muscle myosin II are different in important ways. For example, the motors attach and detach in different phases of their chemical cycles, and whereas kinesin's cycle is intrinsically two-headed, myosin's is not. However, there are myosins that are processive (e.g., myosin V, Mehta et al., 1999) and kinesins that are not (e.g., Ncd, deCastro et al., 1999). Does this mean that myosin V has a hydrolysis cycle like conventional kinesin and Ncd has one like myosin II? The answer is clearly no. Myosin V has the same hydrolysis cycle path as myosin II—just the rates are changed to ensure that the motor has a high duty ratio (De La Cruz et al., 1999). There is probably coordination between the heads as well (Walker et al., 2000). And Ncd has the same hydrolysis cycle as conventional kinesin (Lockhart and Cross, 1994); however, Ncd may not have coordinated heads like kinesin. The hydrolysis cycles of the dyneins are not known in sufficient detail to compare the processive cytoplasmic dyneins with the nonprocessive axonemal ones.

CHAPTER

15

Steps and Forces

In the last chapter I showed that sequential transitions between different chemical states—ATP binding, hydrolysis, and product release—alter the affinity of the motor domain for the filament and lead to the alternation between attached and detached states. It is hypothesized that movement results from a directed conformational change of the crossbridge while it is attached. Following detachment, the motor recovers to its initial conformation, whereupon it is ready to attach to its next binding site on the filament. In this chapter, I describe high-resolution mechanical experiments on single-motor molecules that directly measure the sizes of the conformational changes, the distances between consecutive binding sites on the filament, and the forces that the motors can generate.

Distances that Characterize a Motor Reaction

There are several distances that characterize a motor reaction (Figure 15.1).

1. The **working distance**, δ, is the distance that a crossbridge moves during the attached phase of its hydrolysis cycle (Chapters 12 and 13). It is the distance along the axis of the filament that the distal, load-bearing region of the crossbridge moves while the proximal, filament-binding region is attached. It is also sometimes called the **powerstroke distance** (see Chapter 16 for the distinction).

2. The **distance per ATP**, Δ, is the distance that each motor domain moves during the time it takes to complete one ATP-hydrolysis cycle. As described in Chapter 13, this distance is measured by dividing the speed of movement by the ATPase rate per head.

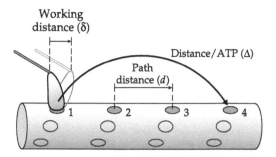

Figure 15.1 Distances associated with the motor reaction
In this example, the crossbridge follows a path parallel to the filament axis and skips two potential binding sites during a hydrolysis cycle. This motor has a low duty ratio.

3. The **path distance**, d, is the distance between the consecutive binding sites on the path that the crossbridge follows as it moves along the filament. It is the distance between the "stepping stones," and arises from the discrete sites on the filament to which the crossbridges can bind. It is not well defined if the motor takes an irregular, nonperiodic course.

The working distance and the distance per ATP were introduced in Chapters 12 and 13. The path distance is a new term. Another term that has been used extensively in the literature is the **step size**. It has been used to denote myosin's working distance (Harada et al., 1990; Uyeda et al., 1990; Bagshaw, 1993), myosin's distance per ATP (Pate et al., 1993), and kinesin's path distance (Svoboda et al., 1993; Howard, 1996; Hua et al., 1997; Kojima et al., 1997). Because of this ambiguity, I will only use the term in a loose sense.

Single-Motor Techniques

The techniques used to measure the distances moved by single molecules, as well as the forces that they generate, are modifications of the in vitro motility assays introduced in Chapter 13. In the filament assay, the actin filament or microtubule is held in a force transducer and presented to a motor that is fixed to a surface (Figure 15.2A). In the bead assay, the motor is attached to a bead held in a force transducer, and the bead is presented to a filament that is fixed to a surface (Figure 15.2B). The force transducers could be cantilevered glass rods, atomic force microscopes (AFMs), or optical tweezers, and they must be able to produce and monitor forces in the piconewton range. The motion produced when the motor interacts with the filament is measured using photodiode detectors capable of sensing subnanometer displacements with millisecond temporal resolution.

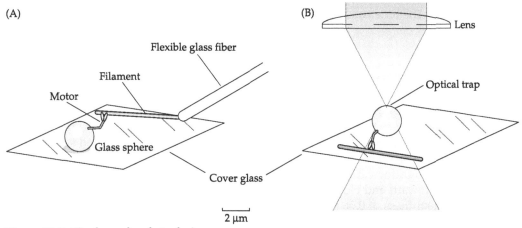

Figure 15.2 Single-molecule techniques
(A) Filament assay in which the filament is held by a glass fiber. (B) Bead assay in which the bead is held in an optical trap.

Forces

A slender glass fiber whose base is held in a micromanipulator can be used as a spring to exert known forces on a protein attached to its tip (Figure 15.3). This is the same principle used in the AFM. The stiffness of a cantilevered fiber of length L, and radius r, is

$$\kappa_f = \frac{3\pi}{4} \frac{Er^4}{L^3} \qquad (15.1)$$

where E is the Young's modulus (Equation 6.5); stiffnesses in the range 0.001 to 1 pN/nm can therefore be achieved with glass rods ($E \sim 70$ GPa) of lengths 100 to 200 μm and radii 100 to 250 nm. Deflection of such fibers by 1 nm to 1 μm leads to forces in the piconewton range. The stiffness of AFM probes is on the order of 10 pN/nm to 10,000 pN/m (0.01 to 10 N/m); the forces are larger than with glass fibers, but still in the piconewton range when the softer probes are used.

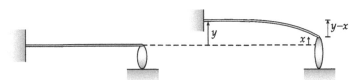

Figure 15.3 Cantilever force transducer
The force equals $F = \kappa_f (y - x)$, where κ_f is the stiffness of the fiber, y is the displacement of the base of the fiber, and x is the displacement of the tip of the fiber, which is also equal to the displacement of the attached point of the protein. $y - x$ is the flexion in the fiber.

The time constant of a cantilevered spring is

$$\tau \cong 0.2 \frac{\eta \cdot L^4}{E \cdot r^4} \tag{15.2}$$

where η is the viscosity of the solution (Example 6.8). The glass fibers considered above have time constants of about 1 ms in aqueous solution, which makes them fast enough to resolve the individual transitions within a hydrolysis cycle. An important design feature is that by varying the length and radius independently it is possible to manufacture a fiber with any desired stiffness and time constant (Crawford and Fettiplace, 1985). For example, it is reasonably straightforward to make glass fibers of length 150 μm and radius ~0.2 μm (Howard and Hudspeth, 1987, 1988; Meyhöfer and Howard, 1995); this gives a stiffness of 0.08 pN/nm and a time constant of ~0.9 ms. The time constant decreases when the relatively stiff motor attaches to the filament (see Example 6.8).

Optical tweezers can also be used to exert forces in the piconewton range. In this method a laser beam is focused down to a diffraction-limited spot using a high-numerical-aperture lens. A dielectric, nonabsorbing particle such as a glass or plastic bead will experience a force that tends to move it into the region of highest light intensity, namely the focus of the laser (Ashkin et al., 1986). There are two ways of thinking about this force. If the particle is much larger than the wavelength of light, then the force can be viewed as arising from the change in momentum of the photons that are refracted by the particle (Figure 15.4A). The force can be calculated using ray optics (Ashkin, 1992). If the particle is much smaller than the wavelength of light, then the force can be viewed as arising from the polarization of the particle induced by the electric field component of the light wave (Figure 15.4B). The polarized particle is then attracted to the focal point where the field is strongest. This is analogous to the electrostatic attraction of a small piece of tissue paper to a comb that has been charged by grooming one's hair. In addition to the gradient force there is also a small scattering force, the magnitude of which is directly proportional to the intensity, that pushes the particle just beyond the focal point of the laser.

To judge the force exerted by an optical trap, F_{opt}, it is useful to compare it to the force associated with the absorption of light (photon pressure):

$$F_{opt} = Q \frac{nP}{c} \tag{15.3}$$

where nP/c is the force exerted by light on a perfect absorber (c/n is the speed of light in the medium, P is the power) and Q is a dimensionless constant called the trapping efficiency. For a spherical particle of radius equal to the wavelength of light, the efficiency is ~0.1. The optical force is therefore ~0.5 pN/mW of laser power (Svoboda and Block, 1994a). For particles and molecules much smaller than the wavelength of light, the force is very much smaller because the polarization (and therefore the trapping efficiency) is proportional to the number of atoms, which depends on the volume (i.e., the third power of the diameter).

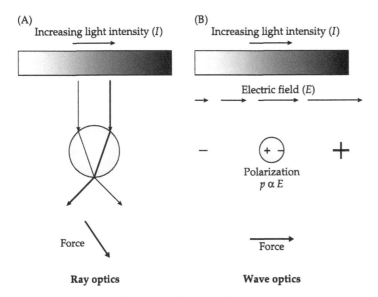

Figure 15.4 Optical gradient forces utilized in laser tweezers
(A) Ray optics view in which the force arises from the change in momentum of the photons that are refracted as the light passes through a high-refractive-index particle. The force is proportional to the intensity gradient, *dI/dx*. (B) Wave optics view in which the force arises from the polarization of the particle by the electric field and its subsequent attraction to the region of highest field strength. The force is proportional to the polarization, *p*, (which is proportional to field strength, *E*) times the field gradient (*dE/dx*). In other words, the force is proportional to *E·dE/dx* ∝ *dI/dx* because the intensity is proportional to the square of the field. This agrees with the ray optics view.

Provided that the particle does not move too far from the center of the trap, the trap will behave approximately like a spring with a spring constant

$$\kappa_{opt} \sim Q \frac{n^2 P}{\lambda c} \quad (15.4)$$

where λ is the wavelength of the light. Spring constants in the range 0.01 to 0.1 pN/nm have been achieved using laser powers of ~100 mW on glass or plastic beads (n = 1.50 or 1.57) of diameter ~1 μm (Svoboda et al., 1993). The physical size of the trap is determined by the wavelength of light and the size of the object. Wavelengths of ~1 μm are used because these longer wavelengths produce less photodamage than shorter, visible wavelengths (Svoboda and Block, 1994a). The traps are linear out to about 200 nm, beyond which the force reaches a maximum, the escape force. The disadvantage of optical traps over cantilever beams and AFMs is that they are softer and produce less force. However, because the damping on a 1-μm-diameter bead is about an order of magnitude smaller than that on a 100-μm-long fiber, an optical trap has a significantly shorter time constant than a fiber of the same stiffness. Thus optical traps have an edge with respect to temporal resolution.

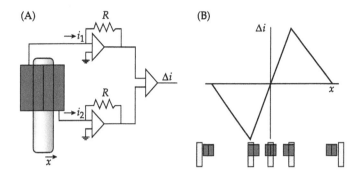

Figure 15.5 A split photodiode detector
(A) The photocurrent from the left photodiode (i_1) is converted to a voltage and subtracted from the photocurrent from the right photodiode (i_2) to provide a signal Δi that depends on the position of the fiber (x) as shown in (B).

Displacements

The molecular motion of a motor held in a force transducer can be measured very precisely by imaging the bead or the fiber onto a photodiode detector (Figure 15.5). Although the resolution of a microscope is limited by the wavelength of light to about $\lambda/2$, or about 250 nm when using green light (Spencer, 1982), there is no such limitation on the precision to which the center of mass of a fiber or a bead can be determined. The reason is that it is possible to estimate the mean of a distribution very precisely even if the distribution itself is quite broad: This is analogous to the standard deviation of the mean being much smaller than the standard deviation when the number of measurements is large. In principle, the precision of a displacement measurement is limited only by the number of photons that are counted (Appendix 15), and noise levels less than 1 nm over a 1000 Hz bandwidth can be achieved routinely (see Example 15.1). Detectors with similar precision are used in AFMs. Provided that the motion of the bead or the fiber faithfully follows that of the motor, the motor can be tracked very precisely.

Example 15.1 Sensitivity of a photodetector Consider a fine glass fiber illuminated by an arc lamp and imaged by darkfield microscopy using a water-immersion objective. The intensity of the light scattered off a segment whose length is a few microns long is $\sim 10^{10}$ photons/s, the diffraction-limited width is ~ 1 μm, and the quantum efficiency is ~ 0.5. The displacement noise is therefore $\sim 10^{-4}$ nm^2/Hz (Appendix 15), in agreement with experiment (Meyhöfer and Howard, 1995). Thus over a 1000 Hz bandwidth, the variance of the equivalent photon shot noise is ~ 0.1 nm^2, corresponding to a standard deviation of ~ 0.3 nm.

Calibration

The detectors are sensitive enough to measure the thermal fluctuations of the force transducers: A spring of stiffness $\kappa = 0.1$ pN/nm has a displacement variance equal to 40 nm^2 ($= kT/\kappa$), corresponding to a standard deviation of ~6 nm, well within the sensitivity of the instrument (Crawford and Fettiplace, 1985). Thus measurement of the thermal fluctuations provides a method for calibrating the stiffness of a force transducer in situ (Howard and Hudspeth, 1988; and see Example 4.7).

Single-Molecule Fluorescence

Recent refinement of the fluorescence microscope allows one to see individual fluorescent molecules in aqueous solution (Funatsu et al., 1995) and this has paved the way for the development of a whole arsenal of single-molecule techniques for studying proteins and other macromolecules. Though single molecules had been visualized by laser-excited fluorescence in solids at low temperature (Moerner and Kador, 1989), the high background light levels in the conventional fluorescence microscope made single-molecule detection impossible. However, by using a total-internal-reflection microscope (Spenser, 1982; Axelrod, 1989) or a scanning confocal microscope (e.g., Zhuang et al., 2000), to reduce out-of-focus background light, it is now possible to image single fluorophores. The fluorophore can be a small dye covalently attached to a protein, a fluorescently labeled ligand (such as ATP) that binds to the protein (Funatsu et al., 1995), or the green fluorescent protein, a naturally occurring fluorescent protein that can be fused to the protein of interest using molecular biology techniques (Dickson et al., 1997). The limited lifetime of the fluorophores (the total number of emitted photons is quite low, generally <10^6) has limited the spatial and temporal resolution of the technique. However, the detection of fluorescence resonance energy transfer (FRET) (Stryer and Haugland, 1967) at the single-molecule level permits the measurement of the distance of two fluorophores separated by less than 5 nm with an accuracy of ~1 nm (Deniz et al., 1999). By coupling one fluorophore to an amino acid on one domain and another on an amino acid in an adjacent domain, the relative positions of the domains can be followed in real time (Deniz et al., 2000).

Steps, Paths, and Forces

Optical tweezers, glass fibers, and AFMs have all been used to measure the mechanics of motor proteins. The displacement sensitivity of these techniques, ~1nm, is sufficiently high to resolve the stepwise movement of conventional kinesin along a microtubule and the working stroke of skeletal muscle myosin. The force sensitivity, ~1pN, is sufficiently high to measure the force generated by a single motor molecule. When taken together, the displacement and force measurements allow one to characterize the energetics of force generation by motor proteins, and permit an estimate of the efficiency of these biological engines.

Conventional Kinesin

Single-molecule techniques confirm that conventional kinesin is processive. A single kinesin molecule can pull a 1-μm-diameter glass bead hundreds of nanometers along a microtubule, even when an optical trap is used to put the motor under load (Figure 15.6; Visscher et al., 1999). At high force, the motion is not smooth: Filtering the traces to remove thermal motion reveals that the motion is stepwise, with the individual steps having an amplitude of 8 nm. Stepwise motion is also seen at low ATP concentration. Similar-sized steps have been obtained in several different labs, using photodiode detectors (Kojima et al., 1997; Crevel et al., 1999; and see Figure 15.8 and the cover of this book) as well as video image processing (Hua et al., 1997).

The 8 nm steps suggest that kinesin moves from one tubulin dimer to the next along a protofilament (the length of the dimer is 8 nm). This has been confirmed using a number of approaches. For example, most cytoplasmic microtubules have 13 protofilaments that run parallel to the axis of the microtubule (see Chapter 7). These microtubules do not rotate as they glide across a kinesin-coated surface. On the other hand, specially synthesized 12- and 14-protofilament microtubules whose protofilaments supertwist about the microtubules' axes do rotate with the pitch and handedness of the supertwist (Figure 15.7). This shows that kinesin follows a path parallel to the protofilaments. The fidelity is high: When the surface density of motors is low and the gliding is driven by a single (two-headed) kinesin molecule, the motor has less than a 1% chance of switching to a neighboring protofilament at each step (Ray et al., 1993). We can therefore think of there being 13 lanes for kinesin motors to travel along on a typical cytoplasmic microtubule. This conclusion is confirmed by

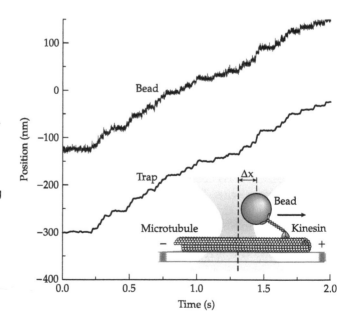

Figure 15.6 The stepping of a single kinesin molecule along a microtubule
The position of a bead (to which a single kinesin molecule is bound) is measured by imaging it onto a photodiode detector (top trace). An opposing force is applied by positioning an optical trap (inset) just behind the bead; the force is kept constant, in this case at 6.5 pN, using a feedback circuit that adjusts the position of the trap (bottom trace) so as to keep the bead a fixed distance from the center of the trap. The amplitude of the steps, 8 nm, corresponds to the spacing of the tubulin dimers along the protofilament. (After Visscher et al., 1999.)

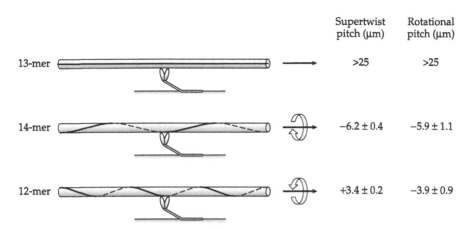

Figure 15.7 Kinesin's path on the microtubule
Experiment showing that kinesin moves parallel to the microtubule's protofilaments. Microtubules with different numbers of protofilaments have different supertwists: The protofilaments of 13-mers run parallel to the microtubule's axis, while those of 14-mers follow a shallow left-handed helix around the surface of the microtubule (note that the width of the microtubule is greatly exaggerated). Values are given at right for the supertwist pitch measured by cryoelectron microscopy and the pitch of the rotation of the microtubules as they move in a gliding assay across a kinesin-coated surface. (After Ray et al., 1993; Howard, 1996.)

tracking kinesin-coated beads using high-resolution video image processing: The bead does not change lanes (Berliner et al., 1995). Interestingly, beads coated with one-headed kinesin molecules wander erratically on the surface of the microtubule, switching protofilaments at least ten times more frequently than two-headed molecules (Berliner et al., 1995). Thus the coordination of two heads within a motor molecule is necessary for precise tracking along a protofilament.

These experiments suggest that kinesin's path distance is 8 nm, as it steps from one tubulin dimer to the next. This is supported by biochemical binding experiments (Huang and Hackney, 1994; Lockhart et al., 1995) and electron micrographs of microtubules decorated with kinesin (Harrison et al., 1993) that show that there is just one binding site for kinesin per tubulin dimer.

The maximum force that kinesin can work against is ~6 pN. This can be measured by allowing kinesin to walk away from the center of a trap: As the distance increases, the load increases, causing the speed to decrease. Eventually, the motor stalls (Figure 15.8). Stall forces in the range of 4 to 8 pN have been measured in several labs using optical tweezers (average 5.8 pN, Svoboda and Block, 1994b; 7.2 pN, Kojima et al., 1997; 7 pN, Visscher et al., 1999; 7.3 pN, Kawaguchi and Ishiwata, 2000), glass fibers (5.4 pN, Meyhöfer and Howard, 1995), microtubule bending (6 pN, Gittes et al., 1996), and viscous loads (4.2 pN, Hunt et al., 1994). The speed of movement decreases approximately linearly as the opposing force is increased, and the maximum force is independent of the ATP concentration, though some small deviations from nonlinearity and ATP-independence were observed in one study (Visscher et al., 1999).

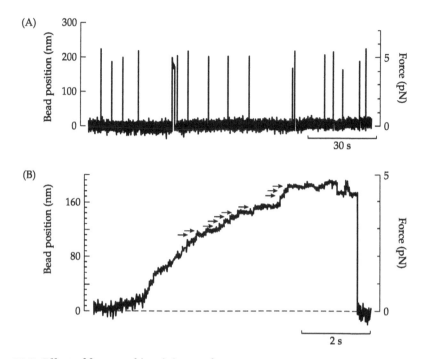

Figure 15.8 Effect of force on kinesin's speed
(A) Record showing 17 examples of a Nkin molecule attached to a bead interacting with a microtubule. The maximum force developed is between 4 and 6 pN. (B) A blowup of one trace shows that as the kinesin moves away from the center of the trap (0 nm) and the force increases, the rate of stepping decreases. Arrows indicate the 8 nm steps. (After Crevel et al., 1999.)

The maximum force is also independent of the temperature (Q_{10} < 1.05), though the speed (Kawaguchi and Ishiwata, 2000) and ATPase rate (Coy and Howard, unpublished) of kinesin increase with a Q_{10} = 2. Optical tweezers give slightly higher stall forces than glass fibers and viscous loads. One possible reason is that optical tweezers also impart forces that tend to pull the motor away from the surface of the microtubule; these perpendicular forces, which are absent in the fiber and viscous assays, have been shown to accelerate the motion and increase the stall force (Gittes et al., 1996).

Knowledge of the step size (8 nm) and the maximum force (6 ± 1 pN) allows one to calculate the maximum work done per step. The maximum work is 48 pN·nm. Kinesin does this much work each time it takes a step while pulling against a load of 6 pN; at lower loads, the work per step is smaller. The maximum work of 48 pN·nm is approximately half the free energy derived from the hydrolysis of ATP under cellular conditions (100 pN·nm, Chapter 14). From this we conclude that the maximum possible efficiency of kinesin is ~50%, though the efficiency will be less when the load is smaller. Kinesin will actually realize this maximum efficiency of 50% provided that two conditions are met. The first condition is that each step at high load is associated with the hydrolysis of just one molecule of ATP. This is not known for sure because 1:1 coupling has only been directly confirmed at low load (Coy et al., 1999b). Although the ATP hydrolysis rate has not been measured at high load, indirect statistical arguments do support a 1:1 coupling of stepping to ATP hydrolysis for forces up to 5 pN (Visscher et al., 1999). Thus, this condition is likely to be met. The second condition is that kinesin be able to generate its maximum force of 6 pN under cellular nucleotide and ionic concentrations. This condition is also likely to be met because the main difference between the assays and cells is that the ADP and P_i concentrations are higher in cells, but this difference is not expected to have a large effect on the maximum force.

Another important conclusion can be made regarding the single-molecule force. The maximum force is only 50% of the **thermodynamic force**. This is the force that corresponds to the equilibrium force, or reversal force (see Chapters 5 and 10). The thermodynamic force is the maximum that a fully reversible motor could work against, and it corresponds to the chemical free energy divided by the step size. For kinesin, the thermodynamic force is 12.5 pN (= 100 pN·nm/8 nm). Because the maximum force is only half the thermodynamic force, kinesin is not at thermodynamic equilibrium when it has stalled. Why, then, does kinesin stall? A priori, there are two reasons why a nonequilibrium motor would stall at high load: Either the frequency of forward steps decreases to near zero, or the frequency of backward steps increases until it balances the frequency of forward steps. Single molecule tracking (see Figures 15.6 and 15.8) indicates that the frequency of backward steps at high load is low, much lower than the frequency of forward steps at low load. Thus kinesin stalls for kinetic reasons: The rate of forward stepping (and presumably the rate of ATP hydrolysis) decreases to near zero as the load is increased. Evidently one (or more) step(s) in the cycle is so strongly force dependent that its rate constant is nearly zero at 6 pN.

Myosin II

Myosin II behaves quite differently from conventional kinesin in single-molecule assays. Myosin only interacts transiently with an actin filament, and does not step along it as kinesin steps along a microtubule. For example, if an actin filament suspended between two glass beads (each held in an optical trap) is presented to a single skeletal muscle myosin on a surface (Figure 15.9A), discrete binding events are observed (Figure 15.9B). The binding can be detected as a decrease in the thermal fluctuations of the actin filament. The traps are very soft, so that when myosin binds, the stiffness is greatly increased with a corresponding decrease in the displacement fluctuations (in accordance with the Equipartition principle, Equation 4.4). After a time (which depends on the ATP concentration), the motor unbinds and the filament resumes its large thermal motion. This behavior is strikingly different to that of kinesin; evidently myosin II is not processive.

The binding of myosin often displaces the actin filament (and the attached bead) away from its average position. But the mean displacements can be pos-

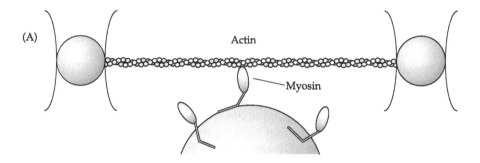

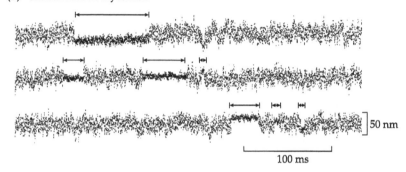

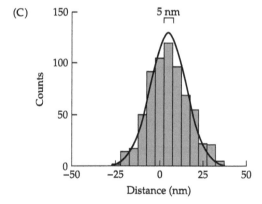

Figure 15.9 Myosin steps
(A) The three-bead assay, in which an actin filament suspended by two beads held in optical traps is lowered down to a third bead to which myosin is bound. The beads are ~1 μm in diameter and are not to scale with the protein. (B) The position of the filament is tracked with high precision by imaging one of the beads onto a quadrant photodiode detector. (C) Histogram of mean displacements associated with binding events. The mean is 5 nm, and corresponds to the amplitude of the working distance. (After Veigel et al., 1998, 1999.)

itive or negative. The interpretation of this finding is that large thermal fluctuations of the actin filament sweep binding sites past the fixed motors; when they are in registration, a motor binds to the filament and holds it in place. Displacements can be positive or negative because the fluctuations are symmetrical. In support of this interpretation, the standard deviation of the histogram of displacements (see Figure 15.9C) is equal to the standard deviation of the position of the free bead; both standard deviations accord with the measured stiffness of the optical traps.

Despite the variability of the displacements, these recordings contain strong evidence that myosin undergoes a conformational change after it binds to an actin filament. The evidence is there are more displacements in one direction than the other (see Figure 15.9C). Furthermore, this directional bias is related to the polarity of the actin filament: If the actin filament is rotated by 180 degrees, the majority of events are now in the other direction relative to the laboratory frame of reference (Molloy et al., 1995a). This suggests that the myosin molecule undergoes a directed conformational change after it binds to the filament. The amplitude of the conformational change is equal to the mean of the displacement histogram, namely 5 nm. This corresponds to myosin's unloaded working distance.

The working distance of myosin II has been very controversial. Earlier work reported just the maximum displacement and therefore an overestimation of the working distance (Finer et al., 1994); subsequent work in this lab has confirmed the smaller working distance (Mehta et al., 1997). Although one lab has reported very large working distances, up to 30 nm, comprising several 5 nm substeps (Kitamura et al., 1999), two other labs have measured the smaller distances (Tyska et al., 1999; Ruff et al., in press). The consensus, therefore, is that the unloaded working distance is 5 nm.

Single-molecule experiments provide strong support for the rotating lever arm model. When the length of the lever is altered by replacing the light-chain binding domain with artificial levers of various lengths, the amplitude of the working stroke is proportional to the length of the lever arm (Figure 15.10; Ruff et al., in press). This suggests a rotation of the lever through 30 degrees, which, for the wild-type lever, corresponds to a working stroke of 5 nm. The single-molecule measurements give a smaller rotation than the 70 degrees inferred from the different crystal structures of myosin (Chapter 12); the most likely interpretation is that the size of the swing depends on whether myosin is bound to actin.

Kinetic measurements from single-molecule experiments on myosin II strongly support much of the biochemical data reported in Chapter 14.

1. The working stroke of myosin II is complete within a few milliseconds, because a transitory displacement corresponding to an initial, prestroke binding is not observed (Veigel et al., 1999). This is consistent with the working stroke occurring during the rapid relaxation of the strained A·M·D state, as argued in Chapter 14.

2. The duration of the attached phase depends on the ATP concentration. The lower the ATP concentration, the longer the binding event, consistent with the attached phase being terminated by the binding of ATP to nucleotide-free myosin in complex with actin.

3. The association of myosin and actin (measured using the three-bead assay) has been correlated with the binding and unbinding of nucleotides (meas-

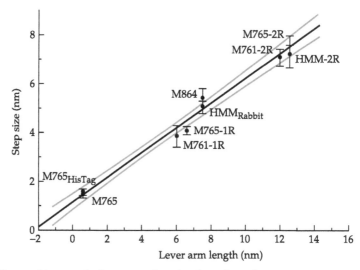

Figure 15.10 The working stroke is proportional to lever length
The length of the light-chain domain of *Dictyostelium* myosin II was modified by genetic engineering. The working stroke, measured using similar assays to that shown in Figure 15.9, is proportional to the lever arm length. M765, M765$_{HisTag}$ have no light-chains; M761-1R and M765-1R have one 6-nm-long α-actinin domain substituting for the light-chain binding domain; M864 and HMM$_{Rabbit}$ have two light chains; M761-2R, M765-2R, and HMM-2R have two α-actinin domains in tandem. (After Ruff et al., in press.)

ured by total-internal-reflection microscopy). As predicted by the crossbridge model, the detachment of myosin S1 from actin is closely coupled to the binding of ATP, and the attachment of myosin is closely coupled to the unbinding of nucleotide, presumably ADP (Ishijima et al., 1998). These experiments confirm that there is a one-to-one coupling between the chemical and mechanical hydrolysis cycles.

4. The duration of the attached phase is increased when myosin S1 is prevented from moving (Finer et al., 1994; Molloy et al., 1995b). This decrease in the detachment rate is consistent with an increase in the duty ratio and a decrease in the ATPase rate as the load on an intact muscle is increased (see Chapter 16).

Myosin II, like kinesin, follows a path that is parallel to the axis of the filament. This has been shown in a number of different ways. Actin filaments rotate little (Sase et al., 1997), if at all (Suzuki et al., 1996), as they glide over surfaces coated with myosin II. This implies that myosin does not move along the two-stranded helix, because this would result in one rotation per 72 nm, the pitch of this helix. Now it is possible that the crossbridge is so flexible that it binds equally well to almost any subunit around the circumference of the filament. However, if a filament is "bowed" quickly past a fixed myosin head (so there is not enough time for the beads and the filament to rotate in the trap), binding is only observed at multiples of approximately 36 nm (Molloy et al., 1995b).

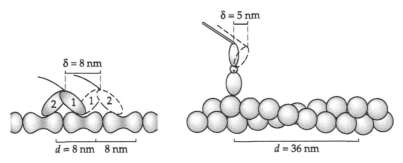

Figure 15.11 Working distances and path distances for conventional kinesin (left) and skeletal muscle myosin (right)

The path distance of myosin is therefore 36 nm, the half-pitch of the actin filament (Figure 15.11).

That myosin follows a path parallel to the actin filament is surprising. Although some earlier models assumed a parallel path (Eisenberg et al., 1980), it has more commonly been assumed that myosin follows the two-stranded helix of actin (e.g., Rayment et al., 1993a), presumably because the consecutive binding sites would be considerably closer (only ~5.5 nm). However, the new conclusions, based on in vitro data, do accord with other biological and structural considerations. If a motor treads upon an angled path, a torque will be generated; this is shown clearly in vitro in the case of kinesin, which rotates supertwisted microtubules (Ray et al., 1993; see Figure 15.7). Such torque would be deleterious for motility: It would disrupt the arrays of the filaments found in muscle and it would cause transported organelles to spiral about their filaments, creating serious steric problems, especially if there are two filaments within an organelle diameter of each other, as is often the case in the cytoplasm.

A lower bound on the force exerted by a single myosin crossbridge has been obtained using the three-bead assay. The assay is modified using a feedback circuit that rapidly moves the trap so as to clamp the position of the bead. The motor force is the force required to keep the bead fixed (Finer et al., 1994). Forces in the range of 1 pN to 10 pN have been measured (Finer et al., 1994; Molloy et al., 1995a; Veigel et al., 1998; Tyska et al., 1999). These forces are lower limits due to compliant connections between the actin filaments and the beads. The maximum force exerted by a myosin crossbridge is therefore ≥10 pN.

Other Motors

Single-molecule techniques are being used to study a number of motor proteins.

1. Myosin I has an additional component of its working stroke that corresponds to the dissociation of ADP (Veigel et al., 1999). This is consistent with cryoEM studies showing that myosin I, upon dissociation of ADP,

undergoes a large structural change (Jontes et al., 1995) like that of smooth muscle myosin (see Figure 12.8).

2. Myosin V takes 36 nm steps as it moves processively along actin filaments (Mehta et al., 1999; Rief et al., 2000; Sakamoto et al., 2000). This step size suggests that myosin V moves parallel to the axis of the actin filament like myosin II (Figure 15.12). This is confirmed by electron microscope images of double-headed myosin V binding to actin: The two heads are spaced 36 nm apart (Walker et al., 2000). The large stride is facilitated by myosin V's six light chains, which create a lever three times longer than that of myosin II (see Figure 12.4).

3. Single-molecule experiments show that *Neurospora* kinesin Nkin is processive (Crevel et al., 1999), while Ncd is not (deCastro et al., 1999, 2000).

4. Recent single-molecule work indicates that the murine kinesin–related protein KIF1A is processive: On average, each encounter of KIF1A with a microtubule results in a net displacement of about 1 μm toward the plus end (Okada and Hirokawa, 1999). This discovery, which is still controversial because it has not been replicated in other labs, is very surprising because KIF1A, a vesicle transporter, has only one head. Thus KIF1A cannot use a hand-over-hand mechanism like conventional kinesin. However, KIF1A has a second microtubule-binding region in its motor domain in addition to the conserved region that is also found in kinesin and all other kinesin-related proteins. This unique binding region is a lysine-rich loop (the K-loop), absent

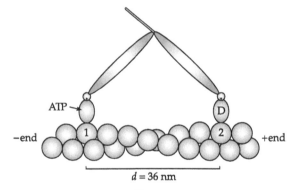

Figure 15.12 Movement of myosin V along an actin filament
Myosin V is a processive motor that is able to make several 36-nm steps along an actin filament before dissociating. Processivity is possible because the light-chain binding domain of myosin V is 24-nm long (myosin V binds six light chains compared to myosin II which binds only two; see Figure 12.4); a single myosin V dimer can therefore span between the consecutive binding sites on the actin filament as shown. The two heads of myosin V are thought to move in a coordinated, hand-over-hand fashion like those of kinesin's (Walker et al., 2000), even though the hydrolysis cycle for myosin (see Figure 14.3) is shifted with respect to that of kinesin (see Figure 14.4). At low ATP concentration, both heads of myosin V are bound to the actin filament as shown. The binding of ATP to the rear head causes it to dissociate. The ATP will then be hydrolyzed and this head will then bind to the next site on the actin filament (another 36 nm past the other head towards the plus end); binding then catalyzes phosphate release in this head. Release of ADP from the other head then completes the cycle.

in kinesin, that interacts with the glutamate-rich C-terminus of tubulin (Okada and Hirokawa, 1999; Kikkawa et al., 2000; Okada and Hirokawa, 2000). Thus it is possible that KIF1A uses its unique binding loop as a tether to maintain contact with the microtubule; this is distinct from kinesin, which uses the conserved microtubule-binding region in the second head as its tether. A possible mechanism for the processivity of KIF1A is that the binding of ATP to the attached KIF1A head biases the K-loop to bind to the next tubulin dimer, and, once it has bound, the conserved region can detach without KIF1A completely dissociating from the microtubule. Such a mechanism may explain the highly diffusive motion of KIF1A, whose velocity fluctuates wildly and whose coupling is very loose (several ATP molecules are hydrolyzed per 8 nm step, even in the absence of load).

5. Single-molecule assays show that outer-arm dynein is not processive at high ATP concentrations, but is processive at low ATP concentrations (Hirakawa et al., 2000). The likely interpretation of this finding is that the working stroke of dynein is large enough to reach the next binding site (8 nm away), but at high ATP concentrations, the dissociation rate is very high and so the duty ratio is small. By contrast, inner-arm dynein appears to be processive even at high ATP concentrations (Shingyoji et al., 1998; Sakakibara et al., 1999). This is surprising because inner-arm dynein is one-headed.

6. Single-molecule techniques have been applied to several other ATPases. ATP synthase rotates in discrete 120-degree steps and generates torques as high as 20 to 40 pN·nm (Yasuda et al., 1998; Soong et al., 2000; and see Problem 6.2). RNA polymerase can generate forces up to 25 pN (Wang et al., 1998), and DNA polymerase can generate forces as high as 34 pN (Wuite et al., 2000).

The Structural Basis for the Duty Ratio

The single-motor distances of myosin and kinesin are summarized in Table 15.1 and Figure 15.11.

The distance per ATP and the path distance are related by:

$$\Delta = n \cdot d \tag{15.5}$$

where n is an integer ≥ 0. $n = 0$ corresponds to a futile ATP hydrolysis event with no associated displacement, as might occur at high load. $n > 1$ if the motor jumps over one or more of the stepping stones as occurs for myosin at low load (Table 15.1). Because the duty ratio is equal to the working distance divided by the distance per ATP, $r = \delta/\Delta$ (Equation 13.5), we obtain the following structural expression for the duty ratio

$$r = \frac{\delta}{n \cdot d} \tag{15.6}$$

(Howard, 1997). This equation represents the crucial steric constraint that is present in a moving motor-filament system but absent when isolated heads are freely diffusing in solution (in which case the path has no meaning). It can

Table 15.1 Motor distances in vitro

Motor Distance	Parameter	Skeletal muscle myosin	Conventional kinesin
Working distance	δ	5 nm[a]	≥ 8 nm
Path distance	d	36 nm	8 nm
Distance/ATP per head[a]	Δ	200–400 nm	16 nm

[a]Low load

also be deduced from first principles by considering the gliding of a filament over a fixed array of motor proteins. If the working distance, δ, is smaller than the path distance, d, then each individual crossbridge must spend a significant time detached while other (attached) crossbridges move the filament and the next binding site forward: The duty ratio must be less than one, and continuous motility will require an assembly of crossbridges. In general, the duty ratio is equal to the fraction of the distance to the next binding site that the working stroke moves the crossbridge, and Equation 15.6 follows.

Summary

Single-molecule experiments have directly confirmed a key prediction of the rotating crossbridge model, namely that motors undergo a directed conformational change (the working stroke) while attached to their filament. The size of the working stroke of rabbit psoas muscle myosin (~5 nm in the absence of load) accords well with the maximum gliding speed of filaments in this muscle (~5000 nm/s; see Table 13.1), given that the attached time is ~1 ms (Chapter 14). Furthermore, because the working stroke is smaller than the path distance (equal to the 36 nm periodicity of the actin filament), skeletal muscle myosin must spend an appreciable fraction of its time detached from the filament: The duty ratio can be no more than ~0.14 (= 5/36 from Equation 15.6) and is likely to be even smaller at high contraction speeds when the motor skips over neighboring binding sites. Single-molecule experiments have also directly confirmed that kinesin is processive and that its path distance is 8 nm, the length of the tubulin dimer along the protofilament. Kinesin's working distance has not been directly measured, but it must be at least 8 nm given that the distance per ATP is 16 nm (see Table 15.1) and the duty ratio is at least 0.5 in order for the kinesin dimer to maintain its attachment to the microtubule.

In addition to confirming key assumptions underlying the rotating crossbridge cycle (such as one mechanical step per ATP hydrolyzed), the single-molecule techniques have provided new data about the operation of motor proteins, data that cannot be obtained using structural or biochemical techniques. These data include the force generated by a single motor as well as the dependence of the chemical and mechanical steps on load. In the next chapter, I show that these single-molecule thermodynamic data account well for the thermodynamic properties of complex motile systems such as muscle tissue.

CHAPTER

16

Motility Models: From Crossbridges to Motion

The goal of this chapter is to bring together the structural, biochemical, and mechanical properties of crossbridges into models that explain how intact motile systems work. We will consider two systems: the contraction of skeletal muscle driven by an ensemble of myosin II heads and the processive movement of kinesin driven by just two heads. These two motile systems offer complementary views of how proteins generate force. The models clearly demonstrate that the mechanics and chemistry of the intact systems can be understood in terms of the molecular properties of the motors.

Macroscopic and Microscopic Descriptions of Motility

There are two ways of viewing a motile system. The first is to treat it as a "black box" and measure its "macroscopic" properties. The second is to take it apart and measure the "microscopic" properties of the individual molecules. The challenge is to explain the macroscopic in terms of the microscopic.

For muscle contraction, the macroscopic properties are the speed, the force, the stiffness, and the rate of ATP hydrolysis per unit volume of muscle tissue. From these parameters the power and efficiency can be calculated. Because we know that contraction is due to the sliding of filaments, we can convert shortening speed into sliding speed and therefore calculate the molecular speed (v) at which myosin crossbridges move past the adjacent actin filaments (see Figure 1.1 and Appendix 13). Furthermore, because we know the number of crossbridges per thick filament and the number of thick filaments per cross-sectional area, we can calculate the **time-average force** per crossbridge ($\langle F \rangle$) from the force per cross-sectional area of the muscle. Likewise, we can calculate the average stiffness per crossbridge. Also, because we know the number of crossbridges, we can calculate the ATPase rate per molecule ($1/T_c$) from the whole-tissue ATPase activity. By dividing the speed by the ATPase rate, we can calculate the distance per ATP, Δ. Similar arguments apply to kinesin: We know the force and ATPase of the dimer, so we can calculate the average force and ATPase per head.

These parameters are macroscopic in the sense that we do not know what the individual crossbridges are doing, just how many there are that contribute to motility. But to know how the individual crossbridges work, we need an additional piece of information: the fraction of the time that the crossbridges are attached. Only when we know this can we calculate the distance moved while the crossbridge is attached, the crossbridge stiffness, and the force per attached crossbridge. This fraction is of course the duty ratio, r, which was introduced in Chapter 13. We have already seen that the duty ratio is equal to the ratio of the working distance to the distance per ATP: $r = \delta/\Delta$. It is also equal to the ratio of the average force ($\langle F \rangle$) to the force per attached crossbridge ($\langle F \rangle_{on}$):

$$\langle F \rangle = \langle F \rangle_{\text{all times}} = \frac{\tau_{on}\langle F \rangle_{on} + \tau_{off}\langle F \rangle_{off}}{\tau_{on} + \tau_{off}} = \frac{\tau_{on}\langle F \rangle_{on}}{T_c} = r\langle F \rangle_{on} \quad (16.1)$$

because the crossbridge generates no force while it is detached ($\langle F \rangle_{off} = 0$). By a similar argument, the duty ratio is the ratio of the average stiffness to the molecular stiffness. Thus, the duty ratio is the key to relating the macroscopic and microscopic descriptions of motility. These relations are summarized in Table 16.1. The duty ratio has an effect on a motor system that is analogous to a mechanical lever: It multiplies the distance and divides the force. Because work is equal to the product of the force and the distance, the duty ratio cancels out so that the macroscopic work per ATP, W, is equal to the microscopic work produced while the crossbridge is attached, w, as expected, since no work is done while a crossbridge is detached.

These considerations are quite general: They apply equally well to an ensemble of motors driving muscle contraction or the swimming of sperm as they do to just two heads of kinesin working within a processive dimer.

Table 16.1 Connections between macroscopic and microscopic parameters

Parameter	Macroscopic (per cycle)	Microscopic (attached)	Connection	Relationships
Speed	V	v	$V = v$	$V = \Delta/T_c = \delta/\tau_{on} = v$
Time	T_c	τ_{on}	$T_c = \tau_{on}/r$	$r = \tau_{on}/T_c$; $T_c = \tau_{on} + \tau_{off}$
Displacement	Δ	δ	$\Delta = \delta/r$	—
Force	$\langle F \rangle$	$\langle F \rangle_{on}$	$\langle F \rangle = r\langle F \rangle_{on}$	—
Stiffness	K	κ	$K = r\kappa$	—
Work	W	w	$W = w$	$W = \langle F \rangle \Delta = \langle F \rangle_{on}\delta = w$

Powerstroke Model

We need a new and final concept in order to explain how a protein conformational change can generate force: A motor protein has an elastic element, a spring, which can store mechanical energy. The idea is that the conformational change strains the spring: The tension in the spring is the motor force, and the relief of the strain is the driving force for the motion. This idea was originally formulated by A. F. Huxley (1957), and it forms the basis of what we call the **powerstroke model** (Figure 16.1). For neither myosin nor kinesin has the spring been identified definitively at the structural level. For myosin, elas-

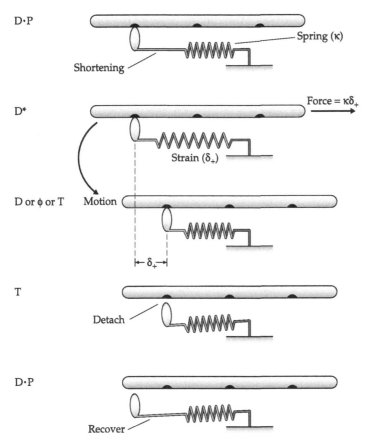

Figure 16.1 Powerstroke model
A conformational change in the crossbridge produces strain in an elastic element which becomes stretches beyond its resting length. The strain generates tension, which is the driving force for motion. The maximum force is the distance associated with the strain, the powerstroke distance (δ_+), times the spring constant (κ). If the crossbridge unbinds when the strain is zero, the average force per attached crossbridge ($\langle F \rangle_{on}$) is the average strain distance ($\frac{1}{2}\delta_+$) times the stiffness. On the left are the nucleotide states for myosin: T = A + ATP, D = ADP, D* = ADP strained, P = P_i.

ticity may correspond to flexion of the light-chain domain—the lever may not be completely rigid (Howard and Spudich, 1996). Alternatively, flexibility could reside in the converter domain that couples the hydrolysis to rotation of the lever; in this case, the elastic element might be analogous to the spiral balance spring found in a watch. For kinesin, the spring could be in the neck or the dimerization domain. The spring could even be hinged coiled coils or relatively unstructured linker peptides acting as entropic springs (Example 6.10).

We call the size of the conformational change associated with the strain the **powerstroke distance**, and denote it by δ_+. It was denoted by "h" in the A. F. Huxley 1957 model. It is equal to the working distance, δ, only if the crossbridge detaches when the strain is zero. However, it is possible that in a slowly contracting muscle a crossbridge might detach before the strain is fully relieved. On the other hand, in a rapidly contracting muscle, a crossbridge might be dragged into a region of negative strain (i.e., compression). To take these possibilities into account, we define δ_- to be the magnitude of the compression in the crossbridge when it detaches. The sign of δ_- is chosen so that its value is positive if the crossbridge is in compression when it detaches. We call δ_- the **drag-stroke distance** (Pate et al., 1993). With these definitions, the working stroke, δ, is $\delta_+ + \delta_-$. When many kinesin molecules participate in the motion of an organelle (or a microtubule in a gliding assay), we also expect that some attached heads will be dragged into negatively strained conformations by other heads. For single-molecule motility, we expect there to be a drag stroke if the duty ratio is greater than 0.5 and there is movement of the load while both heads are bound.

We can now calculate the force per crossbridge and the maximum speed. During steady motion, the average strain in a crossbridge corresponds to an average elongation of the elastic element by a distance $(\delta_+ - \delta_-)/2$; if the stiffness of the elastic element, κ, is constant, then the force per attached crossbridge is

$$\langle F \rangle_{on} = \tfrac{1}{2}\kappa(\delta_+ - \delta_-) \tag{16.2}$$

In view of the relation between the average force and the force per attached crossbridge (Equation 16.1), we have

$$\langle F \rangle = r \langle F \rangle_{on} = \frac{\delta}{\Delta} \cdot \tfrac{1}{2}\kappa(\delta_+ - \delta_-) = \frac{\kappa}{2\Delta}\left(\delta_+^2 - \delta_-^2\right) \tag{16.3}$$

We can think of the force as being due to a positive, elastic force, F_+, and a negative, drag force, F_-, where

$$F_+ = \frac{\kappa}{2\Delta}\delta_+^2 \quad \text{and} \quad F_- = \frac{\kappa}{2\Delta}\delta_-^2 \tag{16.4}$$

The drag force arises from protein friction caused by crossbridges that are dragged into a region of negative strain (compression) as discussed in Chapter 3. The speed of motion of the crossbridges relative to the filament, v, is equal to δ_-/t_-, where t_- is the **drag time** that the crossbridge spends negatively strained; we think of t_- as being a constant that is determined by the rate of detachment of these compressed crossbridges. Equation 16.4 can then be rewritten

$$F_- = \left(\tfrac{1}{2}\frac{\delta_-}{\Delta}\kappa t_-\right)v \tag{16.5}$$

where the bracketed term represents the drag coefficient and can be compared to the drag coefficient for protein friction (see Equation 3.7; $p \sim \delta_-/\Delta$, $\tau_{on} \sim t_-$, and the factor of 2 arises from the different statistics of release between this model and the protein friction model of Chapter 3).

According to the powerstroke model, the speed of muscle contraction is limited by the drag of negatively strained crossbridges. This was a key conclusion of the A. F. Huxley 1957 model. Another way of saying this is that the speed is ultimately limited by protein friction between the motors and the filament. The maximum speed can be obtained from Equation 16.3 by setting $\langle F \rangle = 0$: The drag stroke is then equal to the power stroke and

$$v_{max} = \frac{\delta_{-,max}}{t_-} = \frac{\delta_+}{t_-} \qquad (16.6)$$

These expressions for force and speed are approximations that should be correct within a factor of two: More precise calculations, which require knowledge of the statistical distribution of the working strokes, are made in the next sections of this chapter, where we consider a more detailed crossbridge model.

The crossbridge parameters determined experimentally for myosin and kinesin are summarized in Table 16.2.

Table 16.2 Summary of experimental data on skeletal muscle myosin II and conventional kinesin

Parameter	Description	Myosin	Kinesin
Macroscopic parameters			
F_{max}	Force/head (at $V = 0$)	1.5 pN	3 pN
V_{max}	Speed ($F = 0$, high ATP)	6000 nm/s[a]	800 nm/s
W_{max}	Work (intermediate speeds)	50 pN·nm	50 pN·nm
T_c	Cycle time (at V_{max})	40 ms	20 ms
Δ_{max}	Distance/ATP ($F = 0$)	~240 nm	16 nm
K_{max}	Stiffness (at $V = 0$)	0.6 pN/nm	~1 pN/nm[b]
Single-crossbridge parameters (independent of speed)			
δ_+	Powerstroke distance	5 nm	8–16 nm
d	Path distance	36 nm	8 nm
κ	Crossbridge stiffness	5 pN/nm[c]	~0.5 pN/nm[b]
t_-	Drag time	0.6 ms[a]	
Single-crossbridge parameters (dependent on speed)			
δ_-	Drag-stroke distance	0 to 3.5 nm	
r	Duty ratio	0.035–0.14	0.5–1.0
n	Multiple of binding sites	1–7	2

[a]Depends on the muscle type and temperature (rabbit psoas muscle, 25°C; Pate et al., 1994)
[b]Kojima et al., 1997; Kawaguchi and Ishiwata, 2001
[c]Inferred from Equation 16.9

Relations: $\quad r = \dfrac{\tau_{on}}{T_c} = \dfrac{\delta}{\Delta} = \dfrac{\langle F \rangle}{\langle F \rangle_{on}} = \dfrac{K}{\kappa} \qquad v = \dfrac{\Delta}{T_c} = \dfrac{\delta}{\tau_{on}} = \dfrac{\delta_+}{t_+} = \dfrac{\delta_-}{t_-}$

Role of Thermal Fluctuations in the Power Stroke

What exactly is the power stroke? How does a protein get into a high-energy mechanical state? These questions go to the very core of the force-generation problem. From a physical viewpoint, we expect thermal fluctuations to play crucial roles in the motor reaction because thermal forces at the molecular scale (see Table 2.1) are very large compared to the average directed forces that motor proteins generate (see Table 16.2). Indeed, it is the noisy, diffusive environment in which they operate that distinguishes molecular machines from the man-made machines of our everyday world. For example, an unconstrained protein will diffuse through a distance equal to the working stroke in tens of microseconds (see Example 4.7), three orders of magnitude faster than the millisecond timescale of the ATP hydrolysis cycle (see Table 16.2). From a chemical viewpoint, we also expect thermal fluctuations to play crucial roles. All chemical reactions require thermal energy in order for a molecule to enter the transition state (Chapter 5), and the chemistry of the hydrolysis reaction is no exception. The challenge is to merge the physical and chemical views.

An extreme model is that movement of a motor to its next binding site on the filament is purely diffusive, and the role of ATP hydrolysis is to somehow rectify diffusion so that movement in the wrong direction is blocked (Braxton, 1988; Braxton and Yount, 1989; Vale and Oosawa, 1990). This is reminiscent of the pawl and ratchet discussed by Feynman et al. (1963): If a pawl (the motor) and ratchet (filament) are at different temperatures (analogous to ATP, ADP, and P_i being out of equilibrium; Zhou and Chen, 1996), then they can do work without violating the second law of thermodynamics. Perhaps motor proteins operate analogously to such a heat engine. (But motor proteins cannot literally be heat engines because the diffusion of heat is so rapid over molecular dimensions that thermal gradients will dissipate within picoseconds [see Problem 4.9], much faster than the timescale of the biochemical reactions.)

These ideas about diffusion have inspired a number of models in the physics literature (Astumian and Bier, 1994; Magnasco, 1994; Rousselet et al., 1994). However, purely diffusive, "thermal ratchet" models fail in two respects. First, the maximum force that they can generate against viscous loads is small, only $2kT/d$, where d is the distance between binding sites (Hunt et al., 1994; and see Equation 10.5 for an analogous model for polymerization forces). But for kinesin, this force is only 1 pN, much less than the measured value of 4 to 5 pN (Hunt et al., 1994). And second, because the motor diffuses in the right direction only half the time, it is expected that two molecules of ATP would be hydrolyzed on average for each forward step. But for kinesin, there is one step per ATP hydrolyzed (Coy et al., 1999; Iwatani et al., 1999).

A less extreme model is that, rather than the entire motor protein diffusing to the next binding site, a flexible domain diffuses into the strained, pre-powerstroke state. In other words, the motor spring picks up its mechanical energy via thermal fluctuations; once it is cocked, it can then bind to the filament and generate force. This was the picture in the original A. F. Huxley 1957 model, and it forms the basis of other ratchet models (Cordova et al., 1992). If the cocking of the spring requires a *global* conformational change, then the time will be limited by the stiffness of the spring and the damping from the fluid (as well as possible internal damping within the protein). We call this a Kramers-like mechanism (see Figure 5.9) after Kramers who first solved the diffusion equation for this case (Equation 4.18). This diffusion mechanism was subsequently abandoned by Huxley and Simmons (1971) and by Eisenberg and Hill (1978; Eisenberg et al., 1980) in favor of a mechanism in which a *local* conformational change, presumably localized to the nucleotide-binding pocket, drives the subsequent global conformational change. This is called an Eyring-like mechanism to emphasis its chemical nature (Chapter 5), though even a localized conformation in a protein involves the breaking and formation of many weak bonds. The Kramers-like mechanism was abandoned because it was argued that it would take too long for myosin to pick up an appropriate amount of mechanical energy by diffusion, though I argued above that if a more reasonable mechanical energy is assumed (~12 kT corresponding to an efficiency of ~50%), then the diffusion time is not prohibitive (Example 5.2).

The powerstroke model is equally compatible with the global (physical) or local (chemical) mechanisms. Thermal energy is important in both cases. In the local, Eyring-like model, thermal fluctuations are still needed to get the molecule into the transition state for the localized structural change (see Figure 15.6). The advantage of the local mechanism is that it is faster. Because a short lever is stiffer than a longer one (Appendix 8.1) and because the damping on a small domain is less than the damping on a large domain, high-energy states can be reached much more quickly through localized conformational changes. For example, if the lever ratio is 10, then a local conformational change occurs 1000 times faster than a global change that is associated with the same mechanical potential energy (Figure 16.2). But whether force-generating protein conformational changes are due to diffusive global conformational changes that are locked in by chemical changes or the global changes are driven by more localized chemical changes is still an open question. The development of single-molecule fluorescence-resonance-energy-transfer techniques that can accurately measure intramolecular distances (Deniz et al., 1999) offers one fruitful approach.

Crossbridge Model for Muscle Contraction

In this section we will consider a specific version of the powerstroke model that predicts many of the mechanical properties of muscle in terms of the molecular properties of the crossbridges. The model is illustrated schematically in Figure 16.3 and the equations (Appendix 16.1) follow those in the original A. F. Huxley 1957 paper. The original A. F. Huxley model has just two states (on and off), as does the present model, though the on- and off-rate constants used here, which follow Pate et al. (1993), are somewhat different from those

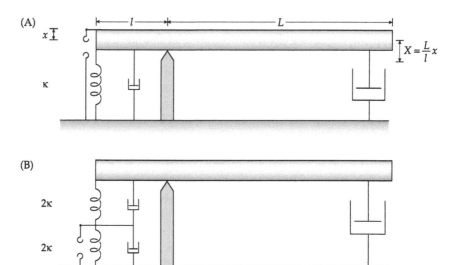

Figure 16.2 Comparison of global and local mechanisms for generating mechanical strain
(A) In the Kramers-like mechanism, the conformational change (represented by the linking of the two hooks) requires the end of the lever to fluctuate through a large distance. (B) By contrast, in an Eyring-like mechanism, the conformational change can be achieved through a localized fluctuation of the spring, which then drives the global conformational change (the rotation of the lever). The local fluctuations are much faster than the global ones. The displacement at the right end of the lever, X, is L/l times the displacement at the left end, x. The stiffness measured at the right end of the lever is the same in both (A) and (B). It is equal to K where

$$U = \tfrac{1}{2}\kappa x^2 = \tfrac{1}{2}\kappa\left(\frac{l}{L}\right)^2 X^2 = \tfrac{1}{2}\left(\frac{\kappa l^2}{L^2}\right)X^2 = \tfrac{1}{2}KX^2$$

The time constant, T, for the thermal ratchet mechanism (A) is much greater than the time constant, τ, for the powerstroke mechanism (B):

$$T = \frac{\Gamma}{K} \sim \frac{L\eta}{\kappa l^2/L^2} = 2\left(\frac{L}{l}\right)^3 \frac{l\eta}{2\kappa} \sim 2\left(\frac{L}{l}\right)^3 \frac{\gamma}{2\kappa} = 2\left(\frac{L}{l}\right)^3 \tau$$

In other words, a local conformational change coupled to a global conformational change leads to *kinetic acceleration*. For example, a 10-nm-long lever arm ($L = 10$ nm) with stiffness 1 pN/nm has a time constant of ~20 ns using the drag expression from Table 6.3. Thus it would take ~10 s to pick up 20 kT of energy in a global conformational change (see Example 5.2). However, the time would be reduced to ~10 ms using a local mechanism with $l = 1$ nm.

used in the Huxley model (where they were denoted "f" and "g"). The rate constants used in this model are improvements over the Huxley rate constants in the sense that they more simply relate to the biochemistry and structure of the crossbridges, as explained below. Yet, like the Huxley model, this model can be readily solved.

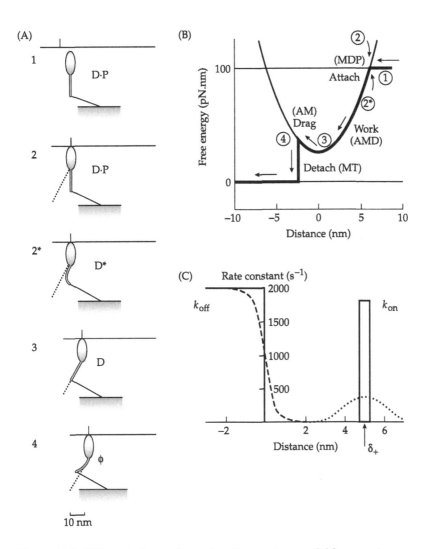

Figure 16.3 Different views of a contracting-spring model for myosin
(A) Rotating lever model in which the spring resides in the light-chain domain. The motion is driven by a rotation of the lever through ~30 degrees with respect to the motor domain (post-powerstroke angle is dashed). The spring could equally well be in the converter domain. (B) The parabola. As the filament moves to the right, the crossbridge binds and becomes strained. The energy of the attached state is the potential energy in the spring and has a parabolic dependence on strain. The crossbridge moves down the parabola as it does work on the filament. At low speed, the crossbridge detaches near the bottom of the well, and a maximum work is done (62 pN·nm in this example). However, at high speed the crossbridge will be dragged up the other side of the parabola where it will become negatively strained (compressed) and will exert negative force and result in less net work being done. The numbers refer to the drawings in (A). (C) Strain-dependent rate constants: The solid curves correspond to the idealized model solved in the text. Dashed and dotted curves correspond to physical models that approximate the idealized curves.

The model assumes that a crossbridge can bind only in a restricted region of positions relative to the filament, and, immediately following binding, its elastic element becomes strained by an amount δ_+ (Figure 16.3A). The strain then decreases as the filament slides past the crossbridge. Because the potential energy of a spring has a square dependence on strain, this part of the cycle can be depicted as sliding down the parabola in Figure 16.3B. The model further assumes that the crossbridge is unable to detach until the strain drops to zero (i.e., until the bottom of the parabola is reached), whereupon the detachment rate constant rises abruptly to $1/t_-$ in the negative-strain region. The attachment and detachment rate constants, k_{on} and k_{off}, respectively, therefore depend on displacement (and thus strain), as shown by the solid curves in Figure 16.3C.

The strain dependence in the model is supported by the following experimental results. First, Fenn (1924) showed that the sum of the rates of heat liberation from and work performed by a muscle increases when the muscle is allowed to shorten: Because this sum is proportional to the rate of ATP hydrolysis (the proportionality constant is ΔH), the **Fenn effect** implies that the overall hydrolysis time, $T_c = 1/k_{on} + 1/k_{off}$, decreases as the load (and strain) decreases. Second, single-molecule recordings show directly that a high load imposed by an optical trap (that prevents the strain from being relieved) decreases the detachment rate (Finer et al., 1994; Molloy et al., 1995b). These data argue for a low detachment rate when the crossbridges are under tension. And third, the stiffness of muscle fibers decreases as the speed of contraction increases (Ford et al., 1985; Haugen, 1988): The simplest explanation is that as binding sites move past at higher and higher speeds, the crossbridges have a lower and lower probability of binding.

The idealized strain dependence of the rate constants (solid curves in Figure 16.3C) can be related to the structure and biochemistry of myosin as follows. The M·D·P crossbridge attaches to the actin filament and immediately becomes strained by an amount δ_+, the powerstroke distance, as the phosphate is released. The crossbridge then detaches in a two-step process. The first step is the release of ADP from the A·M·D state, which is assumed not to occur until the strain in the elastic element has decreased to zero. This could be realized if the transition state for ADP release is highly negatively strained. The second step is the rapid, strain-independent binding of ATP to A·M and the subsequent dissociation of A·M·T. This gives a detachment rate (dashed line in Figure 16.3C) that approximates the solid line. The crossbridge, now in the M·T state, must recover to its initial, pre-powerstroke conformation and move to the next binding site, 36 nm away from the previous one. The model assumes that the rebinding of M·D·P occurs only in a narrow range of displacements, and that at high speed each crossbridge has only a small chance of binding during the short time that it is in the "strike zone." This low chance of binding could be due to an inherently low rate of attachment in the strike zone (as in Figure 16.3C), or it could be due to the hydrolysis step (M·T → M·D·P) being rate limiting (Tesi et al., 1990), or to a refractory M·D·P state that cannot bind to actin; all lead to the same equations.

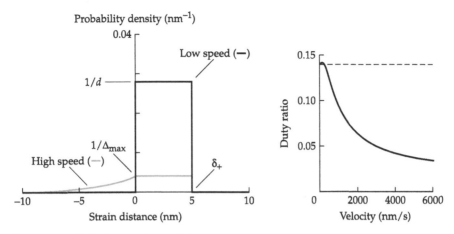

Figure 16.4 Solution to the model
(A) The distribution of strains at low speed (F_{max}) and at high speed (v_{max}). The duty ratios for these two conditions correspond to the areas under the curves. (B) The duty ratio is plotted against the sliding speed.

Comparison of the Model to Muscle Data

Despite its simplicity, the crossbridge model can account for many of the macroscopic mechanical and chemical properties of muscle in terms of the measured molecular properties of the crossbridges.

The model predicts that when the muscle is shortening at a steady speed, the strain in the crossbridges will be distributed as shown in Figure 16.4A. The distribution depends on the speed. At low speed, each crossbridge has a high probability of attaching to the next binding site that comes past, staying on until the strain drops to zero, and accumulating little negative strain before detaching. At high speed, the distribution changes in two respects. First, the attached crossbridges will be dragged into a region of negative strain before they can detach: As a consequence of this increased drag stroke, the average force generated by an attached crossbridge will fall (Equation 16.3). Second, each crossbridge will have only a small chance of attaching during the short time that it is in the "strike zone." This leads to a low duty ratio as the crossbridges jump over potential binding sites (i.e., $n > 1$ in Equation 15.6). The duty ratio is plotted against speed in Figure 16.4B; as noted above, the decrease in duty ratio mirrors the decrease in muscle stiffness as the speed of contraction is increased.

Force–Velocity Curve and the Efficiency of Muscle Contraction

The model predicts a maximum speed

$$v_{max} = \frac{1}{\sqrt{2}} \frac{\delta_+}{t_-} \tag{16.7}$$

The proportionality between v_{max} and δ_+/t_- is a general feature of powerstroke models (Equation 16.6); the prefactor of $1/\sqrt{2}$ arises from the particular strain dependences of this crossbridge model. A maximum speed of 6000 nm/s is consistent with a powerstroke distance $\delta_+ = 5$ nm (Chapter 15) and a drag time (t_-) of 0.6 ms (corresponding to a detachment rate of 1000 to 2000 s^{-1} measured at zero strain, Pate et al., 1993; Ma and Taylor, 1994; and see Chapter 14). The drag time is primarily limited by the dissociation of ADP: It is the rate of ADP release that determines of the speed of contraction of different muscles (Siemankowski et al., 1985). The increase in the rate of release of ADP with increasing temperature, and the corresponding decrease in t_-, accounts for the temperature dependence of the contraction of muscle (Siemankowski et al., 1985).

The total working distance is the powerstroke distance, 5 nm, plus the dragstroke distance. If the elastic element has the same stiffness in compression as it has in tension (as we are assuming), then in a rapidly contracting muscle the drag-stroke distance has an average value of ~3.5 nm (= $5/\sqrt{2}$ nm), and it may reach 7 nm if the attached time is longer than average. Thus the model predicts that the total working distance in a contracting fiber could be as high as 12 nm. This large working stroke can be accommodated by the crystallographic and modeling studies on the myosin head (Chapter 12). However, a powerstroke distance of 13 nm and a total working distance of up to ~25 nm, as postulated by some workers (e.g., Ford et al., 1977; Huxley, 1998), cannot be accommodated by the crystallographic data.

The maximum force is related to the work and the distance per ATP according to

$$W_{max} = F_{max}\Delta(0) \tag{16.8}$$

where $\Delta(0)$ is the distance per ATP at load speed ($V = 0$). Equation 16.8 does not depend on the molecular details of the model (see Table 16.1). For skeletal muscle, the work/ATP is equal to 60 to 70 pN·nm, corresponding to a little over half the free energy obtained from the hydrolysis of ATP under cellular conditions (Figure 16.5; Woledge et al., 1985; Cooke, 1997). $F_{max} = 1.7$ pN (see Figure 16.5). This gives a distance/ATP at low speed, $\Delta(0)$, equal to ~38 nm. This is close to the 36 nm path distance inferred from in vitro experiments (Chapter 15), and so provides in vivo support for myosin II following a path parallel to the axis of the actin filament.

In molecular terms, the maximum work is given by

$$w_{max} = \tfrac{1}{2}\kappa\delta_+^2 \tag{16.9}$$

With a powerstroke distance of 5 nm, and a maximum work of 60 pN·nm, the crossbridge stiffness, κ, is equal to 5 pN/nm (Equation 16.7). This high value is consistent with mechanical measurements on single molecules showing that the stiffness is ≥ 2 pN/nm (Veigel et al., 1998). The high stiffness of a crossbridge also reinforces the notion that it lacks the flexibility to bind anywhere along the two-stranded actin helix: A stiffness of 5 pN/nm gives a r.m.s. fluctuation of <1 nm (=$\sqrt{(kT/\kappa)}$), supporting the assumption that the crossbridges bind in only a narrow window (Figure 16.3C). The high stiffness and small powerstroke distance imply that the maximum single-motor force is 25 pN (attained

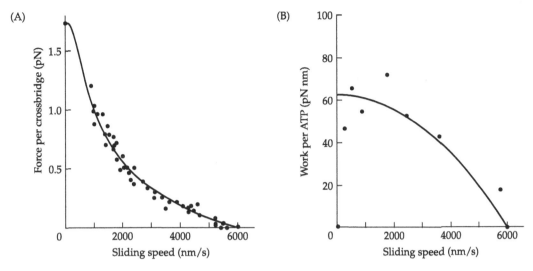

Figure 16.5 Dependence of force and work on speed
(A) Force–velocity curve for a myosin crossbridge in skeletal muscle. The circles are data from rabbit psoas muscle at 25°C recorded using a setup like that illustrated in Figure 1.1A, except that the membrane surrounding the muscle fiber was removed and the fiber activated by immersing it in a solution containing Ca^{2+} and ATP. The speed of contraction was measured at several different loads, as shown in Figure 1.1B. The contraction speed was converted to sliding speed according to Appendix 13.1, and the force per crossbridge was calculated from the whole-muscle tension (the maximum tension was 246 kPa, corrected to 25°C), the number of thick filaments in a cross-section (500 μm^{-2}), and the number of crossbridges per half thick filament (300). The filled circles are data from Pate et al. (1994), and the solid curve is the prediction of the powerstroke model. (B) Mechanical work per ATP hydrolyzed. The circles are data from Kushmerick and Davies (1969), adjusted by Woledge et al. (1985) to account for the ATP hydrolysis by the calcium pump and scaled horizontally to give a maximum speed of 6000 nm/s for comparison to the force–velocity curve. The solid curve is the powerstroke model.

at the beginning of the power stroke). Again, this is consistent with measurements showing that the single-molecule force exceeds 10 pN (Finer et al., 1994; Veigel et al., 1998).

The powerstroke model predicts both the maximum speed and the maximum force of muscle. It also accounts for the characteristic nonlinear shape of the force–velocity curve (Figure 16.5A). In this case, though, the shape of the predicted curve depends strongly on how strain affects the attachment and detachment rates. The strain dependences of the rate constants in Figure 16.3C were chosen to give a good fit to the force–velocity curve and are quite strongly constrained by this data. Models that have strain-independent rate constants such as the Leibler and Huse model (1993), predict linear force–velocity curves and therefore fail to provide an explanation for the concave shape of the force–velocity curve, one of the characteristic features of muscle contraction.

A problem with the present model is that it predicts that the ATPase rate decreases to zero at zero speed. But the ATPase rate of isometrically contracting muscle is actually ~3/s per head (e.g., Ma and Taylor, 1994; He et al., 1997). Because no work is done by an isometrically contracting muscle, the efficiency is zero at zero velocity. The nonzero ATPase rate means that crossbridges must be able to detach without completing their working stroke (k_{off} nonzero at positive strain) and must be able to reattach over a range of displacements (k_{on} nonzero outside the strike zone). If a crossbridge does detach under partial strain, say 2.5 nm, then when the crossbridge recovers it will not be correctly aligned with the binding site. In order to rebind, the crossbridge must first undergo a considerable fluctuation, and so we expect that a Kramers-like mechanism must operate (see Chapter 5). The original Huxley model had nonzero attachment and detachment rate constants to account for the nonzero ATPase rate of isometrically contracting muscle, and the inclusion of small nonzero rate constants could circumvent this limitation of the present model.

Mechanical Transients

The model accounts for some but not all of the transient mechanical behavior of muscle fibers. Experiments show that there is a comparatively fast, partial recovery of the tension in an isometrically contracting muscle whose length is rapidly decreased (Figure 16.6A; Huxley and Simmons, 1971). The model predicts such a recovery (Figure 16.6B). It is due to the rapid detachment of crossbridges that have been brought into their negatively strained region as a result of the decrease in length; the time constant of the recovery is t_- ($\cong 0.6$ ms). However, the experimental recovery can be nearly 100% (for small enough shortenings), whereas the model predicts at most a 50% recovery. But this is not a serious weakness of the model because it is expected that shortening will bring "fresh" crossbridges into striking distance of their target binding sites on the actin filaments, and the rapid binding of these new crossbridges will augment the tension recovery. The arrival of fresh crossbridges has not been taken into account in Figure 16.6B. The mechanical experiments also show that there is a delayed decrease in tension following rapid lengthening of muscle: To account for this, the model would need to be modified so that there is a nonzero detachment rate when the strain is greater than δ_+.

The interpretation of the force recovery here differs markedly from that of Huxley and Simmons (1971). These workers proposed that there was a rapid reequilibration of attached crossbridges between two (or more) conformational states. Our interpretation follows that of Eisenberg and Hill and coworkers (1978; Eisenberg et al., 1980), and it presupposes that only a small fraction of the crossbridges are attached (i.e., the duty ratio is small), even in an isometrically contracting muscle (so that there are "fresh" crossbridges available for binding). A small duty ratio was initially supported by electron paramagnetic spin experiments (Cooke et al., 1982), and has now been quite clearly established by in vitro motility assays and single-molecule experiments (Howard, 1997, 1998; but see Huxley, 1998).

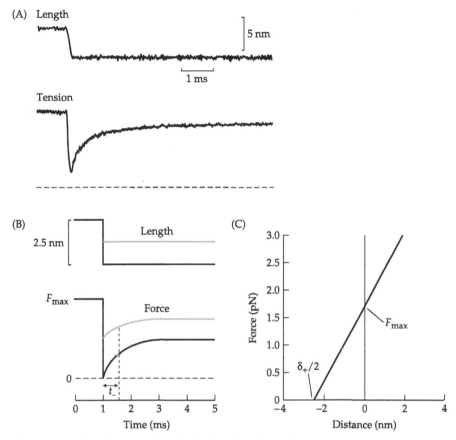

Figure 16.6 Transient mechanical behavior of muscle
(A) Tension change (lower trace) in response to a rapid change in length of a muscle fiber (upper trace). The length scale refers to the length of a half-sarcomere; because of the compliance of the actin filaments (see Figure 8.1), the distance that the thick filament moves relative to the actin filament is only about one-half the change in the length of the half-sarcomere. (B) Predicted tension changes due to the detachment of negatively strained crossbridges. Here the length refers to the change in length of the crossbridge (i.e., relative displacement of the base of the crossbridge, where it attaches to the thick filament with respect to the actin filament, to which the motor domain binds). (C) The change in force (ordinate) just after the change in length of the crossbridge (abscissa), as predicted by the model. (A after Ford et al., 1977.)

Powerstroke Distance, Path Distance, and the Duty Ratio

We have seen that the molecular parameters—powerstroke distance, path distance, and duty ratio—when incorporated into a model, can account for the mechanical and chemical properties of intact muscle. The converse of this statement is also true: The data from muscle actually provide strong independent support for the parameters that have been measured by single-molecule techniques.

The mechanical transients provide strong evidence that the powerstroke distance in muscle is 5 nm, in agreement with the single-molecule data. The evidence comes from plotting the instantaneous force against the amplitude of the shortening step (Figure 16.6C). After correction for the compliance of the actin filament (Chapter 8), it is found that the tension drops to zero if the overlap between the thin and thick filaments is increased by 2.5 nm (Huxley and Simmons, 1971; Lombardi et al., 1992). This is expected if the powerstroke distance is 5 nm and the strain in the crossbridges is uniformly distributed between 0 and 5 nm (see Figure 16.4A).

The large distance per ATP at low velocity (38 nm) inferred from the work and force (Equation 16.8) lends strong independent support for a 36 nm path distance. Such a path distance of 36 nm is consistent with three other experiments on intact muscle.

1. When 600 µM ATP is photoreleased in a muscle fiber held at tensions between 42% and 7% of maximum tension, the filaments slide through 120 to 240 nm (Higuchi and Goldman, 1991). Because the concentration of the crossbridges is 150 µM, this gives an average distance per ATP per head of 30 to 60 nm, comparable to the path distance.

2. When an active muscle fiber is shortened in a "staircase" protocol, the tension recovers rapidly at each step as long as the total extent of shortening is less than ~36 nm (Lombardi et al., 1992). This is consistent with the present model, which predicts that the rapid recovery due to the binding of fresh crossbridges (and unbinding of negatively strained crossbridges) will fail when the total shortening is 36 nm, as this will bring in crossbridges attached to the previous turn of the actin filament—these must complete their ATP hydrolysis cycle before they can come on line again (Chen and Brenner, 1993).

3. The ATPase rate of muscle increases when its length is oscillated; the rate saturates when the peak-to-peak amplitude reaches ~40 nm (Ebus and Stienen, 1996), consistent with each crossbridge having only one binding site per half-pitch of actin filament.

The in vivo and in vitro mechanical experiments therefore all suggest that the path distance is 36 nm.

Muscle data also provide strong support for a low duty ratio, even at high load.

1. The measured, whole-muscle stiffness is 0.7 pN/nm (the rapid release of isometrically contracting muscle by 2.5 nm per half-sarcomere reduces the tension from 1.7 pN to zero, Figure 16.6C), whereas the work (at high load) suggests that the molecular stiffness is ~5 pN/nm (Equation 16.9). This gives a duty ratio of ~0.14 (= 0.7/5), for an isometrically contracting muscle in agreement with the geometric duty ratio of $\delta_+/d = 5/36 \cong 0.14$.

2. If the duty ratio were close to unity, as implied by some workers (e.g., Higuchi and Goldman, 1991), then the high work per ATP would require a powerstroke distance of ~13 nm (w_{max} = 60 pN·nm, κ = 0.7 pN/nm in Equa-

tion 16.9). But this poses two problems: First, 13 nm is still much smaller than the path distance of 36 nm, so it is not consistent with a duty ratio of 1 anyway; and second, as mentioned above, a powerstroke distance of 13 nm is hard to reconcile with the structural data given that a drag stroke of approximately equal magnitude is expected at high shortening speeds.

3. A low duty ratio of ~0.14, limited by the ratio of the powerstroke distance to the path distance, explains why the force of a muscle fiber does not increase as the ATP concentration is lowered (Ferenczi et al., 1984). If there were no steric limitations, then it would be expected that decreasing the ATP concentration would increase the attached time, increasing the fraction of attached crossbridges and therefore the total force per crossbridge (see Equation 16.1). But the force actually decreases a little. Thus, in a muscle fiber only a small fraction of the myosin heads can be attached at the one time (and generating force), even at very low ATP concentrations, consistent with there being a steric constraint to crossbridge binding.

4. A low duty ratio also explains why fluorescent probes attached to the light-chain binding domains of myosin in intact muscle fibers rotate only a few degrees (on average) even when the muscle is shortened so that the attached crossbridges should rotate by ~20% (Corrie et al., 1999). Evidently, only about 10% of the crossbridges are attached.

One Step per ATP

An important, general point is that all the mechanical data—from single molecules, through motility assays, to muscle contraction—are consistent with there being just one powerstroke per ATP hydrolysis cycle. Moreover, the size of the power stroke, ~5 nm, is modest compared to the size of the myosin head domain. In other words, with a small duty ratio, it is not necessary to have a large powerstroke or to have a working stroke made up over several smaller strokes, as has been suggested by Huxley and Simmons (1971), Yanagida et al. (1985), Harada et al. (1990), Higuchi and Goldman (1991), Lombardi et al. (1992) and Kitamura et al. (1999). One step per ATP accords with the measurement of single steps in the optical tweezers assays (Finer et al., 1994; Molloy et al., 1995a; Mehta et al., 1997), and the most recent results of Ishijima et al. (1998), showing that there is approximately a one-to-one ratio of attachment events and ADP release.

Conclusion

In summary, the molecular crossbridge model, which uses values measured by single-molecule (δ_+, κ), structural (d) and biochemical experiments (t_-), provides an excellent explanation for a large body of mechanical experiments on muscle, and in particular, accounts quantitatively for the speed (V), the force ($\langle F \rangle$), the stiffness (K), and the work (W) of contracting muscle.

A Crossbridge Model for Kinesin

When there are many kinesin molecules moving a microtubule, the motion is smooth (Howard, unpublished) and the type of model that we have been discussing for myosin ought to be applicable. However, when there is only one kinesin molecule driving the motility, the motor makes discrete steps along the microtubule (see Figures 15.6 and 15.8); because the speed is not constant, we must consider the force (rather than the speed) to be the dependent variable, and formulate a different type of model.

The model must have a number of properties if it is to account for the motility of kinesin. First, the rate-limiting step at low load must be force sensitive because even a small force slows the movement (Figure 16.7). In order to be force sensitive, this step must be associated with a displacement; that is, there must be a conformational change that brings the load toward the plus end of

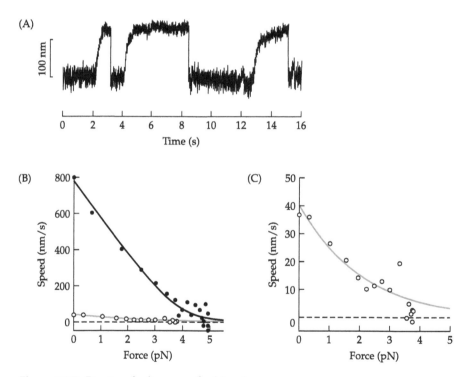

Figure 16.7 Force–velocity curve for kinesin
(A) Position of a kinesin molecule pulling a microtubule attached to a force fiber using the setup illustrated in Figure 15.2A. Shown are three events recorded at high ATP concentration. (B) The speed as a function of the load force exerted by the glass fiber is calculated from recordings at high [ATP] like that shown in (A). The speed at zero force is measured in an in vitro gliding assay. (C) The speed as a function of force is calculated from recordings made at a low [ATP]. (Data are from Meyhöfer and Howard, 1995, and the solid curves are the predictions of the hand-over-hand model, solved in Appendix 16.2.)

Figure 16.8 **Hand-over-hand model with two strain-dependent steps**
(After Schief and Howard, 2001.)

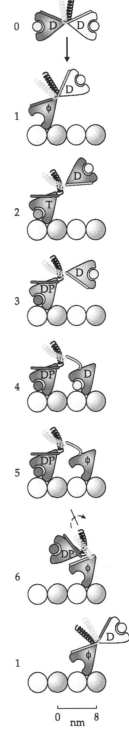

the microtubule. Because the load is carried by kinesin's dimerization domain (see Figure 12.7), this domain must move toward the plus end of the microtubule during this step. Second, at higher load another step must become rate limiting in order that the force–velocity curve be approximately linear (or slightly concave-down), as measured in several labs (Svoboda and Block, 1994; Meyhöfer and Howard, 1995; Coppin et al., 1997; Kojima et al., 1997; Visscher et al., 1999; Kawaguchi and Ishiwata, 2000). A linear force–velocity curve is not consistent with a single force-sensitive rate constant (see Figure 5.10A). Combining these two points, we conclude that the working stroke must be distributed between two or more different chemical steps. Third, in order to account for processivity, the rate of binding of the detached head must be much faster than the rate of unbinding of the attached head, because otherwise there is a good chance of both heads becoming detached and the motor dissociating from the filament. The ratio of the two rate constants must be ~200 in order to account for the finding that kinesin has only a 1% chance of dissociating during each 8 nm step (Howard et al., 1989; Block et al., 1990; Vale et al., 1996; Schnitzer et al., 2000). One way to achieve this is if the binding of the second head accelerates the normally slow detachment of the first head from the microtubule; this is the hand-over-hand model shown in Figure 16.8. This model was discussed in Chapters 14 and 15 and is supported by experiments showing that one-headed kinesin dissociates only very slowly from microtubules (Hancock and Howard, 1998, 1999).

A specific hand-over-hand model is formulated and solved in Appendix 16.2. This model is used in preference to others in the literature (Leibler and Huse, 1993; Peskin and Oster, 1995; Duke and Leibler, 1996; Fisher and Kolomeisky, 1999; Schnitzer et al., 2000) because it incorporates much of the biochemical and structural data, yet can be solved analytically. The predictions are plotted as the solid curves in Figure 16.7. The model considers just three of the six states in Figure 16.8: state 1, occupied just after phosphate release; state 2, occupied just after ATP binding; and state 5, occupied just after ADP release. The binding of ATP is assumed to be fast and reversible and results in a displacement of the load (the distal end of the dimerization domain) 1 nm toward the plus end of the microtubule. The release of ADP is assumed to be the rate-limiting step at high [ATP], and the transition state for this step (perhaps state 3 in Figure 16.8) is a further 1 nm in the plus direction. Together, these two steps provide the force dependence of the speed at low loads seen in Figure 16.7B, C. In order to model the force–velocity curve at high [ATP], it is assumed that the transition state for the release of phosphate (perhaps state 6) is a further +6 nm in the plus direction. The sum of the displacements is 8 nm (though this need not be the case if the transition states are intermediate between the initial and final states; see Chapter 5). The release of phosphate is assumed to be very fast when the motor is unloaded, but rate limiting at high load due to the large displacement associated with the transition state; this linearizes the force–velocity curve at high [ATP].

In addition to accounting for the force–velocity curves, the 3-state hand-over-hand model predicts that kinesin will move with 8 nm steps at low [ATP] (in which case ATP binding to the 1-head-bound motor is rate limiting) and at high [ATP] (in which case release of the rear head and subsequent phosphate release is rate limiting). Such stepwise motion accords with the experimental data (see Figures 15.6 and 15.8). It also explains a puzzling observation made by Gittes et al. (1996): In the presence of a high load parallel to the axis of the microtubule, a perpendicular load that pulls the motor away from the surface of the microtubule accelerates the speed. This observation is explained if the release of the rear head is associated with movement of the load away from the surface of the microtubule, as expected from the structural model (compare state 5 and state 6). Interestingly, the model predicts that at a particular high load and low [ATP], the steps will occur at the same slow rate, and that 2 nm and 6 nm substeps would be observed. This prediction has not yet been tested.

Summary: Comparison between Motile Systems

In this last section, I will compare the two motile systems considered in this chapter. The most fundamental difference between the two systems is that the motors move along different filaments. This statement might seem facile to the reader, but the crucial point, which has not been well appreciated until now, is that the path distances associated with the different filaments are strikingly dissimilar. In the case of kinesin, the tubulin dimer repeat distance is only 8 nm, sufficiently small that the two-headed kinesin dimer can reach its next binding site on its own. As a result, the motor is processive—a single molecule (with two heads) can move continuously for long distances along a microtubule. In the case of myosin II, the binding sites are much more widely separated, 36 nm, due to the structure of the actin filament. As a consequence, a single molecule is not sufficient for motility, and skeletal muscle myosin works only when it is in an ensemble of motors as found in the sarcomere. It has a low duty ratio. This view is reinforced by the finding that myosin V is processive (Mehta et al., 1999; Rief et al., 2000; Sakamoto et al., 2000); its light-chain

Table 16.3 Mechanical amplification of a human biceps muscle

	Phosphate diameter		Working distance		Path distance		Muscle shortening		Hand movement
Distance	0.5 nm	→	5 nm	→	36 nm	→	1.5 mm	→	15 mm
Amplification		10		7		4×10^4		10	
Structural basis		Light-chain domain		Duty ratio		Half sarcomeres per muscle length		Anatomy of the arm	

Note: The release of a phosphate molecule from the active site of each myosin crossbridge, corresponding to shortening of the nucleotide by ~5 Å, leads to the movement of the hand through ~15 mm, a mechanical amplification of ~30 million.

binding domain is three times longer than that of myosin II, allowing the two heads of myosin V to span the 36 nm half-period of the actin filament (see Figure 15.12).

A low duty ratio confers a number of benefits on muscle. First, a high sliding speed can be reached with molecules that have only a modest ATPase rate. Second, muscle can respond quickly when the load is decreased: As the muscle is shortened by movement of the attached crossbridges, "fresh" crossbridges become available to continue the contraction. If all the crossbridges were attached in isometric muscle, there would be a long delay in order that they all complete their ATP hydrolysis cycles before they could enter new high-force states. And third, a low duty ratio contributes an amplification of ~10 toward the overall mechanical amplification of muscle, as a subnanometer structural change around the gamma phosphate in the nucleotide pocket is transduced into the tens of millimeters of displacement at the end of a limb (Table 16.3).

Afterword

What has been learned, and where do we go from here? Over the last ten years, great progress has been made toward understanding the molecular mechanism for muscle contraction. This is a classic problem in physiology and anatomy (Holmes, 1998), and the overall framework of the mechanism has been established through a combination of single-molecule and structural approaches. The mechanism is remarkably similar to that suggested by Hugh Huxley shortly after he discovered the crossbridges using the electron microscope: "How can the crossbridges cause contraction? One can imagine that they are able to oscillate back and forth, and to hook up with specific sites on the actin filament. Then they could pull the filament a short distance (say 100 angstroms) and return to their original configuration, ready for another pull. One would expect that each time a bridge went through such a cycle, a phosphate group would be split from a single molecule of ATP; this reaction would provide the energy for the cycle." (Huxley, 1958, p. 78)

In many ways, the detailed structural, biochemical, and mechanical measurements described in the last five chapters provide a confirmation and elaboration of these ideas, and I hope that I have made a convincing case that the molecular mechanism does indeed provide a satisfactory explanation for the mechanical and biochemical properties of whole muscle.

The microtubule-based motor kinesin has provided crucial insight into the myosin mechanism. Single-molecule techniques for kinesin and myosin were developed in parallel: Kinesin first revealed the steps (between consecutive binding sites on the filament), while myosin first revealed the power stroke (the conformational change within the motor itself). Together, they give a complementary picture of the separate roles of the filament and the motor in motility. In addition, the striking functional differences between conventional kinesin and muscle myosin, the former being processive and the latter operating in large arrays, reinforce the utility of the duty ratio in describing a motor system. The duty ratio also proves useful for generalizing the results from these two motors to the many members of the large kinesin- and myosin-related families of proteins.

Study of the mechanics of motor proteins has given us a deeper understanding of biological force generation. I think that we now know what force is at the molecular level, and we have a good idea about how mechanical, chemical, and thermal forces act on proteins. Indeed, whenever proteins, RNA, and DNA interact with one another, it is the mutual action of forces that drives conformational changes, so the principles learned from motors and the cytoskeleton should have wide application to other macromolecular machines. Central to the operation of a molecular machine is the diffusion and thermal agitation of its moving parts as the protein moves along the reaction coordinate associated with the conformational changes. It is likely that single-molecule techniques will become increasingly useful for defining what a reaction coordinate is for a protein, and ideas from protein folding will probably prove useful.

Where do we go from here? It would be wrong to say that the mechanism of motor proteins is solved, because there are many outstanding structural questions: How does actin trigger the release of phosphate from myosin? How, exactly, does a kinesin head reach the next binding site? What is the structure and mechanism of dynein? Understanding the mechanism of polymerization motility is still at a rudimentary level: How can a filament pull while disassembling, and how can it push while still leaving room for assembly at its end? The questions multiply when one asks about the operation of other macromolecular machines, such as polymerases, ribosomes, and signaling complexes (Alberts, 1998). It is especially fascinating to see how different members of protein families are adapted to their particular cellular roles: For example, some motor proteins go fast and some go slow; some go forward and some go backward. Study of the structural and kinetic specializations of related proteins provides tremendous insight into the biology and the mechanism, and there is still much to find out since many of the motor subfamilies have not been studied in any detail.

In the last decade, the focus has been on molecules. This has been highlighted by the sequencing of whole genomes, the solution of a large number of protein structures, and the manipulation and monitoring of individual proteins. However, we must not lose sight of the bigger question of how ensembles of proteins conspire to produce cells. For, after all, the cell is the minimum unit for sustaining life. Now that we know what force is at the single-molecule level, we should be in a good position to tackle much more complex mechanical questions: How does the cilium beat and the sperm swim? How do cells divide? What determines the shape and internal organization of cells? Answering these questions remains a key challenge for biologists and biophysicists.

Appendix

2.1 Mass and Spring with Damping

The solutions to the mechanical system comprising a mass, a spring, and a dashpot (see Figure 2.4A and Equation 2.10) are as follows (Jaeger and Starfield, 1974). They can be verified by differentiating and substituting back into the equation. The solutions all satisfy the initial conditions $x = 0$ and $dx/dt = 0$ when $t = 0$. At very long times $x \to F/\kappa$.

Underdamped Motion ($\gamma^2 < 4m\kappa$)

$$x(t) = \frac{F}{\kappa}\left[1 - \exp\left(-\frac{t}{\tau}\right)\frac{\sin(\omega t + \phi)}{\sin \phi}\right] \quad (A2.1)$$

where $\quad \tau = \dfrac{2m}{\gamma}, \; \omega^2 = \omega_0^2 - \dfrac{1}{\tau^2}, \; \omega_0^2 = \dfrac{\kappa}{m}, \; \tan \phi = \omega \tau$

Overdamped Motion ($\gamma^2 > 4m\kappa$)

$$x(t) = \frac{F}{\kappa}\left[1 - \frac{\tau_1}{\tau_1 - \tau_2}\exp\left(-\frac{t}{\tau_1}\right) + \frac{\tau_2}{\tau_1 - \tau_2}\exp\left(-\frac{t}{\tau_2}\right)\right] \quad (A2.2)$$

where $\quad \tau_1 = \dfrac{\gamma + \sqrt{\gamma^2 - 4m\kappa}}{2\kappa} \quad$ and $\quad \tau_2 = \dfrac{\gamma - \sqrt{\gamma^2 - 4m\kappa}}{2\kappa}$

Both τ_1 and τ_2 satisfy $(m/\tau) + \kappa\tau = \gamma$. When the motion is highly overdamped ($\gamma^2 \gg 4m\kappa$), the time constants become $\tau_1 = \dfrac{\gamma}{\kappa}$ and $\tau_2 = \dfrac{m}{\gamma}$, where $\tau_1 \gg \tau_2$.

Critically Damped Motion ($\gamma^2 = 4m\kappa$)

$$x(t) = \frac{F}{\kappa}\left[1 - \left(1 + \frac{t}{\tau}\right)\exp\left(-\frac{t}{\tau}\right)\right] \quad (A2.3)$$

where

$$\tau = \frac{2m}{\gamma} = \frac{\gamma}{2\kappa} = \sqrt{\frac{m}{\kappa}}$$

This solution is monotonic, like that in the overdamped case. Note that there is a lag, of duration $\sim\tau/2$, before the displacement starts to rise quickly.

Inertial and Viscous Forces

If the motion is overdamped, then the viscous force, γv, is always larger than the inertial force, ma. This can be shown by calculating the velocity and acceleration from Equation A2.2 and verifying that $\gamma v - ma > 0$. Conversely, if the viscous force is always larger than the inertial force, then the displacement must change monotonically, which is inconsistent with underdamped motion.

2.2 Conservation of Energy

Integration of the equation of motion (Equation 2.10) with respect to x gives

$$\int_0^x m\frac{d^2x'}{dt^2}\,dx' + \int_0^x \gamma\frac{dx'}{dt}\,dx' + \int_0^x \kappa x' \cdot dx' = \int_0^x F \cdot dx'$$

With appropriate changes of variables, this becomes:

$$\int_0^v mv' \cdot dv' + \int_0^t \gamma v^2 \cdot dt' + \int_0^x \kappa x' \cdot dx' = \int_0^x F \cdot dx'$$

or, equivalently,

$$\tfrac{1}{2}mv^2 + \int_0^t \gamma v^2 \cdot dt' + \tfrac{1}{2}\kappa x^2 = \int_0^x F \cdot dx'$$

which can be rewritten

$$\text{K.E.} + \text{heat} + \text{P.E.} = \text{work}$$

3.1 van der Waals Rigidity

To estimate the Young's modulus of a material held together by van der Waals forces, we differentiate Equation 3.3 to obtain the pressure (the force per unit area):

$$P(D) = \frac{8}{3}\frac{U_0}{D}\left[\left(\frac{D_0}{D}\right)^2 - \left(\frac{D_0}{D}\right)^8\right]$$

Because the bond is at equilibrium at the resting separation, D_0, the pressure is zero when $D = D_0$. The stiffness per unit area of the bond holding the two surfaces together is found by differentiating again

$$K(D) = -8\frac{U_0}{D^2}\left[\left(\frac{D_0}{D}\right)^2 - 3\left(\frac{D_0}{D}\right)^8\right]$$

and evaluating at the resting separation

$$K(D_0) = 16\frac{U_0}{D_0^2}$$

To estimate the Young's modulus, we assume that the material comprises a series of plates, of thickness D_1 ($> D_0$), each held together by the van der Waals forces described in the above equations. The Young's modulus resulting from all these springs in series is

$$E = 16\frac{U_0 D_1}{D_0^2}$$

as discussed in relation to Figure 3.3. If we model a protein as packed layers of amino acids, then $U_0 = 40$ pN/nm (Israelachvili, 1991), $D_0 = 0.3$ nm (twice the van der Waals radius; Creighton, 1993) and $D_1 = 0.6$ nm (the average diameter of an amino acid), and we obtain a Young's modulus of 4 GPa.

To find the tensile strength, note that the force is a maximum when the stiffness is zero; this occurs at a separation D_{max} given by

$$D_{max} = \sqrt[6]{3} \cdot D_0$$

The associated maximum pressure, the tensile strength, is

$$P_{max} = \frac{16}{9\sqrt{3}}\frac{U_0}{D_0}$$

or 133 MPa using the above values for U_0 and D_0. This pressure corresponds to 133 pN per nm².

3.2 Criteria for Overdamping

The mass and spring with damping is a crude model for the motion of a protein or cell in a viscous fluid. Is the model applicable, and in particular, is it appropriate to use the Stokes' drag when the speed of motion is changing, as occurs during a protein conformational change? In this section I give more general conditions under which Stokes' law is valid and show that a more realistic model for protein movements also predicts that the motion of proteins is overdamped.

The motion of a sphere (diameter L) held by a spring (stiffness κ) in a viscous fluid (viscosity η) can be solved exactly using the Navier–Stokes equation of fluid mechanics (van Netten, 1991). The spring is assumed to be massless and frictionless. The motion is overdamped or underdamped depending on whether the "periodic Reynolds number," R_{ac}, is less than or greater than unity

$$R_{ac} = \frac{\kappa L \rho}{12\pi\eta^2}$$

where ρ is the density of the sphere (assumed equal to that of the fluid). If $R_{ac} \ll 1$ then Stokes' law holds. Taking $\kappa = EL$ as an estimate of the stiffness, the characteristic length for which $R_{ac} = 1$ is ~6 nm (using $\eta^2/\rho E = 1$ nm^2), consistent with the estimate obtained from Equation 3.9.

These arguments can be generalized to bodies of arbitrary shape. If such a body oscillates back and forth at a frequency ω rad/s with an amplitude, x_0, smaller than the body's diameter, L, then it can be shown that the inertial forces can be neglected provided

$$\omega \ll \frac{4\eta}{\rho L^2}$$

(Landau and Lifshits, 1987, p. 86). In this case $Re < 1$ and Stokes' law holds, because $\omega = v_0/x_0$ where v_0 is the maximum speed. But the converse is not necessarily true: $Re < 1$ does not imply that the inertial forces can be neglected and Stokes' law does not hold for small motions that have a high enough frequency. However, provided that the time constant of the motion is greater than 6 ps, the inertial forces are very small. Therefore, on the microsecond to millisecond timescales of the transitions between different chemical and structural states, the inertial forces can be neglected. The ratio of inertial to viscous forces is

$$\frac{F_{inertial}}{F_{viscous}} = \frac{3}{2}\frac{x(1+2x/9)}{1+x} \qquad x^2 = \frac{L^2}{8}\frac{\rho\omega}{\eta}$$

where L is the diameter (Landau and Lifshits, 1987, p. 89, Equation 3), and where we have used the virtual mass (Landua and Lifshits, 1987, p. 31). Because the resonance frequency is $\sim(\kappa/m)^{0.5} \cong (E/\rho)^{0.5}/L$, it follows that the characteristic length below which the viscous forces are larger than the inertial forces is ~8 nm (using $\eta^2/\rho E = 1$ nm^2), in agreement with the earlier calculations.

3.3 Drag Force on an Oscillating Sphere

Consider a sphere of radius R undergoing a periodic expansion and contraction in an incompressible fluid with viscosity η at low Reynolds number. The velocity field, $v(r, t)$, and the pressure, $p(r, t)$, given by

$$v_r(r,t) = \frac{a}{r^2}\sin(\omega t) \quad \text{and} \quad p(r,t) = \frac{2}{3}\frac{\eta}{r}v_r$$

are solutions to the Navier–Stokes equation

$$\nabla \cdot v = 0 \quad \text{and} \quad \rho\frac{\partial v}{\partial t} = \nabla p + \eta\nabla^2 v$$

(Landau and Lifshits, 1987) under the condition $\omega \ll \eta/\rho R^2$ (see the Appendix section on Criteria for Overdamping, above). The drag force is found by multiplying the pressure at the surface of the sphere (when $r = R$) by the area of the sphere ($4\pi R^2$). Its maximum is $(8\pi/3)\eta R v_r$, similar to the drag force acting on a sphere translating at constant velocity v_r.

3.4 Elongated Proteins and Filaments

An elongated protein is more highly damped than a globular protein of the same molecular mass. The reason is that as the aspect ratio increases, the damping increases and the stiffness decreases. The aspect ratio is defined by $f = L/\sqrt{A}$, where L is the length of the long axis and A is the cross-sectional area. Using the drag coefficients for long cylinders given in Table 6.2, the damping criterion for longitudinal motions (i.e., stretching and compressing) is

$$\frac{4m\kappa}{\gamma^2} \cong \frac{4\rho AL \cdot EAL}{\left[\dfrac{2\pi\eta L}{\ln(L/\sqrt{A}) - 0.2}\right]^2} \cong \left(\frac{\ln f}{\pi}\right)^2 \frac{\rho E}{\eta^2}\left(\frac{L}{f^2}\right)^2$$

and that for bending motions is

$$\frac{4m\kappa}{\gamma^2} \cong \frac{4\rho AL \cdot \dfrac{4EA^2}{3\pi L^3}}{\left[\dfrac{4\pi\eta L}{\ln(L/\sqrt{A}) + 0.84}\right]^2} \cong \frac{\ln^2 f}{3\pi^3}\frac{\rho E}{\eta^2}\left(\frac{L}{f^3}\right)^2$$

We have used the stiffnesses from Examples 3.1 and 3.2. If the aspect ratio is 10, then the characteristic length for stretching motions is ~136 nm, while that for bending motions is ~4200 nm: The corresponding molecular masses are 21 MDa and 600×10^6 kDa! Thus *the motions of proteins with large axial ratios will always be overdamped.*

Suppose that a protein polymerizes to form a filament of length L and diameter d. The aspect ratio is $f \cong L/d$. In this case $4m\kappa/\gamma^2$ is proportional to d^4/L^2 in stretching mode and d^6/L^4 in bending mode. In other words, the longer the filament, the more highly it is damped. This is opposite to the case of globular proteins, in which the larger proteins are less damped. For the two-stranded actin filament of diameter 6 nm, we expect the motion to be overdamped in both modes for all filament lengths. For a microtubule of diameter 25 nm, stretching motions will be overdamped for lengths greater than 500 nm, while bending motions will be overdamped for lengths greater than 100 nm. Due to the high damping, the bending motions of cytoskeletal filaments are very slow, on the order of seconds to minutes, as will be shown when we consider the hydrodynamics of filaments in more detail in Chapter 6.

4.1 Boltzmann's Law

Equipartition of Energy

Suppose that the energy U of a system depends on the parameter x according to $U = ax^2$, where a is a constant. Then the average energy is

$$\langle U \rangle = \int_{-\infty}^{\infty} U(x) p(x) \cdot dx = \frac{1}{Z} \int_{-\infty}^{\infty} U(x) \exp[-U(x)/kT] \cdot dx$$

$$= \frac{\int_{-\infty}^{\infty} U(x) \exp[-U(x)/kT] \cdot dx}{\int_{-\infty}^{\infty} \exp[-U(x)/kT] \cdot dx} = \frac{\int_{-\infty}^{\infty} ax^2 \exp[-ax^2/kT] \cdot dx}{\int_{-\infty}^{\infty} \exp[-ax^2/kT] \cdot dx}$$

$$= \tfrac{1}{2} kT$$

The last step can be verified by looking up a table of integrals.

Independent Degrees of Freedom

Suppose that a system has two degrees of freedom x and y, meaning that the energy depends on both these parameters, and that Boltzmann's law holds

$$p(x,y) = \frac{1}{Z} \exp\left(-\frac{U(x,y)}{kT}\right)$$

Let $p(x) = \sum_y p(x,y)$ and $p(y) = \sum_x p(x,y)$. Then x and y are independent—that is, $p(x,y) = p(x) \cdot p(y)$—if and only if the energy of the system can be written as the sum of energies associated with each degree of freedom

$$U(x,y) = U(x, y_r) + U(x_r, y) - U(x_r, y_r)$$

where (x_r, y_r) is an arbitrary reference point.

This statement follows from the following arguments. First, suppose that x and y are independent. Then

$$\exp\left(-\frac{U(x,y)}{kT}\right) = Z \cdot p(x,y) = Z \cdot p(x) \cdot p(y)$$

$$= Z \frac{p(x) \cdot p(y_r)}{p(y_r)} \frac{p(x_r) \cdot p(y)}{p(x_r)}$$

$$= Z \frac{p(x, y_r) \cdot p(x_r, y)}{p(x_r, y_r)}$$

$$= \exp\left(-\frac{U(x, y_r) + U(x_r, y) - U(x_r, y_r)}{kT}\right)$$

Conversely, suppose that the energies add. Then for all x_0 and y_0,

$$p(x_0) \cdot p(y_0) = \sum_y p(x_0, y) \sum_x p(x, y_0)$$

$$= \frac{1}{Z} \sum_y \exp\left(-\frac{U(x_0, y)}{kT}\right) \frac{1}{Z} \sum_x \exp\left(-\frac{U(x, y_0)}{kT}\right)$$

$$= \frac{1}{Z} \sum_y \exp\left(-\frac{U(x_0, y_r) + U(x_r, y) - U(x_r, y_r)}{kT}\right)$$

$$\cdot \frac{1}{Z} \sum_x \exp\left(-\frac{U(x, y_r) + U(x_r, y_0) - U(x_r, y_r)}{kT}\right)$$

$$= \frac{1}{Z} \exp\left(-\frac{U(x_0, y_r) + U(x_r, y_0) - U(x_r, y_r)}{kT}\right)$$

$$\cdot \frac{1}{Z} \sum_{x,y} \exp\left(-\frac{U(x, y_r) + U(x_r, y) - U(x_r, y_r)}{kT}\right)$$

$$= \frac{1}{Z} \exp\left(-\frac{U(x_0, y_0)}{kT}\right) \frac{1}{Z} \sum_{x,y} \exp\left(-\frac{U(x, y)}{kT}\right)$$

$$= p(x_0, y_0) \cdot \sum_{x,y} p(x, y)$$

$$= p(x_0, y_0)$$

The Law of Detailed Balance

Suppose there are two independent degrees of freedom, x and y, and suppose that we "freeze" one of them out by locking the system so that y is always equal to y_0. Then the new system, with one remaining degree of freedom, remains at equilibrium at the same temperature. In other words, if a system is at equilibrium, then each degree of freedom is separately at equilibrium. This follows from the following argument. Define $P(x) = p(x,y_0)/p(y_0)$ and $U(x) = U(x, y_0)$. Then

$$P(x) = \frac{p(x, y_0)}{p(y_0)} = \frac{1}{p(y_0)} \frac{1}{Z} \exp\left(-\frac{U(x, y_0)}{kT}\right) \equiv \frac{1}{Z \cdot p(y_0)} \exp\left(-\frac{U(x)}{kT}\right)$$

which is Boltzmann's law with partition function $Z_x = Z \cdot p(y_0)$.

Vibrational Frequencies of Proteins

For proteins, the vibrational energy levels are much smaller than kT. This follows because the vibrational frequency, ω, is

$$\omega \sim \sqrt{\frac{\kappa}{m}} \sim \sqrt{\frac{EA/L}{\rho A L}} = \frac{1}{L}\sqrt{\frac{E}{\rho}} \leq 10^{12} \text{ radians/s}$$

where we have taken $E = 1$ GPa, $\rho = 10^3$ kg/m^3, and $L \geq 1$ nm. Thus the energy quanta have magnitude

$$h\nu = \frac{\hbar\omega}{2\pi} \leq 0.1 \times 10^{-21} \text{ J} \ll 4 \times 10^{-21} \text{ J} = kT$$

This calculation assumes that the protein has the maximum rigidity of structural proteins; if it were more flexible, as I have argued to be the case for protein machines, then the vibrational frequency and the associated energy would be even smaller. A vibrational frequency of 10^{12} radians/s corresponds to ~5 cm^{-1}, which is 100 to 1000 times smaller than the vibrational frequencies associated with chemical bonds. The reason why the vibrational frequencies of proteins are so small is that proteins are held together by van der Waals bonds, which are much softer than covalent or metallic bonds. These arguments rule out very-high-frequency quantized resonances of proteins, and of the cytoskeleton in particular, which are envisaged by Penrose to play a role in long-distance information transfer and processing (1994).

4.2 Diffusion

Random Walk, Fick's Law, and the Diffusion Equation

Suppose that in a time interval, Δt, each molecule in a solution steps, with equal probability, to the left or to the right, through a distance Δx. In other words, there is a 50% chance of moving to the left and a 50% chance of moving to the right. From Figure A4.1, the number of molecules that move from the left through the plane at $x + \Delta x$ is $\frac{1}{2} c(x) \Delta x$, while the number that move from right to left is $\frac{1}{2} c(x + \Delta x) \Delta x$. The flux at $x + \Delta x$, $J(x + \Delta x)$, times Δt, is therefore equal to

$$J(x + \Delta x) \cdot \Delta t = \frac{1}{2}[c(x) - c(x + \Delta x)] \cdot \Delta x$$

Dividing both sides by Δt, and rearranging, we obtain

$$J(x + \Delta x) = -\frac{\Delta x^2}{2\Delta t} \frac{c(x + \Delta x) - c(x)}{\Delta x}$$

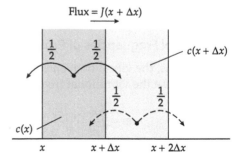

Figure A4.1 Derivation of Fick's law

Taking the limit as Δt and Δx get very small, the left side approaches $J(x)$ while the rightmost term approaches the concentration gradient. Thus we obtain Fick's law (Equation 4.6) provided that

$$D = \frac{\Delta x^2}{2\Delta t}$$

Thus a random walk produces diffusive motion, with the amplitude of the diffusion coefficient depending on the size and frequency of the steps.

Gradients in the flux lead to changes in concentration. If less material leaves a thin slab to the right than enters it from the left, then the concentration will increase (Figure 4.4B). After a time Δt, $A \cdot J(x) \cdot \Delta t$ molecules will enter from the left while $A \cdot J(x + \Delta x) \cdot \Delta t$ will leave toward the right, where A is the area of the slab and Δx its thickness. Provided there are no sinks or sources, the total number of molecules in the slab will increase by $A \cdot J(x) \cdot \Delta t - A \cdot J(x + \Delta x) \cdot \Delta t$. Dividing by the volume, $A \cdot \Delta x$, yields the change in concentration

$$\Delta c = \frac{A \cdot J(x) \cdot \Delta t - A \cdot J(x + \Delta x) \cdot \Delta t}{A \cdot \Delta x} = -\frac{[J(x + \Delta x) - J(x)] \cdot \Delta t}{\Delta x} \rightarrow -\frac{dJ}{dx} \Delta t$$

Dividing both sides by Δt and taking the limit $\Delta t \rightarrow 0$, we obtain Equation 4.7.

Einstein Relation

Suppose that a molecule, with drag coefficient γ, is diffusing, with diffusion coefficient D, in the presence of an external force $F(x)$. If the system is in equilibrium, then the probability, $p(x)$, of finding the molecule at position x does not change with time. The Fokker–Planck equation then reduces to

$$-D\frac{dp}{dx}(x) + \frac{F(x)}{\gamma} p(x) = j_0 = \text{constant}$$

This can be rewritten

$$\frac{d}{dx}\left\{p(x)\exp\left[\frac{U(x)}{D\gamma}\right]\right\} = -\frac{j_0}{D}\exp\left[\frac{U(x)}{D\gamma}\right]$$

where $F(x) = -dU/dx$. Integration gives

$$p(x)\exp\left[\frac{U(x)}{D\gamma}\right] = A - \frac{j_0}{D} B(x)$$

where A is a constant and $dB/dx = \exp[U(x)/D\gamma]$. Upon rearrangement, this becomes

$$p(x) = \left\{A - \frac{j_0}{D} B(x)\right\}\exp\left[-\frac{U(x)}{D\gamma}\right]$$

The only way that this expression can be made to agree with Boltzmann's law (Equation 4.1) is to set $A = 1/Z$, $j_0 = 0$, and $D\gamma = kT$. In this way we obtain the Einstein relation (Equation 4.11). We also conclude that *at equilibrium the flux is zero*. Thus the equilibrium condition is stronger than the steady-state condition, which merely requires that the flux be constant.

First-Passage Times

Consider the box shown in Figure A4.2A in which there is a reflecting wall at $x = 0$ and an absorbing wall at $x = x_0$. If a molecule is placed at $x = 0$, how long will it take to reach $x = x_0$ for the first time? In other words, how long, on average, will it take for the absorbing wall at x_0 to catch the molecule? This is the first-passage time, t_0, sometimes called the **mean time to capture** (Berg, 1993). One way to solve this problem is to imagine that there are many molecules in the box, and that every time a molecule hits the absorbing wall it is instantly placed back at the origin. After a while, depending on the initial conditions, the distribution of molecules will reach steady state—in other words, they will not change over time—and the first-passage time will equal the inverse of the rate at which molecules are removed at x_0. This rate is equal to the flux at x_0. In fact, it will also equal the flux at the origin, and indeed it will be equal to the flux throughout the box because at steady state the flux is a constant (Equation 4.7). Thus the first-passage time is

$$t_0 = \frac{1}{j(x_0)} = \frac{1}{j(x)}$$

In the absence of an external force, it is quite easy to show that the probability $p(x) = -2x/x_0^2 + 2/x_0$ is the solution to the diffusion equation, which satisfies

1. $\partial^2 p/\partial x^2 = \partial p/\partial t = 0$ (i.e., steady state).
2. $p(x_0) = 0$ (the concentration is zero at the absorbing wall).
3. The integrated probability is unity (Equation 4.13).

This solution is plotted in Figure A4.2B. The first-passage time is equal to

$$t_0 = \frac{1}{j(x_0)} = \frac{1}{-D\,dp/dx} = \frac{x_0^2}{2D}$$

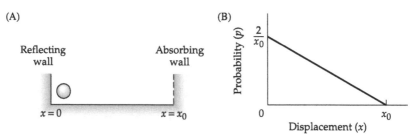

Figure A4.2 First-passage time
(A) The first-passage time is the average time taken for a molecule released at the origin to reach, for the first time, the position x_0. (B) The solution to the diffusion equation that satisfies $p(x_0) = 0$ (absorbing wall). The first-passage time is the flux, which equals $2x_0^2$.

as stated in Equation 4.16. Note that this is an example of a system at steady state that is *not* at equilibrium, as it requires the unseen hand of Maxwell's demon to move molecules from x_0 back to the origin.

The general solution for the first-passage time in the presence of an external force $F(x)$ is more difficult to derive. We start with the second equation in the derivation of the Einstein relation above, and integrate it from x to x_0 to obtain

$$p(x_0)\exp\left[\frac{U(x_0)}{D\gamma}\right] - p(x)\exp\left[\frac{U(x)}{kT}\right] = -p(x)\exp\left[\frac{U(x)}{kT}\right] = -\frac{j_0}{D}\int_x^{x_0}\exp\left[\frac{U(x)}{kT}\right]\cdot dx$$

where we have used $p(x_0) = 0$. Multiplying both sides by $-\exp[-U(x)/kT]$ and integrating from 0 to x_0 then gives

$$\frac{j_0}{D}\int_0^{x_0}\exp\left[-\frac{U(x)}{kT}\right]\left\{\int_x^{x_0}\exp\left[\frac{U(y)}{kT}\right]\cdot dy\right\}\cdot dx = \int_0^{x_0} p(x)\cdot dx = 1$$

Upon rearrangement we obtain

$$t_0 = \frac{1}{j_0} = \frac{1}{D}\int_0^{x_0}\exp\left[-\frac{U(x)}{kT}\right]\left\{\int_x^{x_0}\exp\left[\frac{U(y)}{kT}\right]\cdot dy\right\}\cdot dx$$

In the case of a constant force, $U(x) = -F\cdot x$, it is reasonably straightforward to evaluate the integral to obtain Equation 4.17. In the case where the force is harmonic—that is, when $U(x) = \frac{1}{2}\kappa x^2$—the integral is not straightforward, and for it to be algebraically tractable, we assume that the height of the energy barrier is high—that is, $U_0 \equiv U(x_0) \gg kT$. Substituting the expression for $U(x)$ into the previous equation and rearranging, we obtain

$$t_0 = \frac{1}{D}\int_0^{x_0}\exp\left[-\frac{\kappa x^2}{2kT}\right]\left\{\int_0^{x_0}\exp\left[\frac{\kappa y^2}{2kT}\right]\cdot dy - \int_0^x\exp\left[\frac{\kappa y^2}{2kT}\right]\cdot dy\right\}\cdot dx$$

$$\cong \frac{1}{D}\int_0^{x_0}\exp\left[-\frac{\kappa x^2}{2kT}\right]\left\{\int_0^{x_0}\exp\left[\frac{\kappa y^2}{2kT}\right]\cdot dy\right\}\cdot dx$$

since $\exp[-\kappa x^2/2kT]$ is very small unless x is close to zero, in which case the second integral in the curly bracket can be neglected. We now have

$$t_0 = \frac{1}{D}\left\{\int_0^{x_0} \exp\left[-\frac{\kappa x^2}{2kT}\right]\cdot dx\right\}\cdot\left\{\int_0^{x_0} \exp\left[\frac{\kappa y^2}{2kT}\right]\cdot dy\right\}$$

$$\cong \frac{1}{D}\left\{\int_0^{\infty} \exp\left[-\frac{\kappa x^2}{2kT}\right]\cdot dx\right\}\cdot\left\{\int_0^{x_0} \exp\left[\frac{\kappa x^2}{2kT}\right]\cdot dx\right\}$$

$$\cong \frac{1}{D}\left\{\sqrt{\frac{\pi kT}{2\kappa}}\right\}\cdot\left\{\frac{kT}{\kappa x_0}\exp\left[\frac{\kappa x_0^2}{2kT}\right]\right\}$$

where the first integral can be found in tables of integrals, and the second integral can be solved using the change of variable $z = x_0 - x$. The required first-passage time, or Kramers time, Equation 4.18, is obtained by substituting $U_0 = \frac{1}{2}\kappa x_0^2$ and $D = kT/\gamma$ and $\tau = \gamma/\kappa$.

4.3 Fourier Analysis

Autocorrelation

The autocorrelation function of the motion of a diffusing particle satisfies the Langevin equation. This is shown as follows. First, multiply both sides of Equation 4.19 by $x(t-\tau)$ and take the time average

$$m\left\langle\frac{d^2x}{dt^2}(t)\cdot x(t-\tau)\right\rangle + \gamma\left\langle\frac{dx}{dt}(t)\cdot x(t-\tau)\right\rangle + \kappa\langle x(t)\cdot x(t-\tau)\rangle = \langle F(t)\cdot x(t-\tau)\rangle$$

The right-hand side is zero for the following reason. First, the thermal F must satisfy the property $\langle F\rangle = 0$, because otherwise there would be a net flux of molecules and the system would not be at equilibrium. Second, the thermal force is independent of the molecule's past history; in particular, the force at time t is independent of the molecule's past position. These two properties imply that

$$\langle F(t)\cdot x(t-\tau)\rangle = \langle F(t)\rangle\cdot\langle x(t-\tau)\rangle = 0\cdot\langle x(t-\tau)\rangle = 0 \qquad \text{for } \tau > 0$$

The above differential equation further simplifies by observing that

$$\left\langle\frac{dx}{dt}(t)\cdot x(t-\tau)\right\rangle = \frac{dR_x}{d\tau}(\tau)$$

which follows because

$$\frac{dR_x}{d\tau}(\tau) = \frac{1}{T} \int_{-T/2}^{T/2} x(t) \cdot \frac{d}{d\tau} x(t-\tau) \cdot dt$$

$$= \int_{-T/2}^{T/2} x(t) \cdot \frac{d}{dt} x(t-\tau) \cdot dt$$

$$= -\frac{1}{T}[x(t) \cdot x(t-\tau)]_{-T/2}^{T/2} + \frac{1}{T} \int_{-T/2}^{T/2} \frac{dx}{dt}(t) \cdot x(t-\tau) \cdot dt$$

$$\to \frac{1}{T} \int_{-T/2}^{T/2} \frac{dx}{dt}(t) \cdot x(t-\tau) \cdot dt \text{ as } T \to \infty$$

$$= \left\langle \frac{dx}{dt}(t) \cdot x(t-\tau) \right\rangle$$

Likewise, $\left\langle \frac{d^2x}{dt^2}(t) \cdot x(t-\tau) \right\rangle = \frac{d^2R_x}{d\tau^2}(\tau)$, and the result, Equation 4.21, follows.

Parseval's Theorem

As expected from its definition, the integral of the power spectrum is equal to the total variance of the signal. In other words, the sum of all the variances at each frequency is equal to the total variance. This follows from the following argument. The variance of the signal in terms of the Fourier coefficients is found by multiplying the Fourier series of Equation 4.22 by $x(t)$ and averaging

$$\langle x^2(t) \rangle = \left\langle x(t) \sum_{n=1}^{\infty} \left[a_n \cos \frac{2\pi nt}{T} + b_n \sin \frac{2\pi nt}{T} \right] \right\rangle$$

$$= \sum_{n=1}^{\infty} \left[a_n \left\langle x(t) \cos \frac{2\pi nt}{T} \right\rangle + b_n \left\langle x(t) \sin \frac{2\pi nt}{T} \right\rangle \right]$$

$$= \frac{1}{2} \sum_{n=1}^{\infty} \left[a_n^2 + b_n^2 \right]$$

The power spectrum at frequency f is equal to

$$G_x(f) \cdot \Delta f \equiv \langle x_{f,\Delta f}^2(t) \rangle$$

$$= \left\langle \left[a_n \cos \frac{2\pi nt}{T} + b_n \sin \frac{2\pi nt}{T} \right]^2 \right\rangle$$

$$= \frac{1}{2}\left(a_n^2 + b_n^2\right)$$

where we have used the familiar results that

$$\langle \cos^2(2\pi nt/T)\rangle = \langle \sin^2(2\pi nt/T)\rangle = \tfrac{1}{2} \quad \text{and} \quad \langle \cos(2\pi nt/T)\sin(2\pi nt/T)\rangle = 0$$

The integral of the power spectrum is therefore

$$\int_0^\infty G_x(f)\cdot df \equiv \sum_{f>0} G_x(f)\cdot \Delta f = \sum_{n>0} \tfrac{1}{2}(a_n^2 + b_n^2) = \langle x^2\rangle = \sigma_x^2$$

as claimed. This is known as **Parseval's theorem**. Parseval's theorem is of great practical use when actually calculating a power spectrum. This is because the power spectrum can be defined in a number of slightly different ways that differ by multiplicative constants. (For example, we are using the single-sided power spectrum, which only considers positive frequencies, Bracewell, 1986; Bendat and Piersol, 1986.) Furthermore, different numerical algorithms for calculating the power spectrum often use different scaling factors. For example, they may return the power per frequency increment, which will not be the same as the power per Hz unless the frequency increment is 1 Hz. The net result is that the integral of a numerically computed power spectrum may not equal the variance; this power spectrum should therefore be rescaled so that its integral is equal to the independently calculated variance. In this way, Parseval's theorem will be satisfied, and the y-axis of the spectrum will be correct. A second technical point about computing the power spectrum is that most algorithms define the power spectrum at zero frequency to be $G_x(0) = \langle x\rangle^2$ so that the power at zero frequency corresponds to the "D.C.," or background level. However, this means that the power spectrum is discontinuous at $f = 0$, and this can create complications when taking the integral. We will generally assume that the signal has zero mean, and we will ignore the zero-frequency power.

The Power Spectrum Is the Fourier Transform of the Autocorrelation Function

This follows from the following argument.

$$R_x(\tau) = \langle x(t)x(t-\tau)\rangle = \sum_n \sum_m \left[a_n a_m \left\langle \cos\frac{2\pi nt}{T}\cos\frac{2\pi m(t-\tau)}{T}\right\rangle\right.$$

$$+ a_n b_m \left\langle \cos\frac{2\pi nt}{T}\sin\frac{2\pi m(t-\tau)}{T}\right\rangle + b_n a_m \left\langle \sin\frac{2\pi nt}{T}\cos\frac{2\pi m(t-\tau)}{T}\right\rangle$$

$$\left. + b_n b_m \left\langle \sin\frac{2\pi nt}{T}\sin\frac{2\pi m(t-\tau)}{T}\right\rangle\right]$$

This can be simplified by substituting

$$\cos(t-\tau) = \cos t\cos\tau + \sin t\sin\tau$$
$$\sin(t-\tau) = \sin t\cos\tau - \cos t\sin\tau$$

and using

$$\langle\cos^2(2\pi nt/T)\rangle = \langle\sin^2(2\pi nt/T)\rangle = \tfrac{1}{2}$$

$$\langle\cos(2\pi nt/T)\sin(2\pi mt/T)\rangle = 0 \quad \text{all } n, m$$

$$\langle\cos(2\pi nt/T)\cos(2\pi mt/T)\rangle = \langle\sin(2\pi nt/T)\sin(2\pi mt/T)\rangle = 0 \quad n \neq m$$

We then obtain

$$R_x(\tau) = \tfrac{1}{2}\sum_{n=1}^{\infty}(a_n^2 + b_n^2)\cos\frac{2\pi n\tau}{T}$$

Thus the Fourier transform of R_x is

$$4\int_0^{\infty} R_x(\tau)\cos(2\pi f\tau)\cdot d\tau = 2\sum_{n=1}^{\infty}(a_n^2 + b_n^2)\int_0^{\infty}\cos\frac{2\pi n\tau}{T}\cos(2\pi f\tau)\cdot d\tau$$

$$= \tfrac{1}{2}\sum_{n=1}^{\infty}(a_n^2 + b_n^2)\delta(f - \frac{n}{T})$$

using

$$\frac{2}{\pi}\int_0^{\infty}\cos f_0\tau\cos f\tau\cdot d\tau = \delta(f - f_0)$$

where $\delta(f)$ is the delta function. Integrating from f to $f + \Delta f$ gives

$$\int_f^{f+\Delta f}\left\{4\int_0^{\infty}R_x(\tau)\cos(2\pi f\tau)\cdot d\tau\right\}\cdot df = \tfrac{1}{2}(a_n^2 + b_n^2) = G_x(f)\cdot\Delta f \quad f = n/T \quad \Delta f = 1/T$$

This shows that the Fourier transform of the autocorrelation function equals the power spectrum, as required.

Power Spectrum of the Thermal Force

Because we know the power spectrum of the position of a diffusing molecule, it should be possible to calculate the power spectrum of the equivalent force necessary to produce the motion. This is the thermal force. Let us go back to the equation of motion for our molecule, and consider how it moves in response to an external sinusoidal force, $F_f(t)$, of frequency f

$$\gamma\frac{dx_f}{dt}(t) + \kappa x_f(t) = F_f(t) \equiv F_0\sin(2\pi ft)$$

where we are ignoring the mass. One can verify, by differentiation, that the solution is

$$x_f(t) = F_0\frac{\kappa^{-1}}{1 + (2\pi f\tau_0)^2}[\sin(2\pi ft) - 2\pi f\tau_0\cos(2\pi ft)] \quad \tau_0 = \gamma/\kappa$$

The variance of the solution is therefore equal to

$$G_x(f) \equiv \langle x_f^2(t) \rangle = \tfrac{1}{2} F_0^2 \frac{\kappa^{-2}}{1+(2\pi f \tau_0)^2} = \langle F_f^2(t) \rangle \frac{\kappa^{-2}}{1+(2\pi f \tau_0)^2} = G_F(f) \frac{\kappa^{-2}}{1+(2\pi f \tau_0)^2}$$

Comparing this expression to the expression in Example 4.7, we see that the thermal motion of the molecule is identical to that produced by an external force whose variance at each frequency f is

$$G_F(f) = 4kT\gamma$$

Autocorrelation of the Thermal Force

Another way to think of the thermal force is to go back to the collision model. Suppose a stationary molecule is struck by an impulsive force. By this we mean that the molecule experiences a very large force F for a very short time Δt. During this time, the force, which is much greater than any drag or elastic forces, causes the molecule to accelerate with an acceleration equal to F/m. Over the duration of the impulse, the velocity increases to $F\Delta t/m = v_0$, and momentum $mv_0 = F \cdot \Delta t$ is imparted to the molecule. In the limit of very small Δt, the molecule therefore satisfies the Langevin equation (Equation 4.18) for all times $t>0$, with initial conditions $x(0) = 0$ and $v(0) = v_0$. Now suppose that the molecule is struck by impulses at random times and from random directions. The autocorrelation function of this barrage of impulsive forces is

$$R_F(\tau) = \langle F(t)F(t-\tau) \rangle = 2kT\gamma \cdot \delta(\tau)$$

where $\delta(\tau)$ is the delta function

$$\delta(\tau) = \delta(-\tau) \text{ and } \int_{-\infty}^{\infty} \delta(\tau) \cdot d\tau = 1$$

The amplitude of the autocorrelation function is right because it leads to the correct power spectrum

$$G_F(f) = 4 \int_{-\infty}^{\infty} R_F(\tau) \cos(2\pi f \tau) \cdot d\tau = 2 \int_{-\infty}^{\infty} \{2kT\gamma\delta(\tau)\} \cos(2\pi f \tau) \cdot d\tau = 4kT\gamma$$

The trouble with the last equation is that the variance of the force is infinite. The reason is that we are assuming that the impacts are truly instantaneous, when in fact they are not. This problem can be solved by writing

$$\gamma \frac{dx}{dt}(t) + \kappa x(t) = F(t) \quad \langle F(t)F(t-\tau) \rangle = R_F(\tau) = kT\gamma \left\{ \frac{1}{\tau_0} \exp\left(-\frac{|\tau|}{\tau_0} \right) \right\} \quad \tau_0 = \frac{m}{\gamma}$$

which gives the correct value for the power spectrum of the force for frequencies $<< (2\pi\tau_0)^{-1}$. In this case, the variance of the force is finite, and equals

$$\langle F^2 \rangle = R_F(0) = kT\gamma \frac{1}{\tau_0} = m\langle v^2 \rangle \gamma \frac{\gamma}{m} = \gamma^2 \langle v^2 \rangle$$

using the Equipartition principle for one dimension (Equation 4.5). This shows that the root-mean-square thermal force is proportional to the drag coefficient times the root-mean-square velocity, as expected of a drag force. This is another way of thinking about the fluctuation–dissipation theorem.

5.1 Free Energies Associated with Ensembles of Conformational States

Discrete States

Suppose that a molecule can be in any of N different conformational states each with potential energy U_i. The average energy is

$$\langle U \rangle = \sum_{i=1}^{N} U_i \cdot p_i$$

where the probability of being in state i is

$$p_i = \frac{1}{Z} \exp\left[-\frac{U_i}{kT}\right]$$

and the partition function is

$$Z = \sum_{i=1}^{N} \exp\left[-\frac{U_i}{kT}\right]$$

The **entropy** is

$$S = -k \sum_{i=1}^{N} p_i \ln p_i$$

and so the free energy is

$$\begin{aligned} G &\equiv \langle U \rangle - TS \\ &= \sum_{1}^{N} U_i \cdot p_i + kT \sum_{1}^{N} p_i \ln p_i \\ &= \sum_{1}^{N} p_i \{U_i + kT \ln p_i\} \\ &= \sum_{1}^{N} \frac{1}{Z} \exp\left[-\frac{U_i}{kT}\right] \left\{ U_i + kT \ln\left(\frac{1}{Z} \exp\left[-\frac{U_i}{kT}\right]\right)\right\} \\ &= \frac{1}{Z} \sum_{1}^{N} \exp\left[-\frac{U_i}{kT}\right] \{U_i - kT \ln Z - U_i\} \\ &= -kT \ln Z \end{aligned}$$

This equation provides the link between classical thermodynamics, where the free energy is expressed in terms of enthalpy ($\Delta H = \langle U \rangle$) and entropy, and statistical thermodynamics, where it is expressed in terms of probabilities (p_i) and energy levels (U_i). The simple relationship between the free energy and the partition function makes the latter a useful tool for performing statistical calculations.

Now suppose that there are two ensembles of conformational states, E_1 and E_2, with associated energies $\{U_1, ..., U_n\}$ and $\{U_1, ..., U_m\}$ where $n + m = N$. Then

$$P_1 = \text{Prob}\{E_1\} = \sum_{i=1}^{n} p_i = \frac{1}{Z}\sum_{i=1}^{n} \exp\left[-\frac{U_i}{kT}\right] \equiv \frac{Z_1}{Z}$$

with a similar expression for P_2. Z_1 is the partition function for the first ensemble. Thus the ratio of concentrations of E_1 and E_2 is

$$\frac{[E_2]}{[E_1]} = \frac{P_2}{P_1} = \frac{Z_2}{Z_1} = \exp\left[-\frac{\Delta G}{kT}\right]$$

This establishes Equation 5.2. This equation shows that the equilibrium constant depends on the free energy rather than the potential energy. For example, suppose that the potential energies of all the conformational states are the same, but that ensemble 2 has m states whereas ensemble 1 has n states. Then the difference in free energy is $\Delta G = -kT\ln(m/n)$: If ensemble 2 has twice as many states, then its free energy is $kT\ln 2$ lower than that of ensemble 1. Thus the state with the greater entropy is preferred. Likewise, the ensemble that has more closely spaced energy levels has a greater entropy and therefore a lower free energy (assuming the average potential energies are the same).

The potential energy comprises the internal energy, the energies associated with formation of the bonds between atoms (including the solvent), and the external energy, the energy associated with external forces. The free energy of each ensemble depends on the external force according to

$$G_1(F) = -kT \ln Z_1(F) = -kT \ln \sum_{i=1}^{n} \exp\left[-\frac{U_i(F)}{kT}\right]$$

where $U_i(F)$ is the potential energy of state i in the presence of the force. Now

$$\frac{dG_1}{dF} = -\frac{kT}{Z_1}\frac{dZ_1}{dF} = \langle x_i(F)\rangle_1 \quad \text{where} \quad x_i(F) \equiv \frac{dU_i(F)}{dF}$$

The average is taken over the first ensemble of conformational states that comprises structural state 1. Thus for small forces we have

$$\Delta G = G_2 - G_1 \cong \Delta G^0 + F \cdot \Delta x^0 \quad \text{where} \quad \Delta x^0 = \langle x(0)\rangle_2 - \langle x(0)\rangle_1$$

Note that Δx^0 is the length change in the absence of a force. This approximation will work well in a number of situations. First, if the only difference between the two ensembles is that E_2 is displaced a distance Δx relative to E_1 (Figure 5.2A), then the equation holds exactly. The equation also holds exactly if the ensembles correspond to rigid translations of domains such as shown in Figure 5.2B. Furthermore, if the two ensembles have the same compliance—that is, $d\langle x_1\rangle/dF = d\langle x_2\rangle/dF$ (Figure 5.2C)—then again the difference in free energy will be linearly related to the strength of the external force. These three assertions can be verified by thinking about how the changes affect the energy levels.

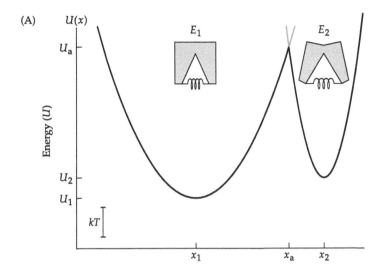

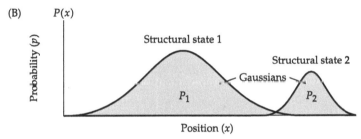

Figure A5.1 Entropy of a spring
(A) Free energy diagram for a molecular spring that can assume a short, soft state (left) or a rigid, long state (right). (B) The probability of being in either of the states.

Free Energy of a Spring

Consider the free energy landscape shown in Figure A5.1. We define state 1 as being all the conformations to the left of the activated state (at x_a) and state 2 as being all the conformations to the right of the activated state. The probabilities of being in states 1 and 2 are

$$P_1 = \frac{1}{Z}\int_{-\infty}^{x_a} \exp\left[-\frac{U(x)}{kT}\right]\cdot dx = \frac{Z_1}{Z} \quad \text{and} \quad P_2 = \frac{1}{Z}\int_{x_a}^{\infty} \exp\left[-\frac{U(x)}{kT}\right]\cdot dx = \frac{Z_2}{Z}$$

where Z_1, Z_2, and Z are the partition functions defined by

$$Z_1 = \int_{-\infty}^{x_a} \exp\left[-\frac{U(x)}{kT}\right]\cdot dx \quad Z_2 = \int_{x_a}^{\infty} \exp\left[-\frac{U(x)}{kT}\right]\cdot dx \quad Z = Z_1 + Z_2$$

The two states have free energies

$$G_1 = -kT \ln Z_1 \qquad G_2 = -kT \ln Z_2$$

and it is easy to show that

$$P_1 = \frac{1}{Z}\exp\left[-\frac{G_1}{kT}\right] \qquad P_2 = \frac{1}{Z}\exp\left[-\frac{G_2}{kT}\right]$$

in accordance with Boltzmann's law.

If state 2 is softer than state 1, then there are more conformations available in state 2 than in state 1 (the entropy of state 2 is greater than the entropy of state 1) and the free energy of state 2 will be correspondingly decreased relative to state 1. This is illustrated in Figure A5.1, where we suppose that the two energy wells are parabolic with stiffnesses κ_1 and κ_2. Then

$$U(x) = \begin{cases} U_1(x - x_1) & x \leq x_a \\ U_2(x - x_2) & x \geq x_a \end{cases}$$

where

$$U_1(x) = U_1^0 + \tfrac{1}{2}\kappa_1 x^2 \quad \text{and} \quad U_2(x) = U_2^0 + \tfrac{1}{2}\kappa_2 x^2$$

Then

$$Z_1 = \int_{-\infty}^{x_a} \exp\left[-\frac{U(x)}{kT}\right] \cdot dx$$

$$= \exp\left[-\frac{U_1^0}{kT}\right] \int_{-\infty}^{x_a} \exp\left[-\frac{\kappa_1 x^2}{2kT}\right] \cdot dx$$

$$\cong \exp\left[-\frac{U_1^0}{kT}\right] \int_{-\infty}^{\infty} \exp\left[-\frac{\kappa_1 x^2}{2kT}\right] \cdot dx$$

$$= \sqrt{\frac{2\pi kT}{\kappa_1}} \exp\left[-\frac{U_1^0}{kT}\right]$$

In this case

$$G_1 = -kT \ln Z_1 = U_1^0 + \tfrac{1}{2} kT \ln \frac{\kappa_1}{2\pi kT}$$

The difference in free energy between the two states is

$$\Delta G = G_2 - G_1 = \Delta U^0 + \tfrac{1}{2} kT \ln \frac{\kappa_2}{\kappa_1}$$

For example, if state two is 4 times stiffer than state 1, then the free energy of state 2 is increased by $kT \ln 2$.

5.2 Chemical Kinetics

Global Conformational Changes Are Limited by the Speed of Sound in Proteins

Suppose that a force F is applied to a protein and that the magnitude of the final deformation of the protein is x_0 (less than the diameter, L, of the protein). From Appendix 2.2 it follows that throughout the deformation

$$\tfrac{1}{2}mv^2 + \tfrac{1}{2}\kappa x^2 = Fx - Q \le Fx = \kappa x_0 x$$

because the heat Q is always positive. Thus

$$\tfrac{1}{2}mv^2 \le \kappa x_0 x - \tfrac{1}{2}\kappa x^2 \le \tfrac{1}{2}\kappa x_0^2$$

Hence

$$v \le x_0\sqrt{\kappa/m} = x_0\sqrt{\frac{EA}{L}\frac{1}{\rho AL}} = \frac{x_0}{L}\sqrt{\frac{E}{\rho}} \le \sqrt{\frac{E}{\rho}} \approx c$$

where c is the speed of sound (Landau and Lifshits, 1976, p. 100). This argument holds whether the motion is underdamped or overdamped.

Q_{10}

$$Q_{10} \equiv \frac{k_{T+10}}{k_T} = \frac{A(T+10)\exp\left[-\dfrac{\Delta U_a(T+10) - T\Delta S_a(T+10)}{k(T+10)}\right]}{A(T)\exp\left[-\dfrac{\Delta U_a(T) - T\Delta S_a(T)}{kT}\right]}$$

$$\cong \exp\left[\frac{\Delta U_a}{kT}\left\{1 - \frac{1}{1+10/T}\right\}\right]$$

$$\cong \exp\left[\frac{\Delta U_a}{kT}\left\{1 - \left(1 - \frac{10}{T}\right)\right\}\right]$$

$$= \exp\left[\frac{\Delta U_a}{kT}\cdot\frac{10}{T}\right]$$

where the approximate equality holds because A, ΔU, and ΔS are not expected to have a strong temperature dependence. Therefore

$$\frac{\Delta U_a}{kT} = \frac{T}{10}\ln Q_{10}$$

Kramers Rate Equation

Consider a molecule in state 1 that contains a spring of stiffness κ_1. Suppose that the spring is free to diffuse in the positive and negative directions but that when it reaches the positive position x_a, corresponding to energy $\Delta U_{a1} = U_a - U_1 = \frac{1}{2}\kappa_1(x_a - x_1)^2$, it has a probability ε_1 of changing into state 2 (Figure A5.1A). This is similar to the problem considered in Figure 4.7. Because the molecule spends about half its time in the negative side of the well, the mean time to escape in the positive direction will be twice the Kramers time t_K (Equation 4.17). The rate constant for the transition between state 1 and state 2 is therefore

$$k_1 = \varepsilon_1 \frac{1}{2t_0} = \varepsilon_1 \frac{1}{\tau_1} \frac{1}{\sqrt{\pi}} \sqrt{\frac{\Delta U_{a1}}{kT}} \exp\left(-\frac{\Delta U_{a1}}{kT}\right)$$

A similar expression holds for k_{-1}. The efficiency factors ε_1 and $\varepsilon_{-1} = 1 - \varepsilon_1$ can be calculated using the equilibrium condition $k_1/k_{-1} = P_2/P_1$. After a little bit of algebra, ε_1 is

$$\varepsilon_1 = \frac{\sqrt{\kappa_2 \Delta U_{a2}}}{\sqrt{\kappa_1 \Delta U_{a1}} + \sqrt{\kappa_2 \Delta U_{a2}}} = \frac{\kappa_2 \Delta x_2}{\kappa_1 \Delta x_1 + \kappa_2 \Delta x_2} \qquad \Delta x_1 = x_1 - x_a \quad \Delta x_2 = x_2 - x_a$$

The ratio of the efficiency factors is the reciprocal of the ratio of the slopes at the transition state.

5.3 Diffusion-Limited Association Rate Constants

Diffusion-Limited Collision Rate Constant

In two or more dimensions, Fick's law is

$$J = -D\nabla c$$

and the diffusion equation is

$$\frac{\partial c}{\partial t} = D\nabla^2 c$$

If there is spherical symmetry, these equations reduce to

$$J(r) = -D\frac{\partial c}{\partial r} \quad \text{and} \quad \frac{\partial c}{\partial t} = D\frac{1}{r^2}\frac{\partial}{\partial r}\left(r^2 \frac{\partial c}{\partial r}\right)$$

The steady-state solution ($dc/dt=0$) that satisfies the boundary conditions $c(R) = 0$ and $c(\infty) = c_\infty$ is

$$c(r) = c_\infty \left(1 - \frac{R}{r}\right)$$

This corresponds to an absorbing sphere of radius R in a solution of particles whose concentration a long way from the sphere is c_∞. The total flux of particles through the surface of the sphere is

$$I = \int_{surface} J(R) \cdot da = 4\pi D R c_\infty$$

Suppose that the particles have radius R_1, that the absorbing sphere (the target) has radius R_2, and that adsorption takes place when the spheres touch. The effective radius is then $R_1 + R_2$. Suppose also that both spheres can freely diffuse, so that the total diffusion coefficient is $D_1 + D_2$. The collision rate constant is

$$k_1 = 4\pi(D_1+D_2)(R_1+R_2) = \tfrac{2}{3}\frac{kT}{\eta}\frac{(R_1+R_2)^2}{R_1 R_2}$$

where we have used the Einstein relation and Stokes' law (see Equation 4.12). The unit is (particles/m^3)$^{-1}\cdot$s^{-1} and to convert it to the usual M$^{-1}\cdot$s^{-1} we need to multiply by $1000N$ (the number of molecules in 1 m^3 when the concentration is 1 M). In the case of two identical spheres, the **diffusion-limited on-rate** is 8×10^9 M$^{-1}\cdot$s^{-1}. If one of the spheres is larger than the other, the on-rate will be even larger. On the other hand, if one of the spheres, the target, is thought of as a stationary point ($D_2 = 0$, $R_2 = 0$), then the diffusion limited on-rate is 2×10^9 M$^{-1}\cdot$s^{-1}.

Effect of Orientation on Collision Rates

To calculate the effect of orientation on collision rates, we adopt a different method for calculating the on-rate (Doi, 1975; Bell, 1978). In this method, the on-rate is calculated as the product of the probability of finding the molecule in a particular volume and in a particular range of orientations, and the rate at which the molecule diffuses out of that particular volume and orientation.

The method works well in the absence of orientation constraints. For example, consider a sphere with a stationary point target. The probability that the target is within the sphere is $(4\pi/3)R^3$. The time to diffuse away is $R^2/6D$. The on-rate calculated by this method is

$$k_1 = \frac{4\pi R^3}{3}\frac{6D}{R^2} = \frac{4\pi R}{3}\frac{6kT}{6\pi\eta R} = \frac{4kT}{3\eta}$$

which differs from the exact solution by only a factor of 2.

Now we consider the effect of orientation on the on-rate. Suppose that there is a limited target radius s (~1 Å) and a limited orientation $\theta \sim s/R$. Now the probability is $(4\pi/3)s^3 \times \theta^2/4\pi \times \theta/2\pi = s^6/6\pi R^3$, where the individual terms correspond to the spatial coordinates, the direction of an axis, and the rotation about that axis. The rate of leaving is the sum of the rates for lateral diffusion $(s^2/6D_{lat})^{-1}$ and rotational diffusion $(\theta^2/6D_{rot})^{-1}$. The leaving rate is therefore

$$k_{-1} = \frac{1}{t_{lat}} + \frac{1}{t_{rot}} = \frac{kT}{\eta}\left(\frac{6}{s^2}\frac{1}{6\pi R} + \frac{6}{\theta^2}\frac{1}{8\pi R^3}\right) = \frac{kT}{\eta}\left(\frac{1}{\pi s^2 R} + \frac{3}{4\pi s^2 R}\right) = \frac{7}{4\pi}\frac{kT}{\eta s^2 R}$$

where the rotational diffusion coefficient for a sphere (D_{rot}) is from Table 6.2. The on-rate is then

$$k_{on} = \frac{s^6}{6\pi R^3}\frac{7kT}{4\pi\eta s^2 R} = k_1\frac{7}{24\pi^2}\frac{kT}{\eta}\left(\frac{s}{R}\right)^4 \cong k_1\left(\frac{s}{2.6R}\right)^4$$

where k_1 is the rate in the absence of orientational constraints.

5.4 Kinetic Equations

The general solution of Equation 5.12 is

$$\frac{d[ES]}{dt} = k_1[E][S] - (k_{-1} + k_2)[ES]$$

$$= k_1([E_t] - [ES])([S_t] - [ES]) - (k_{-1} + k_2)[ES]$$

$$= k_1[ES]^2 - \{k_{-1} + k_2 + k_1([E_t] + [S_t])\}[ES] + k_1[E_t][S_t]$$

$$= k_1([ES] - \alpha)([ES] - \beta)$$

where

$$\alpha = \tfrac{1}{2}\left\{(K_M + [E_t] + [S_t]) - \sqrt{(K_M + [E_t] + [S_t])^2 - 4[E_t][S_t]}\right\}$$

$$\beta = \tfrac{1}{2}\left\{(K_M + [E_t] + [S_t]) + \sqrt{(K_M + [E_t] + [S_t])^2 - 4[E_t][S_t]}\right\}$$

$$K_M = \frac{k_{-1} + k_2}{k_1}$$

and $[E_t]$ and $[S_t]$ are the total concentrations of enzyme and substrate. Rearranging gives

$$\frac{d[ES]}{[ES] - \alpha} - \frac{d[ES]}{[ES] - \beta} = -k_1(\beta - \alpha)dt$$

which can be integrated to give

$$\ln\{[ES] - \alpha\} - \ln\{[ES] - \beta\} = -k_1(\beta - \alpha)t + A$$

or

$$\frac{[ES] - \alpha}{[ES] - \beta} = A'\exp[-k_1(\beta - \alpha)t]$$

This has the steady-state solution

$$[ES] = \alpha$$

for times

$$t \gg \frac{1}{k_1(\beta - \alpha)} = \frac{1}{\sqrt{\{k_{-1} + k_2 + k_1([E_t] + [S_t])\}^2 - 4k_1[E_t][S_t]}}$$

which occurs when
$$t \gg \frac{1}{k_{-1} + k_2 + k_1[[E_t] - [S_t]]}$$
which, in turn, occurs when t is much greater than any of the individual rate constants. If $[S_t] \gg [E_t]$ then
$$[ES] \cong [E_t] \frac{[S_t]}{K_M + [S_t]}$$
which is the Michaelis–Menten equation.

5.5 Free Energies of a Reaction

Consider a mixture initially comprising n_1 molecules of E_1 (with free energy G_1) and n_2 molecules of E_2 (with free energy G_2). The initial free energy is

$$G_i = n_1 G_1 + n_2 G_2 + (n_1 + n_2)kT \left\{ \frac{n_1}{n_1 + n_2} \ln \frac{n_1}{n_1 + n_2} + \frac{n_2}{n_1 + n_2} \ln \frac{n_2}{n_1 + n_2} \right\}$$

If we change one molecule of E_1 into a molecule of E_2, then the mixture has a new final free energy equal to

$$G_f = (n_1 - 1)G_1 + (n_2 + 1)G_2 + (n_1 + n_2)kT \left\{ \frac{n_1 - 1}{n_1 + n_2} \ln \frac{n_1 - 1}{n_1 + n_2} + \frac{n_2 + 1}{n_1 + n_2} \ln \frac{n_2 + 1}{n_1 + n_2} \right\}$$

The difference in free energy is

$$\Delta G_{S \to P} = G_f - G_i$$
$$= -G_1 + G_2 + kT \left\{ n_1 \ln \frac{n_1 - 1}{n_1} + n_2 \ln \frac{n_2 - 1}{n_2} + \ln \frac{n_2 + 1}{n_1 - 1} \right\}$$
$$\cong \Delta G^0 + kT \left\{ -n_1 \cdot \frac{1}{n_1} + n_2 \cdot \frac{1}{n_2} + \ln \frac{n_2}{n_1} \right\}$$
$$= \Delta G^0 + kT \ln \frac{[E_2]}{[E_1]}$$

6.1 Bending and Buckling

Energy of Bending

Consider a short segment of length Δs bent by the force shown in Figure A6.1. For convenience, denote the curvature ($d\theta/ds$) by ρ. The bending moment, $M(\rho)$ equals $EI\rho$, by the beam equation. The potential energy is the integral of the moment with respect to the curvature—this is analogous to the energy being the integral of the force with respect to displacement. Thus the bending energy is

$$\Delta U = \left[\int_0^{\rho_0} EI\rho \cdot d\rho\right]\Delta s = \tfrac{1}{2}EI\rho_0^2 \Delta s = \tfrac{1}{2}EI\left(\frac{d\theta}{ds}(s_0)\right)^2 \Delta s$$

Taking the limit for small segment length gives

$$\frac{dU}{ds} = \tfrac{1}{2}EI\left(\frac{d\theta}{ds}\right)^2$$

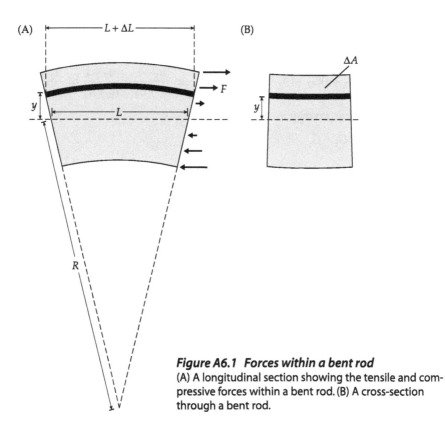

Figure A6.1 Forces within a bent rod
(A) A longitudinal section showing the tensile and compressive forces within a bent rod. (B) A cross-section through a bent rod.

Forces in a Bent Rod

When a rod is bent, the external bending moment is balanced by the bending moments generated by the compressive and tensile forces within the material (see Figure A6.1A). At a distance y above the midplane, the extension in the shaded section is Δl, and from the geometry $\Delta l/l = y/R$. On the other hand, we know that the strain is proportional to the stress (the force per unit area) according to $\Delta F/\Delta A = E \cdot \Delta l/l = E \cdot y/R$, where E is the Young's modulus (Equation 3.2). This stress generates a moment $y\Delta F$. The total moment acting on the cross section is the integral of all these moments over the cross-section of the rod (see Figure A6.1B)

$$M = \int_{area} dM = \int_{area} y \cdot dF = \int_{area} y \frac{dF}{dA} \cdot dA = \frac{E}{R} \int_{area} y^2 \cdot dA$$

This is identical to the beam equation (Equation 6.1) provided that I is given by Equation 6.2.

Principal Axes of Bending

For an arbitrary cross-sectional area, there exist two orthogonal axes such that if I_1 is the moment about one axis, and I_2 is the moment about the other, then the moment about an axis at angle ϕ relative to these principal axes is

$$I(\phi) = I_1 \cos^2 \phi + I_2 \sin^2 \phi$$

To see this, we make the following definitions. The center of mass (x_0, y_0) is

$$x_0 = \frac{\int x \cdot dA}{A} \quad \text{and} \quad y_0 = \frac{\int y \cdot dA}{A}$$

where A is the area. The "variances" are

$$\sigma_x^2 = \int x^2 dA - x_0^2 A \qquad \sigma_y^2 = \int y^2 dA - y_0^2 A \qquad \sigma_{xy}^2 = \int xy \cdot dA - x_0 y_0 A$$

Define one principal axis as passing through (x_0, y_0) at an angle ϕ_0, equal to

$$\phi_0 = \tfrac{1}{2} \tan^{-1}\left(\frac{\sigma_{xy}^2}{\sigma_x^2 + \sigma_y^2}\right)$$

The second moment of inertia with respect to an axis at angle ϕ relative to this axis is

$$I(\phi) = \int [(x - x_0)\cos(\phi + \phi_0) + (y - y_0)\sin(\phi + \phi_0)]^2 dA$$

The moments about the principal axes are $I_1 = I(0)$ and $I_2 = I(90)$, and in general,

$$I(\phi) = \cos^2(\phi+\phi_0)\sigma_x^2 + \sin^2(\phi+\phi_0)\sigma_y^2 + 2\cos(\phi+\phi_0)\sin(\phi+\phi_0)\sigma_{xy}^2$$

$$= \cos^2\phi \cdot \left[\cos^2\phi_0 \cdot \sigma_x^2 + \sin^2\phi_0 \cdot \sigma_y^2 + 2\cos\phi_0\sin\phi_0 \cdot \sigma_{xy}^2\right]$$

$$+ \sin^2\phi \cdot \left[\sin^2\phi_0 \cdot \sigma_x^2 + \cos^2\phi_0 \cdot \sigma_y^2 - 2\cos\phi_0\sin\phi_0 \cdot \sigma_{xy}^2\right]$$

$$+ 2\cos\phi\sin\phi\left[-\cos\phi_0\sin\phi_0 \cdot (\sigma_x^2+\sigma_x^2) + (\cos^2\phi_0 - \sin^2\phi_0)\sigma_{xy}^2\right]$$

$$= \cos^2\phi \cdot I_1 + \sin^2\phi \cdot I_2 + 0$$

as required. Note that symmetry axes are also principal axes because the covariance σ_{xy}^2 about a symmetry axis is zero.

Second Moment of a Helical Rod

A helical rod is one in which the cross-section rotates at a constant rate from one end to the other. An actin filament is an example of a helical rod. The second moment of inertia of a helical rod is the geometric mean of the moments about the rod's principle axes. To see this, let ϕ be the angle that one principal axis makes with respect to the coordinate system. Now suppose that a moment M acts on the rod. Then at each angle ϕ, the curvature $\rho(\phi)$ is $M/EI(\phi)$, where $I(\phi)$ is given in the previous section. If the rod makes one helical turn over a unit length, then the bending energy is

$$U = \int_0^1 \tfrac{1}{2} E \cdot I(s) \cdot \rho^2(s) \cdot ds = \frac{E}{4\pi}\int_0^1 I(\phi) \cdot \rho^2(\phi) \cdot d\phi$$

$$= \frac{M^2}{4\pi E}\int_0^1 \frac{d\phi}{I_1\cos^2\phi + I_2\sin^2\phi} = \frac{M^2}{2E\sqrt{I_1 \cdot I_2}}$$

Comparing this to the expression for a rod of constant second moment I, we deduce that $I = \sqrt{(I_1 \cdot I_2)}$ as required.

Small Buckles

In Figure 6.4A, the force on the right-hand end induces a clockwise, negative moment on the point at (x,y) equal to $-Fy$. The small-angle beam equation becomes

$$\frac{d^2y}{dx^2} = -\frac{F}{EI}y$$

A solution is $y(x) = y_{max}\sin(\pi x/L)$, which satisfies

$$\frac{d^2y}{dx^2} = -\frac{\pi^2}{L^2}y$$

This has the same form as the previous equation provided that we equate the constants. This gives the Euler force of Equation 6.6. Note that the force is independent of y_{max}; this means that all amplitudes are valid solutions to the equation.

Large Buckles

Consider Figure 6.4E. In vector notation, $M = (r_0 - r) \times F$, where $r = (x,y)$, $r_0 = (x_0, y_0)$ and $\times$ is the cross product. The beam equation can be written more simply if we differentiate it to give

$$\frac{d^2\theta}{ds^2} = \frac{1}{EI}\frac{dM}{ds} = \frac{1}{EI}F \times \frac{dr}{ds} = \frac{1}{EI}F \times t = -\frac{|F|}{EI}\sin[\theta(s) - \varphi_0]$$

where $t = (dx/ds, dy/ds) = (\cos\theta, \sin\theta)$ is the tangent vector.

6.2 Hydrodynamics of Slender Rods

Drag on a Bending Filament

In the small-angle approximation, the total bending moment at point x due to the drag forces on the right-hand side of the point is

$$M(x) = \int_x^L f_\perp(x')(x'-x) \cdot dx'$$

where $f_\perp$ is the drag force per unit length (see Figure 6.5). Differentiating twice with respect to x gives

$$\frac{\partial^2 M}{\partial x^2} = f_\perp(x) = -c_\perp v_\perp = -c_\perp \frac{\partial y}{\partial t}$$

where $c_\perp$ is the drag coefficient per unit length. Substitution into the beam equation (Equation 6.4), gives the **hydrodynamic beam equation**:

$$\frac{\partial^4 y}{\partial x^4} = \frac{1}{EI}\frac{d^2 M}{dx^2} = -\frac{c_\perp}{EI}\frac{\partial y}{\partial t}$$

Hydrodynamic Bending Modes

The hydrodynamic beam equation can be solved for appropriate initial conditions and boundary conditions. The boundary conditions for an unconstrained filament are that both $\partial^2 y/\partial x^2$ and $\partial^3 y/\partial x^3$ are zero at both ends ($x = 0$ and $x = L$). The differential equation is solved by separation of variables to give a set of modes

$$y_n(x,t) = e^{-t/\tau_n}\left[\sinh\alpha_n \cos\frac{2\alpha_n}{L}(x-\frac{L}{2}) - \sin\alpha_n \cosh\frac{2\alpha_n}{L}(x-\frac{L}{2})\right] \quad n \text{ odd}$$

$$y_n(x,t) = e^{-t/\tau_n}\left[\cosh\alpha_n \sin\frac{2\alpha_n}{L}(x-\frac{L}{2}) + \cos\alpha_n \sinh\frac{2\alpha_n}{L}(x-\frac{L}{2})\right] \quad n \text{ even}$$

where

$$\tau_n = \frac{c_\perp}{EI}\left(\frac{L}{2\alpha_n}\right)^4 \qquad \tan\alpha_n = (-1)^n \tanh\alpha_n \qquad n = 1, 2, 3, \ldots$$

Approximate values for the constants α_n are

$$\alpha_n \cong (n+\tfrac{1}{2})\frac{\pi}{2} \qquad n = 1, 2, 3, \ldots \quad (\alpha_1 \approx 2.365)$$

Each mode, as well as any linear combination of modes, satisfies the differential equation as well as the boundary conditions.

Dynamics of a Cantilever Spring

Consider a slender fiber clamped at $x = 0$ and either free at $x = L$ or attached to a spring of stiffness κ. For small deflections, the fiber satisfies the small-angle beam equation with boundary conditions $y(0) = 0$, $y'(0) = 0$, $y''(L) = 0$; $y'''(L) = 0$ if the end is free, or $y'''(L) = \kappa y(L)/EI$ if the end is attached to the spring.

The solutions to the hydrodynamic beam equation are

$$y_n(x) = \exp\left(-\frac{t}{\tau_n}\right)\left\{\cos\frac{\beta_n x}{L} - \cosh\frac{\beta_n x}{L} + A_n\left(\sin\frac{\beta_n x}{L} - \sinh\frac{\beta_n x}{L}\right)\right\}$$

where

$$\tau_n = \frac{c_\perp \cdot L^4}{EI \cdot \beta_n^4} \quad \text{and} \quad A_n = -\frac{\cos\beta_n + \cosh\beta_n}{\sin\beta_n + \sinh\beta_n}$$

For the free-end boundary condition ($\kappa = 0$), $\beta_n(0)$ satisfies

$$\cos\beta_n \cdot \cosh\beta_n = -1 \qquad \beta_n \cong (n-\tfrac{1}{2})\pi \qquad n = 1, 2, 3, \ldots \quad \beta_1(0) \cong 1.875$$

This solution correctly predicts the transient motion of a free fiber: For example, if the base is moved very rapidly, then the model predicts that the free end actually begins to move in the opposite direction, just like the whiplash observed for a real probe (Howard, unpublished).

For the attached boundary condition, $\beta_n(\kappa)$ satisfies

$$\frac{(\beta_n)^3}{3}\frac{1+\cos\beta_n \cdot \cosh\beta_n}{\cos\beta_n \cdot \sinh\beta_n - \sin\beta_n \cdot \cosh\beta_n} = \frac{\kappa L^3}{3 \cdot EI} = \frac{\kappa}{\kappa_f}$$

where $\kappa_f = 3EI/L^3$ is the stiffness of the fiber. By differentiating the above equation and using the approximations

$$\beta_1(\kappa) \cong \beta_1(0) + \kappa\frac{d\beta_1}{d\kappa}(0) \quad \text{and} \quad \frac{\tau_1^{\text{free}}}{\tau_1^{\text{attached}}} = \left(\frac{\beta_1(\kappa)}{\beta_1(0)}\right)^4 \cong 1 + 4\frac{\kappa}{\beta_1(0)}\frac{d\beta_1}{d\kappa}(0)$$

it follows that if $\kappa \ll \kappa_f$, then

$$\frac{\tau_1^{free}}{\tau_1^{attached}} \cong 1 + \frac{12}{\beta_1^4(0)}\frac{\kappa}{\kappa_f} \cong 1 + 0.96\frac{\kappa}{\kappa_f} \cong \frac{\kappa_f + \kappa}{\kappa_f}$$

In other words, the fiber time constant is approximately inversely proportional to the "total" stiffness, $\kappa + \kappa_f$, at the unclamped end.

On the other hand, if $\kappa \gg \kappa_f$, then

$$\beta_n(\kappa) \cong (n + \tfrac{1}{4})\pi \qquad n = 1, 2, 3, \ldots$$

This is interesting because it shows that the elastic load has little effect on the kinetics of the higher modes. Even on the first mode, a very large elastic load only speeds up the relaxation about 20-fold ($= \{\beta_1(\infty)/\beta_1(0)\}^4$). Thus when the elastic load is large, our picture of the cantilever acting as a simple spring in parallel with the load breaks down.

Dynamics of Buckling

In the presence of a compressive force, F, the hydrodynamic beam equation becomes

$$EI\frac{\partial^4 y}{\partial x^4} + F\frac{\partial^2 y}{\partial x^2} = -c_\perp \frac{\partial y}{\partial t}$$

Let $z(x) = z_0 \sin(\pi x/L)$ be a static solution to the beam equation

$$\frac{d^2 z}{dx^2}(x) = -\frac{F_c}{EI} z(x)$$

where F_c is the critical force (Equation 6.6). By substitution, it can be shown that $y(x,t) = z(x)\exp(t/\tau)$ is a solution of the hydrodynamic beam equation provided the time constant τ satisfies

$$\tau = \frac{c_\perp EI}{F_c(F - F_c)}$$

This means that if the compressive force exceeds the critical force, F_c, and the rod begins to buckle, as, for example, through a thermal fluctuation, then the buckle will increase exponentially with the stated time constant. On the other hand, if the force is less than the critical force, then the negative time constant means that the buckle will die out, and the rod will straighten out again.

6.3 Thermal Bending

Derivation of the Persistence Length

Let $f(s) = \langle \cos[\theta(s)] \rangle$, where θ is the tangent angle of a filament constrained to lie in a plane. We assume that $\theta(0) = 0$, without loss of generality. Then

$$\frac{df}{ds}\Delta s \cong f(s+\Delta s) - f(s)$$
$$= \langle \cos[\theta(s+\Delta s)] - \cos[\theta(s)] \rangle$$
$$= \langle \cos[\theta(s+\Delta s) - \theta(s) + \theta(s)] - \cos[\theta(s)] \rangle$$
$$= \langle \cos[\Delta\theta + \theta(s)] \rangle - \langle \cos[\theta(s)] \rangle$$

where $\Delta\theta = \theta(s+\Delta s) - \theta(s)$. Now the tangent angles $\theta(s)$ and $\Delta\theta$ are statistically independent because the thermal forces on the segment $(s, s+\Delta s)$ are independent of the thermal forces acting on $(0, s)$. Hence

$$\frac{df}{ds}\Delta s \cong \langle \cos[\theta]\cos[\Delta\theta] - \sin[\theta]\sin[\Delta\theta] \rangle - \langle \cos[\theta] \rangle$$
$$= \langle \cos[\theta] \rangle \cdot \langle \cos[\Delta\theta] \rangle - \langle \sin[\theta] \rangle \cdot \langle \sin[\Delta\theta] \rangle - \langle \cos[\theta] \rangle$$
$$= \langle \cos[\theta] \rangle \cdot \{ \langle \cos[\Delta\theta] \rangle - 1 \}$$

where the last line follows because the average values of the sine terms are zero (the angles are equally likely to be positive or negative). Thus

$$\frac{df}{ds} \cong \frac{\langle \cos[\Delta\theta] - 1 \rangle}{\Delta s} f(s)$$
$$\cong -\tfrac{1}{2} \left\langle \frac{\Delta\theta^2}{\Delta s} \right\rangle f(s)$$
$$= -\tfrac{1}{2} \left\langle \left(\frac{\Delta\theta}{\Delta s}\right)^2 \Delta s \right\rangle \cdot f(s)$$
$$= -\frac{\langle \Delta U \rangle}{EI} \cdot f(s)$$
$$= -\tfrac{1}{2} \frac{kT}{EI} f(s)$$

where we used the energy of bending in the penultimate step and the Equipartition principle in the last step. Solving this differential equation then gives

$$\langle \cos[\theta(s) - \theta(0)] \rangle = \exp\left(-\frac{kTs}{2EI}\right)$$

as required (Equation 6.10). In three dimensions, there are two independent angular degrees of freedom, so the mean energy in the short segment is kT, rather than $\tfrac{1}{2}kT$, and the cosine of the angle decays twice as quickly (Landau et al., 1980).

End-to-End Length

Let R be the root-mean-squared end-to-end length *in three dimensions* of a slender rod of persistence length L_p. Let $t(s)$ be the tangent vector at point s. If R

is the end-to-end vector, then following Landau et al. (1980), the mean-squared end-to-end distance is

$$\langle R^2 \rangle = \langle R \cdot R \rangle = \left\langle \left[\int_0^L t(s) ds \right]^2 \right\rangle = \left\langle \left[\int_0^L t(s) ds \right] \cdot \left[\int_0^L t(s') ds' \right] \right\rangle$$

$$= \int_0^L \int_0^L \langle t(s) \cdot t(s') \rangle ds ds' = \iint_{s>s'} \langle t(s) \cdot t(s') \rangle ds ds' + \iint_{s'>s} \langle t(s) \cdot t(s') \rangle ds ds'$$

$$= 2 \int_{s=0}^L \int_{s'=s}^L \langle t(s) \cdot t(s') \rangle ds' ds = 2 \int_{s=0}^L \int_{s'=s}^L \langle \cos[\theta(s') - \theta(s)] \rangle ds' ds$$

$$= 2 \int_{s=0}^L \int_{s'=s}^L \exp\left(-\frac{s'-s}{L_p}\right) ds' ds = 2 L_p^2 \left\{ \exp\left(-\frac{L}{L_p}\right) - 1 + \frac{L}{L_p} \right\}$$

Thermal Fluctuations in Shape

The shape of a slender filament, $\theta(s)$, can be expressed as the superposition of Fourier modes

$$\theta(s) = \sqrt{\frac{2}{L}} \sum_{n=0}^\infty a_n \cos\frac{n\pi s}{L} \quad \text{where} \quad a_n = \sqrt{\frac{2}{L}} \int_0^L \theta(s) \cos\frac{n\pi s}{L} \cdot ds \quad n \geq 1$$

The energy of bending, U, is a quadratic sum of the mode amplitudes

$$U = \int_0^L \tfrac{1}{2} EI \left(\frac{d\theta}{ds} - \frac{d\theta^0}{ds}\right)^2 \cdot ds = \tfrac{1}{2} EI \sum_{n=1}^\infty \left(\frac{n\pi}{L}\right)^2 \left(a_n - a_n^0\right)^2$$

where $\theta^0(s)$ is the shape in the absence of external forces and a_n^0 is the amplitude of the mode in the absence of an external force. By the Equipartition principle, each mode has $\tfrac{1}{2} kT$ of thermal energy in it, so that

$$\text{Var}(a_n) = \langle a_n^2 \rangle - \langle a_n^0 \rangle^2 = \frac{kT}{EI}\left(\frac{L}{n\pi}\right)^2 = \frac{1}{L_p}\left(\frac{L}{n\pi}\right)^2 \quad \text{for } n \geq 1$$

Thus measurement of the variance of the mode amplitude, together with knowledge of the filament length, provides an estimate of the persistence length (Gittes et al., 1993).

For a cantilevered beam, the Equipartition principle implies that the mean-square deflection of the free end, $\langle y(L)^2 \rangle = kT/\kappa = L^3/3L_p$. On the other hand, the problem could have been analyzed using the modal approach in which the shape is decomposed into fluctuating cosine modes each of which has energy $\tfrac{1}{2} kT$. The two approaches should give the same mean-square deflection, and indeed they do. For, without loss of generality, let $\theta(0)=0$. Then

$$a_0 = -\sum_{n=1}^\infty a_n$$

Furthermore,

$$y(L) = \int_0^L \sin\theta(s) \cdot ds \cong \int_0^L \theta(s) \cdot ds = \sqrt{2L} \cdot a_0$$

Thus

$$\langle y(L)^2 \rangle = 2L\langle a_0^2 \rangle = 2L\Big\langle \sum_{n=1}^{\infty} a_n \sum_{m=1}^{\infty} a_m \Big\rangle = 2L \sum_{n,m=1}^{\infty} \langle a_n a_m \rangle$$

$$= 2L \sum_{n=1}^{\infty} \langle a_n^2 \rangle = \frac{2L^3}{\pi^2 L_p} \sum_1^{\infty} \frac{1}{n^2} = \frac{2L^3}{\pi^2 L_p} \frac{\pi^2}{6} = \frac{L^3}{3L_p}$$

and the two approaches are in agreement.

The Langevin Function

Consider one freely jointed rod whose left-hand end is fixed but still able to swivel freely. In the absence of an external force, the free end of the segment is equally likely to be anywhere on the surface of the sphere. In the presence of a horizontal force, rotation of the rod through an angle θ is associated with a potential energy $U = -Fb\cos\theta$. It follows from Boltzmann's law that the probability of the free end being in the annular strip defined by $(\theta, \theta + d\theta)$ is proportional to the area of the strip, $2\pi\sin\cdot d\theta$, multiplied by the Boltzmann factor $\exp(-U/kT)$. Thus we have

$$p(\theta) \cdot d\theta = \frac{1}{Z} \exp\left(-\frac{U(\theta)}{kT}\right) \sin\theta \cdot d\theta = \frac{1}{Z} \exp\left(\frac{Fb \cdot \cos\theta}{kT}\right) \sin\theta \cdot d\theta$$

where

$$Z = \int_0^\pi p(\theta) \cdot d\theta = \int_{-1}^{1} \exp(xy) dy = \frac{1}{x}\left(e^x - e^{-x}\right)$$

with $x = Fb/kT$. We can now calculate $\langle \cos\theta \rangle$

$$\langle \cos\theta \rangle = \int_0^\pi \cos\theta \cdot p(\theta) \cdot d\theta = \frac{1}{Z}\int_0^\pi \cos\theta \cdot \exp\left(\frac{Fb}{kT}\cos\theta\right) \sin\theta \cdot d\theta$$

$$= \frac{\int_{-1}^{1} y \exp(xy) \cdot dy}{\int_{-1}^{1} \exp(xy) \cdot dy}$$

$$= \frac{e^x + e^{-x}}{e^x - e^{-x}} - \frac{1}{x} \equiv L(x)$$

This is the Langevin function of Figure 6.10.

If there are n segments, then the total extension is $X = nb\langle\cos\theta\rangle$ and Equation 6.14 follows.

Figure A6.2 Mean-squared end-to-end distance

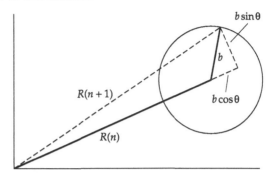

Mean-Squared End-to-End Length of a Freely Jointed Chain

Let $\langle R^2(n) \rangle$ be the mean-squared end-to-end length of a freely-jointed chain with n segments in the absence of a force. With the aid of Figure A6.2 it follows that

$$\langle R^2(n+1) \rangle = \langle [R(n)+b\cos\theta]^2 + [b\sin\theta]^2 \rangle = \langle R^2(n) \rangle + 2b\langle R(n) \rangle \cdot \langle \cos\theta \rangle + b^2$$
$$= \langle R^2(n) \rangle + b^2$$

Because $\langle R^2(1) \rangle = b^2$, it follows by induction that $\langle R^2(n) \rangle = nb^2$.

8.1 A Pivotal Spring

The stiffness of a "pivotal spring," a spring in which a rigid rod rotates about a flexible connection at a fixed point (the pivot point), is equal to

$$K = K_r/L^2$$

where L is the height above the pivot at which the stiffness is measured, and K_r is the pivotal or rotational stiffness (Howard and Ashmore, 1986).

This follows from considering Figure A8.1. If the application of a force F to the right-hand end of the pivoting bar moves it through a distance X, then the stiffness is $K = F/X$. The total energy of the deflection is

$$U = \tfrac{1}{2} K \cdot X^2 = \tfrac{1}{2} \kappa \cdot h^2$$

Hence

$$K = \kappa \frac{h^2}{X^2} = \kappa \frac{l^2}{L^2} = (\kappa \cdot l^2)/L^2 = K_r/L^2$$

A8.1 A pivotal spring

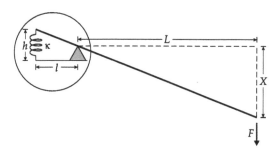

as required. We idealize a pivotal spring by making the circle very small while maintaining a constant K_r (i.e., make κ large as l gets small).

8.2 Stiffness of a Stereocilium

Consider the cantilevered beam shown in Figure A8.2, which has a narrow base of radius a and length l. This is a model for a stereocilium (Crawford and Fettiplace, 1985; Howard and Ashmore, 1986). Suppose that the distal region is much longer (length $L \gg l$) and more rigid than the base. In the basal region we may write the beam equation (Equation 6.4)

$$\frac{d^2 y}{dx^2}(x) = \frac{M(x)}{E_1 I_1} \cong \frac{FL}{E_1 I_1}$$

where $E_1 I_1$ is the flexural rigidity of the base, and using the approximation $L \gg l$. If the distal portion is much more rigid than the base, then from the geometry

$$\frac{X}{L} = \frac{dy}{dx}(l) = \int_0^l \frac{d^2 y}{dx^2} dx = \int_0^l \frac{FL}{E_1 I_1} dx = \frac{FLl}{E_1 I_1}$$

Upon rearrangement, we obtain

$$\kappa = \frac{F}{X} = \frac{E_1 I_1}{l L^2}$$

If the base is composed of n crosslinked filaments of radius r, then the filaments occupy a fraction $f = n \cdot r^2 / a^2$ of the cross-sectional area and the flexural rigidity of the base is

$$E_1 I_1 = f E \cdot I_1 = \frac{n r^2}{a^2} E \frac{\pi a^4}{4} = n E \frac{\pi r^4}{4} \frac{r^2}{a^2} = n E I \frac{r^2}{a^2}$$

where EI is the flexural rigidity of a single filament. If the basal filaments are crosslinked, the stiffness is

$$\kappa = n \frac{EI}{l L^2} \frac{a^2}{r^2}$$

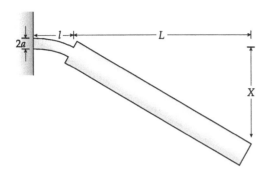

A8.2 A cantilevered beam with a flexible base

If the filaments are not crosslinked, the flexural rigidity of the base is simply n times the flexural rigidity of the individual filaments, or

$$\kappa = n \frac{EI}{lL^2}$$

8.3 Damping on Hair Bundles

It is interesting to ask whether the motion of a hair bundle is overdamped or underdamped. We model the hair bundle as a collection of N stereocilia that pivot about a basal region, each having a pivotal stiffness, κ_r (equal to 0.5×10^{-15} N·m·rad^{-1} for a frog saccular hair bundle, Howard and Ashmore 1986). By analogy to the damped mass on a spring, the equation of motion for rotation about a pivot point is

$$M_r \ddot{\theta} + \Gamma_r \dot{\theta} + K_r \theta = T$$

(Crawford and Fettiplace, 1985), where T is the torque, θ is the angle, $M_r = \frac{1}{3} \rho A N L^3$ is the moment of inertia, $\Gamma_r = \frac{1}{3} c_\perp L^3 = \frac{4}{3} \pi \cdot \eta \cdot c(N, L) \cdot L^3$ is the rotational drag, and $K_r = N\kappa_r$ is the rotational stiffness. ρ is the density ($\sim 10^3$ kg/m^3), A is the cross-sectional area occupied by each stereocilium (0.25 μm^2 corresponding to a 0.53 μm spacing on a hexagonal lattice; Howard and Ashmore, 1986), and $c(N, L)$ is a term that depends only weakly on N and L and is approximately equal to 1. As we saw in Chapters 2 and 3, the motion will be overdamped or underdamped depending on whether the damping ratio is less than or greater than unity. In this case the ratio is

$$\frac{4M_r K_r}{\Gamma_r^2} = \frac{3\rho A \kappa_r}{2\pi^2 \eta^2 c^2} \frac{N^2}{L^3}$$

If the motion is underdamped (i.e., oscillatory), then the resonant frequency in radians/s is

$$\omega \approx \sqrt{\frac{K_r}{M_r}} = \sqrt{\frac{3\kappa_r}{\rho A}} L^{-3/2}$$

There are a number of interesting implications of these equations.

1. The damping ratio increases as the length decreases. This is the opposite to the case of globular materials where the damping ratio decreases as the length decreases (Chapter 3). This raises the possibility that small hair bundles could resonate, even in fluid.
2. Frog saccular hair bundles are expected to be overdamped; with $N = 50$ stereocilia and height $L = 8$ μm (see Figure 8.2), the damping ratio is 0.05. This agrees with experiment (see "Bending Stiffness of Actin in Stereocilia" in Chapter 8).
3. The short hair bundles from mammalian cochleas that respond to high frequencies of sound might be underdamped. For example, the echolocating bats *Tadarida brasiliensis* (Vater and Siefer, 1995) and *Hipposideros bicolor* (Dannhof and Bruns, 1991) have sonar systems operating at 20 to 80 kHz and 150 to 200 kHz, respectively; the hair cells that respond to these sound frequencies have bundles comprising up to 100 stereocilia of length 2 μm (inner hair cells) and ~1 μm (outer hair cells). The corresponding damping ratios are about 10 and 100 (indicating that the motion is expected to be underdamped) and the predicted resonant frequencies are ~135 kHz and 400 kHz (using the other parameters measured in the frog sacculus). Thus hair bundles are possible exceptions to the rule that cellular structures are overdamped.

The hair cells at the basal (high frequency) end of the human cochlea have stereocilia with lengths of 4 μm (inner hair cells) and 2.5 μm (outer hair cells) (Wright, 1984). The predicted damping ratios are 1.6 and 6 (indicating that there is a possibility of resonance), and the predicted resonance frequencies are 50 kHz and 100 kHz (again assuming parameter values measured in the frog). On the other hand, the apical (low frequency) hair cells have stereocilia of height ~7 μm (both inner and outer hair cells) and are very likely to be overdamped like the frog saccular hair bundles.

These calculations show that it is plausible that mechanical properties of hair bundles may play a role in frequency tuning in the mammalian cochlea, at least at high frequencies. The exact frequency of resonance could be tuned by tuning the stiffness or spacing of the stereocilia within the bundle. This analysis undercuts the argument of Gold (1948) that the damping by fluids in the ear is too high for small structures such as hair bundles to undergo "passive" resonance. Thus the sharp mechanical tuning of the basilar membrane (Ruggero et al., 1997) and corresponding sharp frequency tuning of cochlear hair cells (Pickles, 1988) does not necessarily imply the existence of an "active process" that uses metabolic energy to provide highly tuned responses as postulated by Gold. The present analysis does not deny the existence of such an active process, termed the cochlear amplifier; indeed, there are several other arguments in its favor (see, e.g., Camalet et al., 2000; Eguíluz et al., 2000). Instead, the analysis cautions against leaping to conclusions, and it highlights the need for more high-resolution mechanical experiments on the cochlear cells and tissues.

9.1 Single-Stranded Polymer

It is easy to confirm by substitution that the solution to Equation 9.1 is

$$a_n = a_1^n = \exp\left(-\frac{n}{n_0}\right) \quad \text{where } a_n = \frac{[A_n]}{K}, \; n_0 = -\frac{1}{\ln a_1} \quad n \geq 1$$

This means that there is an exponential distribution of polymer lengths. The average length of the polymers (not counting the monomers) is

$$n_{av} = \sum_2^\infty n p_n = \sum_2^\infty n \frac{a_n}{\sum_2^\infty a_n} = 1 + \frac{1}{1-a_1}$$

The total number of subunits is

$$a_t = \sum_1^\infty n a_n = \frac{a_1}{(1-a_1)^2}$$

The concentration of monomers therefore depends on the total concentration of subunits according to

$$a_1 = 1 + \frac{1}{2a_t} - \sqrt{\frac{1}{a_t} + \frac{1}{4a_t^2}}$$

When $a_t \gg 1$, we have

$$a_1 \cong 1 - \frac{1}{\sqrt{a_t}} \qquad n_0 \cong \sqrt{a_t} \qquad n_{av} \cong 1 + \sqrt{a_t}$$

9.2 Two-Stranded Polymer

Length

The two-stranded solution is

$$a_n = \frac{K}{K_1} a_1^n = \frac{K}{K_1} \exp\left(-\frac{n}{n_0}\right) \quad n \geq 3$$

$$a_2^* = \frac{K}{K_1} a_1^2 \qquad a_2^{**} = \frac{K}{K_2} a_1^2$$

The total number of subunits is

$$a_t = a_1 + \frac{K}{K_2} a_1^2 + \frac{K}{K_1} \frac{a_1^2(2-a_1)}{(1-a_1)^2}$$

The average length of the polymers (not counting monomers or the nuclei a_2^* and a_2^{**}) is

$$n_{av} = 2 + \frac{1}{(1-a_1)} \cong \sqrt{\frac{K_1}{K}} \sqrt{a_t}$$

where the approximation holds when $a_t \gg 1$. Under this condition $a_1 \cong 1$ and $a_n \cong \frac{K}{K_1}$ provided $n \ll n_0$.

Annealing and Breakage

Consider the annealing/breakage reaction:

$$A_n + A_m \underset{k_{off,m}}{\overset{k_{on,m}}{\longleftrightarrow}} A_{n+m} \qquad \frac{[A_n]\cdot[A_m]}{[A_{n+m}]} = \frac{K^2}{K_1} = \frac{k_{off,m}}{k_{on,m}}$$

It is expected that the second-order association rate for annealing of an m-mer ($k_{on,m}$) is smaller than the corresponding rate for monomer addition (k_{on}) because the polymer will diffuse more slowly than the monomer. Thus we expect

$$k_{off,m} = \frac{K^2}{K_1} k_{on,m} \leq \frac{K^2}{K_1} k_{on} = \frac{K}{K_1} k_{off}$$

10.1 Reaction–Diffusion Equation for the Brownian Ratchet Model

Let there be a fixed number of nucleation sites for the polymers. Let $p_n(x,t)$ be the probability, at time t, that a polymer is an n-mer with a gap between its tip and the object equal to x. The probability, $p_n(x,t)$, satisfies the following differential equation:

$$\frac{\partial p_n(x,t)}{\partial t} = -\frac{\partial J_n}{\partial x}(x,t) - k_{on}[A_1]\left[p_n(x,t)H(x-\delta) - p_{n-1}(x+\delta,t)\right]$$
$$+ k_{off}\left[p_{n+1}(x-\delta,t) - p_n(x,t)\right] \qquad n \geq 1 \quad \text{(A10.1)}$$

$$\frac{\partial p_0(x,t)}{\partial t} = -\frac{\partial J_0}{\partial x}(x,t) - k_{on}[A_1]\left[p_0(x,t)H(x-\delta)\right] + k_{off}\left[p_1(x-\delta,t)\right]$$

$$J_n(x,t) = -D\frac{\partial p_n}{\partial x}(x,t) - \frac{FD}{kT}p_n(x,t) \qquad n \geq 0$$

for $x \geq 0$ (Peskin et al., 1993). The case $n = 0$ corresponds to an empty polymerization site. This reaction–diffusion equation provides the link between the chemical and physical forces. The flux component, J, arises from diffusion of the object. The two reaction terms arise from the transitions associated with subunit addition and subtraction. The wrinkle is that we have included an x dependence because in our picture the addition reactions can only take place provided that the gap is larger than δ. Hence the inclusion of the "Heavyside step function," $H(x - \delta)$, which is 0 when $x < \delta$ and 1 when $x > \delta$. Furthermore, because the addition of a subunit partially fills the gap, subunit addition and subtraction change the gap width: this is why $x + \delta$ and $x - \delta$ appear in the p_{n-1} and p_{n+1} terms. The reaction–diffusion equation is a complete description of the polymerization process in the presence of external and thermal forces. $[A_1]$ is the monomer concentration, assumed to be a constant. F is the force, taken as positive when it opposes polymerization (i.e., when it is compressive).

The equation simplifies considerably because in the Peskin model, the diffusion of the gap is independent of the size of the polymer. Furthermore, we seek a solution in which the average gap is time independent: That is, we consider the steady-state growth of the polymer. This means that we can write:

$$p_n(x,t) = P_n(t) \cdot p(x) \qquad \int_0^\infty p(x) \cdot dx = 1 \qquad \text{(A10.2)}$$

where the gap size, $p(x)$, is independent of n and t. In this case, $P_n(t)$ is the total probability of there being an n-mer and we must have

$$P_n(t) \equiv \int_0^\infty p_n(x,t) \cdot dx \qquad \sum_{n=0}^\infty P_n(t) = 1 \quad \text{and} \quad n_{\text{av}}(t) = \sum_{n=0}^\infty n P_n(t) \qquad \text{(A10.3)}$$

where n_{av} is the average polymer length. A few comments need to be made about this scenario. First, steady state requires that the monomer concentration be constant; otherwise, depletion of monomers would slow the growth. Steady state therefore necessitates either an infinite pool of monomers, or replacement of monomers that add on to the polymers. Second, we assume that there is a fixed (large) number of nucleation sites and that growth occurs equally well on all sites. After the polymers have grown for a while, we expect the number of empty sites—and therefore $P_0(t)$—to be very small.

Substituting Equation A10.2 into Equation A10.1 and integrating with respect to x gives

$$\frac{dP_n}{dt}(t) = [j(0) - j(\infty)]P_n(t) - k_{\text{on}}[A_1][P_n(t) - P_{n-1}(t)] \cdot p_\delta \qquad n \geq 1 \qquad \text{(A10.4)}$$
$$+ k_{\text{off}}[P_{n+1}(t) - P_n(t)]$$

$$\frac{dP_0}{dt}(t) = [j(0) - j(\infty)]P_0(t) - k_{\text{on}}[A_1]P_0(t) \cdot p_\delta + k_{\text{off}} P_1(t)$$

where the gap flux, the rate at which the number of gaps of size greater than x increases, is

$$j(x) = -D\frac{dp}{dx}(x) - \frac{FD}{kT}p(x) \qquad \text{(A10.5)}$$

and p_δ is the probability that the gap is greater than δ:

$$p_\delta \equiv \int_\delta^\infty p(x) \cdot dx \qquad \text{(A10.6)}$$

A subtle argument shows that the gap flux term in Equation A10.4 vanishes; summing Equation A10.4 over n gives:

$$0 = \frac{d}{dt}\sum_{n=0}^{\infty} P_n(t) = \sum_{n=0}^{\infty} \frac{dP_n}{dt}(t) = [j(0) - j(\infty)] \quad \text{(A10.7)}$$

where we have used $P_n \to 0$ as $n \to \infty$. Now, because $j(\infty) = 0$ (because both $p(x)$ and its derivative $p'(x)$ vanish as $x \to \infty$), we must also have $j(0) = 0$. Thus the gap flux is zero at the boundaries (because there are no sinks or sources).

We are interested in finding the rate of polymerization,

$$\frac{dn_{av}}{dt} \equiv \frac{d}{dt}\sum_{n=0}^{\infty} nP_n(t) = \sum_{n=0}^{\infty} n\frac{dP_n}{dt}(t) \quad \text{(A10.8)}$$

This can be evaluated by substituting Equation A10.4 into Equation A10.8 and using Equation A10.7 to obtain

$$\frac{dn_{av}}{dt} = k_{on}[A_1]p_\delta - k_{off} \quad \text{(A10.9)}$$

There is a simple intepretation of this equation: The shrinkage rate (k_{off}) is not affected by the applied force, but the growth rate is weighted by the probability that the gap is larger than δ (i.e., large enough for another monomer to fit in).

Finding the elongation rate now reduces to the problem of finding the probability p_δ. To find p_δ we need an expression for $p(x)$. This is obtained by substituting Equation A10.2 into Equation A10.1 and summing with respect to n:

$$\begin{aligned}0 &= p(x)\frac{d}{dt}\sum_{n=0}^{\infty} P_n(t)\\ &= -\frac{dj}{dx}(x) - k_{on}A_1\big[p(x)H(x-\delta) - p(x+\delta)\big] \\ &\quad + k_{off}\big[p(x-\delta) - p(x)\big]\cdot\big[1 - P_0(t)\big]\end{aligned} \quad \text{(A10.10)}$$

where $P_0(t) \cong 0$.

The Reaction-Limited Case

In the reaction-limited case we assume that the diffusion coefficient, D, is very large. If the flux is finite, then $j(x)/D$ goes to zero. Therefore, by Equation A10.5, $p(x)$ must satisfy

$$-\frac{dp}{dx}(x) - \frac{F}{kT}p(x) = \frac{j(x)}{D} \to 0 \quad \text{(A10.11)}$$

It follows that $p(x)$ is an exponential given by

$$p(x) = \frac{F}{kT}\exp\left(-\frac{Fx}{kT}\right) \qquad F > 0 \qquad (A10.12)$$

In other words, Boltzmann's law holds. In this case

$$p_\delta = \exp\left(-\frac{F\delta}{kT}\right) \qquad (A10.13)$$

which, upon substitution into Equation A10.9 gives the elongation rate of Equation 10.3.

The Diffusion-Limited Case

In the diffusion-limited case we assume that diffusion is slow and that the monomer addition rate, $k_{on}[A_1]$, is very fast, so that almost as soon as a gap opens up, a monomer will drop in. This means that $p(x)$ must drop almost to 0 above δ. If we integrate Equation A10.10 from δ to ∞, we get

$$0 = j(\delta) - j(\infty) + k_{on}A_1 \int_{2\delta}^{\infty} p(x) \cdot dx - k_{on}A_1 p_\delta + k_{off} - k_{off}p_\delta \qquad (A10.14)$$

$$= j(\delta) - \frac{dn_{av}}{dt} - k_{off}p_\delta$$

The second term on the right-hand side of the first line, $j(\infty)$, vanishes according to our earlier discussion, the third term vanishes because the probability that the gap reaches a size of 2δ is extremely small by our assumption that the monomer addition rate is very high, and the fourth and fifth terms together are the negative of the elongation rate. The last term cannot be ignored, even though in the diffusion limit p_δ is 0. This is because if the association rate is very high, then the dissociation rate must also be very high; otherwise, the equilibrium constant would be vanishingly small. Because $k_{off}p_\delta$ is not necessarily small, we write $k_{off} = K_c k_{on}$, in which case

$$k_{off}p_\delta = \frac{K_c k_{on}A_1 p_\delta}{A_1} = \frac{K_c}{A_1}\left(\frac{dn_{av}}{dt} + k_{off}\right) \qquad (A10.15)$$

using Equation A10.9. Substituting this into Equation A10.14 and rearranging gives

$$\frac{dn_{av}}{dt} = \frac{j(\delta)}{1 + K_c/[A_1]} - \frac{K_c/[A_1]}{1 + K_c/[A_1]}k_{off} \qquad (A10.16)$$

If we assume that $[A_1] \gg K_c$ and that the off-rate constant is negligible, then the growth rate equals the flux (the rate at which a gap opens up). These two assumptions are reasonable if the actin monomer concentration is ~30 µM, about 300 times greater than the critical concentration (see the discussion of *Listeria* motility at the beginning of Chapter 14), and $k_{off} = 1.4\ s^{-1}$ (see Table 11.1), much less than the polymerization rate associated with *Listeria* motility.

10.2 Acrosomal Reaction

The diffusion model of acrosome growth pictured in Figure B in Example 10.3, predicts an elongation rate of

$$L(t) = 2z\sqrt{Dt} \qquad z \cdot \text{erf}(z) = \frac{c_0 \delta}{\rho\sqrt{\pi}} = \text{constant}$$

where c_0 is the concentration of actin (molecules/m^3), δ is the monomer addition length, ρ is the density of filaments in the cross-section, and D is the diffusion coefficient of actin monomers in solution. Using the values $c_0 = 4.4 \times 10^6$ μm^{-3}, $\delta = 2.75$ nm, and $\rho = 2800$ μm^{-2} (giving $z \cong 2.4$) from Tilney and Inoue (1982) and using $D = 50$ μm^2/s for the diffusion of a globular protein of the molecular weight of actin (Chapter 3), the "diffusion-limited" growth rate is seen to exceed the measured growth rate (see Example 10.3C).

The model is solved in the following way. The differential equation is the Fokker–Planck equation

$$\frac{\partial c}{\partial t} = D\frac{\partial^2 c}{\partial x^2} - \frac{\partial L}{\partial t}\frac{\partial c}{\partial x}$$

with boundary and initial conditions

$$c(0,t) = c_0, \quad c(L,t) = 0, \text{ and } \quad \frac{\partial L}{\partial t} = \frac{\delta}{\rho} \times \text{flux of monomers} = -\frac{\delta}{\rho}D\frac{\partial c}{\partial x}(L,t)$$

The solution is

$$c(x,t) = c_0 \frac{\text{erf}\left[(L-x)/\sqrt{4Dt}\right]}{\text{erf}(z)} \qquad z = L/\sqrt{4Dt} = \text{constant}$$

where the constant z satisfies the stated condition. Note that depletion of monomer in the acrosomal cup will slow the growth down, whereas narrowing the process will speed growth up.

11.1 Diffusional Length Changes at Equilibrium

The diffusion coefficient follows by considering polymerization to be a random walk with step size δ (the incremental increase in length of the polymer when a subunit is added). In a short time interval Δt, the probability of addition is $k_{on}[A]\Delta t$ and the probability of subtraction is $k_{off}\Delta t$. The variance of the length change is therefore

$$\text{Variance} = \sum_{x=0,+\delta,-\delta} x^2 p(x) = 2\delta^2 k_{on}[A]\Delta t$$

(remembering that $k_{on}[A] = k_{off}$ at equilibrium). Comparing this to the variance associated with diffusion (see the discussion after Equation 4.15), we obtain the effective diffusion coefficient $D = \delta^2 k_{on}[A]$.

11.2 Dynamic Instability

Let $p_+(n,t)$ be the probability that a polymer is a growing n-mer, and let $p_-(n,t)$ be the probability that it is a shrinking n-mer. Then $p_+(n,t)$ and $p_-(n,t)$ satisfy the following reaction–diffusion equations (Dogterom and Leibler 1993):

$$\frac{\partial p_+}{\partial t}(n,t) = -v_+ \frac{\partial p_+}{\partial n}(n,t) - f_{+-}p_+(n,t) + f_{-+}p_-(n,t)$$

$$\frac{\partial p_-}{\partial t}(n,t) = v_- \frac{\partial p_-}{\partial n}(n,t) + f_{+-}p_+(n,t) - f_{-+}p_-(n,t)$$

In the steady state ($\partial p / \partial t = 0$), the solution is

$$p_+(n) = \frac{P_+}{n_0} \exp\left(-\frac{n}{n_0}\right)$$

$$p_-(n) = \frac{P_-}{n_0} \exp\left(-\frac{n}{n_0}\right)$$

$$p(n) \equiv p_+(n) + p_-(n) = \frac{1}{n_0} \exp\left(-\frac{n}{n_0}\right)$$

where n_0 is the mean polymer length (ignoring monomers and small nuclei) given by

$$n_0 = \frac{v_- v_+}{v_- f_{+-} - v_+ f_{-+}}$$

and P_+ (P_-) is the probability that a polymer is growing (shrinking). P_+ is given by

$$P_+ \equiv \int_{n=0}^{\infty} p_+(n) \cdot dn = \frac{v_-}{v_- + v_+} = 1 - P_-$$

This solution is valid provided that n_0 is positive, which occurs if and only if $f_{+-}v_- > f_{-+}v_+$

13.1 Contraction and Filament Sliding

Muscle

The contraction of muscle is driven by the relative sliding of the thin and thick filaments. If the sliding speed is v, then each sarcomere will contract at a speed $2v$, since there are two actin filaments per sarcomere (see Figure 1.1). Hence the speed of muscle contraction, V, in units of muscle lengths per second is

$$V = \frac{2v}{s}$$

Sperm

The motility of sperm is driven by the sliding of adjacent doublet microtubules in the axoneme (see Figure 8.3B). The sliding, in turn, is driven by the most-radial crossbridges, those formed by outer-arm dynein (as shown in genetic studies; Kurimoto and Kamiya, 1991). The relationship between the sliding speed and the speed of sperm movement is not as straightforward as it is in muscle, and a good deal of background information is needed for the calculation.

The shear between adjacent microtubules on one side of the axoneme leads to bending of the sperm because the doublet microtubules are crosslinked at their minus ends near the sperm head; for bending to occur, the dyneins on the opposite side of the axoneme must not be active in order to allow the microtubules on this opposite side to shear in the other direction. Thus the undulatory bending of sperm is due to a zone of active dyneins on one side and inactive dyneins on the other and the propagation of this zone of activation from the head to the tail. The reason that the undulation leads to forward movement is that the drag coefficients in the directions parallel ($c_{||}$) and perpendicular ($c_\perp$) to the axis of the sperm are not equal: Because the perpendicular drag coefficient is larger than the parallel one (by about a factor of 2; see Tables 6.2 and 6.3), the undulation of the sperm meets less resistance if the sperm moves forward (Lighthill, 1975). The microtubule sliding speed, v, equals the rate of shear between adjacent microtubule doublets:

$$v = b\frac{d\theta}{dt}$$

where $b = 60$ nm is the center-to-center spacing of the doublets and $\theta(s,t)$ is the tangent angle of the sperm tail at position s along its length and time t. If the beating sperm's tail lies in a plane and the shape is sinusoidal, then

$$\theta(s,t) \cong \frac{dy}{dx} = \frac{2\pi h}{\lambda}\cos\frac{2\pi(x - c \cdot t)}{\lambda}$$

where h is the amplitude of the wave, λ is the wavelength, and c is the wave velocity (the rate at which the beat propagates from the sperm's head). The approximation holds provided that h/λ is small (which is approximately true). The speed of swimming is predicted to equal

$$V = \frac{(1-\alpha)(1-\beta)}{1-\beta+\alpha\beta}c \qquad \alpha = \frac{c_{||}}{c_\perp} \qquad \beta = \left(1 + \frac{4\pi^2 h^2}{\lambda^2}\right)^{-\frac{1}{2}}$$

(Brennen and Winet, 1977), ignoring the drag contributed by the head (Example 6.5).

The sperm of the sea urchin swim at ~150 μm/s: The planar beats have $h \cong 5$ μm, $\lambda \cong 30$ μm, $c \cong 550$ μm/s (Bray, 1992), giving a maximum shear rate between adjacent doublets that lie in a plane parallel to the plane of bending equal to ~7 μm/s.

15.1 Photon Shot Noise

Because light is quantized and the absorption of a photon by a detector is a random process, optical signals are inherently noisy. The statistical properties of this so-called photon shot noise can be calculated as follows (Rice, 1954). Suppose that the absorption of a photon at time $t = 0$ gives rise to a photocurrent

$$g(t) = \frac{ze}{\tau_0} \exp\left(-\frac{t}{\tau_0}\right)$$

where τ_0 is the time constant of the detector, e is the electronic charge, and z is the total number of charges displaced upon absorption of the photon. If an average of n photons is absorbed per second, then the mean photocurrent is

$$\langle i \rangle = n \int_0^\infty g(t) dt = nze$$

If photon absorption occurs randomly in time, then the variance is

$$\sigma_i^2 = \langle i^2 \rangle - \langle i \rangle^2 = n \int_0^\infty g^2(t) dt - \langle i \rangle^2 = \frac{n(ze)^2}{2\tau_0} - n^2(ze)^2 \cong \frac{n(ze)^2}{2\tau_0}$$

where the last approximate equality holds provided that the time constant is very short ($\tau_0 n \ll 1$). The autocorrelation function is

$$R(\tau) = \sigma_i^2 \exp\left(-\frac{|\tau|}{\tau_0}\right)$$

(compare to Example 4.7), and the power spectrum is

$$\sigma_i^2(f) = \frac{4\sigma_i^2 \tau_0}{1 + (2\pi f \tau_0)^2} \cong 4\sigma_i^2 \tau_0 = 2n(ze)^2$$

where the approximate equality holds provided that $2\pi f \tau_0 \ll 1$. In other words, the power spectrum is frequency independent provided that the duration of the single-photon events is very much briefer than the reciprocal of the frequency.

15.2 Sensitivity of a Split Photodiode Detector

Photon shot noise limits the precision with which an instrument can measure position. For illustrative purposes, we calculate the sensitivity of a dual photodiode detector to the movement of the tip of a luminous fiber (as shown in Figure 15.5A). Suppose that on average, n_1 photons strike one of the detectors per second, that the detector has an efficiency q ($q \cong 0.5$ for a photodiode), and that z electrons are produced for each photon absorbed ($z = 1$ for a photodiode, $z \sim 10^6$ for a photomultiplier). The mean photocurrent, $\langle i_1 \rangle$, is $\langle i_1 \rangle = qn_1 ze$, where e is the charge on the electron. Likewise, the mean photocurrent in the other detector is $\langle i_2 \rangle = qn_2ze$. The differential signal is therefore

$$\langle \Delta i \rangle = \langle i_2 \rangle - \langle i_1 \rangle = q(n_2 - n_1)ze$$

This is shown in Figure 15.5B. The gain of the detector is

$$\frac{d\langle \Delta i \rangle}{dx} = q\left(\frac{dn_2}{dx} - \frac{dn_1}{dx}\right)ze = 2\frac{qnze}{d}$$

for an object of width d and uniform cross-section; it has units amps/meter. Note that if the width of the object is smaller than the wavelength of light, then its image will be broadened by diffraction—it is the broadened width that is relevant. $n = n_1 + n_2$ is the total light intensity in photons.

The sensitivity of the detector depends on the amplitude of the noise. One source of noise is Johnson noise due to thermal fluctuations in the feedback resistors. Johnson noise is analogous to the thermal fluctuations of a damped spring, and its amplitude is $\sigma_i^2(f) = 4kT/R$ in each of the feedback resistors of resistance R (Sigworth, 1995). The variance of the current noise due to both feedback resistors in a frequency bandwidth B is $8kTB/R$; it can be made arbitrarily small by choosing a large enough feedback resistor.

The other, limiting source of noise is photon shot noise due to the random arrival of photons at the detector. From Appendix 15.1, the power spectrum of the current fluctuations due to photon shot noise is

$$\sigma_i^2(f) = 2qn(ze)^2$$

To evaluate the significance of the current noise, we must compare it to the detector's gain. What we really want is the power spectrum of the "equivalent displacement noise," which is the current noise divided by the gain squared:

$$\sigma_x^2(f) = \frac{\sigma_i^2(f)}{\left(d\langle \Delta i \rangle/dx\right)^2} = \frac{1}{2}\frac{d^2}{qn}$$

The equivalent displacement noise tells us how the photon shot noise limits the resolution. There are a number of interesting points: The resolution is independent of the electronic charge, e, the instrument amplification, z, and, if we think about it for a bit, it is also independent of the magnification. It depends on the total number of photons collected by the detector, as well as on the spatial width of the object.

Had we considered a circular luminous object with diameter D imaged onto the center of a quadrant detector, the equivalent displacement noise would be

$$\sigma_x^2(f) = \frac{\pi^2}{32}\frac{D^2}{qn}$$

16.1 Crossbridge Model for Myosin

Consider an actin filament moving past a crossbridge at a constant, positive speed v. The probability, $p_{on}(x,t)$, that a crossbridge is attached to the actin filament and at a distance, x, (defined in Figure 16.3) at time, t, is given by

$$\frac{\partial p_{on}}{\partial t}(x,t) = k_{on}(x)p_{off}(x,t) - k_{off}(x)p_{on}(x,t) + v\frac{\partial p_{on}}{\partial x}(x,t)$$

The steady-state solution, $p_{on}(x)$, satisfies

$$v\frac{dp_{on}}{dx}(x) = -k_{on}(x)p_{off}(x) + k_{off}(x)p_{on}(x)$$

For the rate constants shown in Figure 16.3C, the solution is

$$p_{on}(x) = \begin{cases} \frac{1}{\Delta}\exp\left(\frac{x}{\delta_-}\right) & x \leq 0 \\ \frac{1}{\Delta} & 0 \leq x \leq \delta_+ \\ 0 & x \geq \delta_+ \end{cases}$$

where $\delta_- = vt_-$. In this model, in which the crossbridge immediately becomes strained upon binding, x corresponds to the strain in the crossbridge. To facilitate the solution of the equation, it has been assumed that attachment occurs at a very high rate (k_0) over a very small range of positions (x_0). Integration over all positions gives a total probability of being attached of

$$P_{on} = \int_{-\infty}^{\infty} p_{on}(x) \cdot dx = \frac{1}{\Delta}(\delta_- + \delta_+) = \frac{\delta}{\Delta} = r$$

as required. The higher the speed, the lower the probability of attaching when moving through the "strike zone." The probability of hitting is

$$\int_0^{t_0} k_0 \exp(-k_0 t_0) \cdot dt = 1 - \exp(-k_0 t_0) = 1 - \exp\left(-\frac{k_0 x_0}{v}\right)$$

where $t_0 = x_0/v$ is the time it takes to cross the "plate" of width x_0. Given that the next opportunity to connect will occur after a distance d, the path distance, the average distance traveled relative to the filament before the crossbridge attaches is

$$\Delta(v) = \frac{d}{1-\exp\left(-\frac{k_0 x_0}{v}\right)} = \frac{d}{1-\exp\left(-\frac{k_{max}d}{v}\right)}$$

where k_{max} is the maximum ATPase rate at high speed (~25 s^{-1}; Ma and Taylor, 1994, and see Chapter 14). $k_{max} = 1/T_c$. The equality $k_0 x_0 = k_{max}d$ is made so that $\Delta_{max} = v_{max}/k_{max}$, as required by the definition of Δ. The average force is

$$\langle F \rangle = r \cdot \kappa \cdot \langle x \rangle = \frac{\delta}{\Delta} \kappa \left(p_+ \langle x_+ \rangle - p_- \langle x_- \rangle \right) = \frac{\delta}{\Delta} \kappa \left(\frac{\delta_+}{\delta} \frac{\delta_+}{2} - \frac{\delta_-}{\delta} \delta_- \right)$$

$$= \tfrac{1}{2} \frac{\kappa \delta_+^2}{d} \left[1 - \exp\left(-\frac{k_{max} d}{v} \right) \right] \left[1 - \frac{2 v^2 t_-^2}{\delta_+^2} \right]$$

Thus we obtain an explicit formula for the force–velocity curve in terms of the measurable parameters. The maximum speed given in Equation 16.7 is obtained by setting the force to zero. It is also straightforward to calculate the work ($\langle F \rangle \Delta$), working distance (δ), and the duty ratio (δ/Δ) as a function of speed.

16.2 Hand-Over-Hand Model for Dimeric Kinesin

We consider a 3-state model, with states K, K·T, and K·P, corresponding to species 1, 2, and 5 in Figure 16.8. For simplicity, we assume that the ADP and P_i concentrations are zero (as is approximately the case in the in vitro assays). This gives the following kinetic equation

$$K \underset{k_{-1}}{\overset{k_1[ATP]}{\rightleftharpoons}} K \cdot T \overset{k_2}{\rightarrow} K \cdot P \overset{k_3}{\rightarrow} K$$

The solution to this equation is

$$v = \delta k_{ATPase} = \delta k_{cat} \frac{[ATP]}{K_M + [ATP]} \qquad k_{cat} = \frac{k_2 k_3}{k_2 + k_3} \qquad K_M = \frac{k_3(k_2 + k_{-1})}{k_1(k_2 + k_3)}$$

where v is the average speed of movement, k_{ATPase} is the ATPase rate, and δ is the distance per ATP (8 nm for kinesin).

If a cyclic reaction is coupled to motion, as is the case here, then the rate constants will in general depend on force as discussed in Chapter 5. We can describe this in the following way. If the transition $E_i \rightleftharpoons E_{i+1}$ is associated with a structural change through a distance δ_i, then the rates will depend on a load, F, in the opposite direction as δ_i, according to

$$k_{+i} = k_{+i}^0 \exp\left[-f_i \frac{F \delta_i}{kT} \right]$$

$$k_{-i} = k_{-i}^0 \exp\left[(-f_i) \frac{F \delta_i}{kT} \right]$$

where f_i is the location of the transition state as a fraction of the distance between the two states E_i and E_{i+1}, and k^0_{+i} and k^0_{-i} are the rate constants in the absence of load. Note that $\delta_1 + \delta_2 + \delta_3 = 8$ nm. Normally we think that f_i lies between 0 and 1, though this is not an absolute requirement. For the hand-over-hand model, we make the simple assumption that all $f_i = 1$, meaning that the transition state is displaced all the way toward the final state. This puts all the displacement sensitivity in the forward step. The solid curves in Figure 16.7 are generated with $\delta_1 = 1$ nm, $\delta_2 = 1$ nm, $\delta_3 = 6$ nm; $k^0_{+1} = 100\ \mu M^{-1} \cdot s^{-1}$, $k^0_{-1} = 3000\ s^{-1}$, $k^0_2 = 105\ s^{-1}$, and $k^0_3 = 5000\ s^{-1}$.

Bibliography

Adams, M. D., et al. 2000. The genome sequence of *Drosophila melanogaster*. *Science* 287: 2185–2195.

Afzelius, B. 1959. Electron microscopy of the sperm tail. Results obtained with a new fixative. *J. Biophys. Biochem. Cytol.* 5: 269–278.

Afzelius, B. A. 1976. A human syndrome caused by immotile cilia. *Science* 193: 317–319.

Aidley, D. J. 1998. *The Physiology of Excitable Cells*. Cambridge University Press, New York.

Aist, J. R., and C. J. Bayles. 1991. Detection of spindle pushing forces in vivo during anaphase B in the fungus *Nectria haematocca*. *Cell Motil. Cytoskeleton* 19: 18–24.

Alberts, B. 1994. *Molecular Biology of the Cell*. Garland Publishing, New York.

Alberts, B. 1998. The cell as a collection of protein machines: Preparing the next generation of molecular biologists. *Cell* 92: 291–294.

Alberty, R. A., and R. N. Goldberg. 1992. Standard thermodynamic formation properties for the adenosine 5'- triphosphate series. *Biochemistry* 31: 10610–10615.

Alexander, P., and C. Earland. 1950. Structure of wool fibers. *Nature* 166: 396–397.

Alexander, S. P., and C. L. Rieder. 1991. Chromosome motion during attachment to the vertebrate spindle: initial saltatory-like behavior of chromosomes and quantitative analysis of force production by nascent kinetochore fibers. *J. Cell Biol.* 113: 805–815.

Allen, R. D., N. S. Allen, and J. L. Travis. 1981. Video-enhanced contrast, differential interference contrast (AVEC-DIC) microscopy: A new method capable of analyzing microtubule-related motility in the reticulopodial network of *Allogromia laticollaris*. *Cell Motil.* 1: 291–302.

Allen, R. D., D. G. Weiss, J. H. Hayden, D. T. Brown, H. Fujiwake, and M. Simpson. 1985. Gliding movement of and bidirectional transport along single native microtubules from squid axoplasm: Evidence for an active role of microtubules in cytoplasmic transport. *J. Cell Biol.* 100: 1736–1752.

Amos, L., and A. Klug. 1974. Arrangement of subunits in flagellar microtubules. *J. Cell Sci.* 14: 523–549.

Amos, L. A. 1995. The microtubule lattice—20 years on. *Trends Cell Biol.* 5: 48–51.

Amos, L. A. 2000. Focusing-in on microtubules. *Curr. Opin. Struct. Biol.* 10: 236–241.

Amos, L. A., and W. B. Amos. 1991. *Molecules of the Cytoskeleton*. Guilford Press, New York.

Ansari, A., C. M. Jones, E. R. Henry, J. Hofrichter, and W. A. Eaton. 1992. The role of solvent viscosity in the dynamics of protein conformational changes. *Science* 256: 1796–1798.

Armstrong, C. M., and F. Bezanilla. 1974. Charge movement associated with the opening and closing of the activation gates of the Na channels. *J. Gen. Physiol.* 63: 533–552.

Arnal, I., and R. H. Wade. 1998. Nucleotide-dependent conformations of the kinesin dimer interacting with microtubules. *Structure* 6: 33–38.

Arnal, I., E. Karsenti, and A. A. Hyman. 2000. Structural transitions at microtubule ends correlate with their dynamic properties in *Xenopus* egg extracts. *J. Cell Biol.* 149: 767–774.

Ashburner, M. 1989. *Drosophila*. Cold Spring Harbor Laboratory, Cold Spring Harbor, NY.

Ashkin, A. 1992. Forces of a single-beam gradient laser trap on a dielectric sphere in the ray optics regime. *Biophys. J.* 61: 569–582.

Ashkin, A., J. M. Dziedzic, J. E. Bjorkholm, and S. Chu. 1986. Observation of a single-beam gradient force optical trap for dielectric particles. *Opt. Lett.* 11: 288–290.

Ashkin, A., K. Schutze, J. M. Dziedzic, U. Euteneuer, and M. Schliwa. 1990. Force generation of organelle transport measured in vivo by an infrared laser trap. *Nature* 348: 346–348.

Astumian, R. D., and M. Bier. 1994. Fluctuation driven ratchets: Molecular motors. *Phys. Rev. Lett.* 72: 1766–1769.

Atkins, P. W. 1986. *Physical Chemistry*. W. H. Freeman and Company, New York.

Axelrod, D. 1989. Total internal reflection fluorescence microscopy. *Methods Cell Biol.* 30: 245–270.

Bagshaw, C. R. 1993. *Muscle Contraction*. Chapman & Hall, New York.

Bagshaw, C. R., and D. R. Trentham. 1973. The reversibility of adenosine triphosphate cleavage by myosin. *Biochem. J.* 133: 323–328.

Bagshaw, C. R., J. F. Eccleston, F. Eckstein, R. S. Goody, H. Gutfreund, and D. R. Trentham. 1974. The magnesium ion-dependent adenosine triphosphatase of myosin. Two-step processes of adenosine triphosphate association and adenosine diphosphate dissociation. *Biochem. J.* 141: 351–364.

Bagshaw, C. R., D. R. Trentham, R. G. Wolcott, and P. D. Boyer. 1975. Oxygen exchange in the gamma-phosphoryl group of protein-bound ATP during Mg2+-dependent adenosine triphosphatase activity of myosin. *Proc. Natl. Acad. Sci. USA* 72: 2592–2596.

Ban, N., P. Nissen, J. Hansen, M. Capel, P. B. Moore, and T. A. Steitz. 1999. Placement of protein and RNA structures into a 5 A-resolution map of the 50S ribosomal subunit. *Nature* 400: 841–847.

Barany, M. 1967. ATPase activity of myosin correlated with speed of muscle shortening. *J. Gen. Physiol.* 50 Suppl: 197–218.

Barton, N. R., and L. S. Goldstein. 1996. Going mobile: Microtubule motors and chromosome segregation. *Proc. Natl. Acad. Sci. USA* 93: 1735–1742.

Batiza, A. F., I. Rayment, and C. Kung. 1999. Channel gate! Tension, leak and disclosure. *Structure Fold. Des.* 7: R99–R103.

Bayley, P. M., M. J. Schilstra, and S. R. Martin. 1990. Microtubule dynamic instability: numerical simulation of microtubule transition properties using a Lateral Cap model. *J. Cell Sci.* 95: 33–48.

Bell, G. I. 1978. Models for the specific adhesion of cells to cells. *Science* 200: 618–627.

ben-Avraham, D., and M. M. Tirion. 1995. Dynamic and elastic properties of F-actin: A normal-modes analysis. *Biophys. J.* 68: 1231–1245.

Bendat, J. S., and A. G. Piersol. 1986. *Random Data: Analysis and Measurement Procedures*. Wiley, New York.

Benoit, M., D. Gabriel, G. Gerisch, and H. E. Gaub. 2000. Discrete interactions in cell adhesion measured by single-molecule force spectroscopy. *Nat. Cell Biol.* 2: 313–317.

Berg, H. C. 1993. *Random Walks in Biology*. Princeton University Press, Princeton, NJ.

Berg, O. G., and P. H. von Hippel. 1985. Diffusion-controlled macromolecular interactions. *Annu. Rev. Biophys. Biophys. Chem.* 14: 131–160.

Berger, B., D. B. Wilson, E. Wolf, T. Tonchev, M. Milla, and P. S. Kim. 1995. Predicting coiled coils by use of pairwise residue correlations. *Proc. Natl. Acad. Sci. USA* 92: 8259–8263.

Berliner, E., E. C. Young, K. Anderson, H. K. Mahtani, and J. Gelles. 1995. Failure of a single-headed kinesin to track parallel to microtubule protofilaments. *Nature* 373: 718–721.

Blanchoin, L., K. J. Amann, H. N. Higgs, J. B. Marchand, D. A. Kaiser, and T. D. Pollard. 2000. Direct observation of dendritic actin filament networks nucleated by Arp2/3 complex and WASP/Scar proteins. *Nature* 404: 1007–1011.

Block, S. M., L. S. Goldstein, and B. J. Schnapp. 1990. Bead movement by single kinesin molecules studied with optical tweezers. *Nature* 348: 348–352.

Bloom, G. S., and S. A. Endow. 1995. Motor proteins 1: kinesins. *Protein Profile*. 2: 1105–1171.

Bourne, H. R., D. A. Sanders, and F. McCormick. 1991. The GTPase superfamily: conserved structure and molecular mechanism. *Nature* 349: 117–127.

Bowater, R., and J. Sleep. 1988. Demembranated muscle fibers catalyze a more rapid exchange between phosphate and adenosine triphosphate than actomyosin subfragment 1. *Biochemistry* 27: 5314–5323.

Boyer, P. D. 1997. The ATP synthase—a splendid molecular machine. *Annu. Rev. Biochem.* 66: 717–749.

Bracewell, R. N. 1986. *The Fourier Transform and Its Applications*. McGraw-Hill, New York.

Brady, S. T. 1985. A novel brain ATPase with properties expected for the fast axonal transport motor. *Nature* 317: 73–75.

Brady, S. T., K. K. Pfister, and G. S. Bloom. 1990. A monoclonal antibody against kinesin inhibits both anterograde and retrograde fast axonal transport in squid axoplasm. *Proc. Natl. Acad. Sci. USA* 87: 1061–1065.

Brändén, C.-I., and J. Tooze. 1999. *Introduction to Protein Structure*. Garland Publishing, New York.

Braxton, S. M. 1988. Synthesis and Use of a Novel Class of ATP Carbamates and a Ratchet Diffusion Model for Directed Motion in Muscle. Ph.D. diss., Washington State University, Pullman, WA.

Braxton, S. M., and R. G. Yount. 1989. A ratchet diffusion model for directed motion in muscle. *Biophys. J.* 55: 12a.

Bray, D. 1992. *Cell Movements*. Garland Publishing, New York.

Bray, D. 2001. *Cell Movements: From Molecules to Motility*, 2nd ed. Garland Publishing, New York.

Bray, D., and M. B. Bunge. 1981. Serial analysis of microtubules in cultured rat sensory axons *J. Neurocytol.* 10: 589–605.

Brennen, C., and H. Winet. 1977. Fluid mechanics of propulsion by cilia and flagella. *Annu. Rev. Fluid Mech.* 9: 339–398.

Brenner, B., M. Schoenberg, J. M. Chalovich, L. E. Greene, and E. Eisenberg. 1982. Evidence for cross-bridge attachment in relaxed muscle at low ionic strength. *Proc. Natl. Acad. Sci. USA* 79: 7288–7291.

Brylawski, B. P., and M. Caplow. 1983. Rate for nucleotide release from tubulin. *J. Biol. Chem.* 258: 760–763.

Bustamante, C., J. F. Marko, E. D. Siggia, and S. Smith. 1994. Entropic elasticity of lambda-phage DNA. *Science* 265: 1599–1600.

Byers, T. J., and D. Branton. 1985. Visualization of the protein associations in the erythrocyte membrane skeleton. *Proc. Natl. Acad. Sci. USA* 82: 6153–6157.

Byrd, P. F., and M. D. Friedman. 1971. *Handbook of Elliptic Integrals for Engineers and Scientists*. Springer-Verlag, New York.

Calladine, C. R. 1983. Construction and operation of bacterial flagella. *Sci. Prog.* 68: 365–385.

Camalet, S., T. Duke, F. Julicher, and J. Prost. 2000. Auditory sensitivity provided by self-tuned critical oscillations of hair cells. *Proc. Natl. Acad. Sci. USA* 97: 3183–3188.

Cao, Z., and F. A. Ferrone. 1997. Homogeneous nucleation in sickle hemoglobin: stochastic measurements with a parallel method. *Biophys. J.* 72: 343–352.

Caplow, M., and J. Shanks. 1996. Evidence that a single monolayer tubulin-GTP cap is both necessary and sufficient to stabilize microtubules. *Mol. Biol. Cell* 7: 663–675.

Caplow, M., R. L. Ruhlen, and J. Shanks. 1994. The free energy for hydrolysis of a microtubule-bound nucleotide triphosphate is near zero: All of the free energy for hydrolysis is stored in the microtubule lattice. *J. Cell Biol.* 127: 779–788.

Cardon, J. W., and P. D. Boyer. 1978. The rate of release of ATP from its complex with myosin. *Eur. J. Biochem.* 92: 443–448.

Carlier, M. F., D. Pantaloni, and E. D. Korn. 1984. Evidence for an ATP cap at the ends of actin filaments and its regulation of the F-actin steady state. *J. Biol. Chem.* 259: 9983–9986.

Carrion-Vazquez, M., A. F. Oberhauser, S. B. Fowler, P. E. Marszalek, S. E. Broedel, J. Clarke, and J. M. Fernandez. 1999. Mechanical and chemical unfolding of a single protein: a comparison. *Proc. Natl. Acad. Sci. USA* 96: 3694–3699.

Carslaw, H. S., and J. C. Jaeger. 1986. *Conduction of Heat in Solids*. Oxford University Press, Oxford.

Case, R. B., D. W. Pierce, N. Hom-Booher, C. L. Hart, and R. D. Vale. 1997. The directional preference of kinesin motors is specified by an element outside of the motor catalytic domain. *Cell* 90: 959–966.

Cassimeris, L., N. K. Pryer, and E. D. Salmon. 1988. Real-time observations of microtubule dynamic instability in living cells. *J. Cell Biol.* 107: 2223–2231.

Caterina, M. J., M. A. Schumacher, M. Tominaga, T. A. Rosen, J. D. Levine, and D. Julius. 1997. The capsaicin receptor: A heat-activated ion channel in the pain pathway. *Nature* 389: 816–824.

Caterina, M. J., T. A. Rosen, M. Tominaga, A. J. Brake, and D. Julius. 1999. A capsaicin-receptor homologue with a high threshold for noxious heat. *Nature* 398: 436–441.

C. Elegans Sequencing Consortium, 1998. Genome sequence of the nematode *C. elegans*: A platform for investigating biology. The *C. elegans* Sequencing Consortium. *Science* 282: 2012–2018.

Chalfie, M., and J. N. Thomson. 1982. Structural and functional diversity in the neuronal microtubules of *Caenorhabditis elegans*. *J. Cell Biol.* 93: 15–23.

Chandra, R., E. D. Salmon, H. P. Erickson, A. Lockhart, and S. A. Endow. 1993. Structural and functional domains of the *Drosophila* ncd microtubule motor protein. *J. Biol. Chem.* 268: 9005–9013.

Chang, G., R. H. Spencer, A. T. Lee, M. T. Barclay, and D. C. Rees. 1998. Structure of the MscL homolog from Mycobacterium tuberculosis: A gated mechanosensitive ion channel. *Science* 282: 2220–2226.

Chang, P., and T. Stearns. 2000. Delta-tubulin and epsilon-tubulin: Two new human centrosomal tubulins reveal new aspects of centrosome structure and function. *Nat. Cell Biol.* 2: 30–35.

Chen, Y. D., and B. Brenner. 1993. On the regeneration of the actin-myosin power stroke in contracting muscle. *Proc. Natl. Acad. Sci. USA* 90: 5148–5152.

Cheney, R. E., M. K. O'Shea, J. E. Heuser, M. V. Coelho, J. S. Wolenski, E. M. Espreafico, P. Forscher, R. E. Larson, and M. S. Mooseker. 1993. Brain myosin-V is a two-headed unconventional myosin with motor activity. *Cell* 75: 13–23.

Chik, J. K., U. Lindberg, and C. E. Schutt. 1996. The structure of an open state of beta-actin at 2.65 A resolution. *J. Mol. Biol.* 263: 607–623.

Chothia, C., and J. Janin. 1975. Principles of protein-protein recognition. *Nature* 256: 705–708.

Chretien, D., and R. H. Wade. 1991. New data on the microtubule surface lattice. *Biol Cell* 71: 161–174.

Chretien, D., F. Metoz, F. Verde, E. Karsenti, and R. H. Wade. 1992. Lattice defects in microtubules: Protofilament numbers vary within individual microtubules. *J. Cell Biol.* 117: 1031–1040.

Chretien, D., S. D. Fuller, and E. Karsenti. 1995. Structure of growing microtubule ends: Two-dimensional sheets close into tubes at variable rates. *J. Cell Biol.* 129: 1311–1328.

Clemons, W. M., Jr., J. L. May, B. T. Wimberly, J. P. McCutcheon, M. S. Capel, and V. Ramakrishnan. 1999. Structure of a bacterial 30S ribosomal subunit at 5.5 A resolution. *Nature* 400: 833–840.

Cleveland, D. W., S. Y. Hwo, and M. W. Kirschner. 1977. Purification of tau, a microtubule-associated protein that induces assembly of microtubules from purified tubulin. *J. Mol. Biol.* 116: 207–225.

Close, R. 1964. Dynamic properties of fast and slow skeletal muscles of the rat during development. *J. Physiol. (Lond).* 173: 74–95.

Cole, D. G., S. W. Chinn, K. P. Wedaman, K. Hall, T. Vuong, and J. M. Scholey. 1993. Novel heterotrimeric kinesin-related protein purified from sea urchin eggs. *Nature* 366: 268–270.

Cooke, R. 1997. Actomyosin interaction in striated muscle. *Physiol. Rev.* 77: 671–697.

Cooke, R., and E. Pate. 1985. The effects of ADP and phosphate on the contraction of muscle fibers. *Biophys. J.* 48: 789–798.

Cooke, R., M. S. Crowder, and D. D. Thomas. 1982. Orientation of spin labels attached to crossbridges in contracting muscle fibres. *Nature* 300: 776–778.

Cooke, R., K. Franks, G. B. Luciani, and E. Pate. 1988. The inhibition of rabbit skeletal muscle contraction by hydrogen ions and phosphate. *J. Physiol. (Lond).* 395: 77–97.

Cooper, G. M. 2000. *The Cell: A Molecular Approach*, 2nd ed. Sinauer Associates, Sunderland, MA.

Cordova, N. J., B. Ermentrout, and G. F. Oster. 1992. Dynamics of single-motor molecules: The thermal ratchet model. *Proc. Natl. Acad. Sci. USA* 89: 339–343.

Corey, D. P., and J. Howard. 1994. Models for ion channel gating with compliant states. *Biophys. J.* 66: 1254–1257.

Corey, D. P., and A. J. Hudspeth. 1983. Analysis of the microphonic potential of the bullfrog's sacculus. *J. Neurosci.* 3: 942–961.

Corrie, J. E., B. D. Brandmeier, R. E. Ferguson, D. R. Trentham, J. Kendrick-Jones, S. C. Hopkins, U. A. van der Heide, Y. E. Goldman, C. Sabido-David, R. E. Dale, S. Criddle, and M. Irving. 1999. Dynamic measurement of myosin light-chain-domain tilt and twist in muscle contraction. *Nature* 400: 425–430.

Cortese, J. D., B. D. Schwab, C. Frieden, and E. L. Elson. 1989. Actin polymerization induces a shape change in actin-containing vesicles. *Proc. Natl. Acad. Sci. USA* 86: 5773–5777.

Coue, M., V. A. Lombillo, and J. R. McIntosh. 1991. Microtubule depolymerization promotes particle and chromosome movement in vitro. *J. Cell Biol.* 112: 1165–1175.

Coulombe, P. A., M. E. Hutton, A. Letai, A. Hebert, A. S. Paller, and E. Fuchs. 1991. Point mutations in human keratin. 14 genes of epidermolysis bullosa simplex patients: Genetic and functional analyses. *Cell* 66: 1301–1311.

Coy, D. L., and J. Howard. 1994. Organelle transport and sorting in axons. *Curr. Opin. Neurobiol.* 4: 662–667.

Coy, D. L., W. O. Hancock, M. Wagenbach, and J. Howard. 1999a. Kinesin's tail domain is an inhibitory regulator of the motor domain. *Nat. Cell Biol.* 1: 288–292.

Coy, D. L., M. Wagenbach, and J. Howard. 1999b. Kinesin takes one 8-nm step for each ATP that it hydrolyzes. *J. Biol. Chem.* 274: 3667–3671.

Crawford, A. C., and R. Fettiplace. 1985. The mechanical properties of ciliary bundles of turtle cochlear hair cells. *J. Physiol. (Lond).* 364: 359–379.

Creighton, T. E. 1993. *Proteins: Structures and Molecular Properties*. W. H. Freeman and Company, New York.

Cremo, C. R., and M. A. Geeves. 1998. Interaction of actin and ADP with the head domain of smooth muscle myosin: Implications for strain-dependent ADP release in smooth muscle. *Biochemistry* 37: 1969–1978.

Crevel, I., N. Carter, M. Schliwa, and R. Cross. 1999. Coupled chemical and mechanical reaction steps in a processive *Neurospora* kinesin. *EMBO J.* 18: 5863–5872.

Crick, F. H. C. 1953. The packing of α-helices: simple coiled coils. *Acta Cryst.* 6: 689–697.

Cross, R. A. 1995. On the hand-over-hand footsteps of kinesin heads. *J. Muscle Res. Cell Motil.* 16: 91–94.

Cudmore, S., P. Cossart, G. Griffiths, and M. Way. 1995. Actin-based motility of *vaccinia* virus. *Nature* 378: 636–638.

Cyr, J. L., K. K. Pfister, G. S. Bloom, C. A. Slaughter, and S. T. Brady. 1991. Molecular genetics of kinesin light chains: Generation of isoforms by alternative splicing. *Proc. Natl. Acad. Sci. USA* 88: 10114–10118.

Dabiri, G. A., J. M. Sanger, D. A. Portnoy, and F. S. Southwick. 1990. Listeria monocytogenes moves rapidly through the host-cell cytoplasm by inducing directional actin assembly. *Proc. Natl. Acad. Sci. USA* 87: 6068–6072.

Dannhof, B. J., and V. Bruns. 1991. The organ of Corti in the bat *Hipposideros bicolor*. *Hear. Res.* 53: 253–268.

Dantzig, J. A., Y. E. Goldman, N. C. Millar, J. Lacktis, and E. Homsher. 1992. Reversal of the crossbridge force-generating transition by photogeneration of phosphate in rabbit psoas muscle fibres. *J Physiol.* 451: 247–278.

Dawson, R. M. C. 1986. *Data for Biochemical Research*. Clarendon Press, Oxford.

De La Cruz, E. M., A. L. Wells, S. S. Rosenfeld, E. M. Ostap, and H. L. Sweeney. 1999. The kinetic mechanism of myosin V. *Proc. Natl. Acad. Sci. USA* 96: 13726–13731.

De Lozanne, A., and J. A. Spudich. 1987. Disruption of the *Dictyostelium* myosin heavy chain gene by homologous recombination. *Science* 236: 1086–1091.

deCastro, M. J., C. H. Ho, and R. J. Stewart. 1999. Motility of dimeric ncd on a metal-chelating surfactant: Evidence that ncd is not processive. *Biochemistry* 38: 5076–5081.

deCastro, M. J., R. M. Fondecave, L. A. Clarke, C. F. Schmidt, and R. J. Stewart. 2000. Working strokes by single molecules of the kinesin-related microtubule motor ncd. *Nat. Cell Biol.* 2: 724–729.

Deniz, A. A., M. Dahan, J. R. Grunwell, T. Ha, A. E. Faulhaber, D. S. Chemla, S. Weiss, and P. G. Schultz. 1999. Single-pair fluorescence resonance energy transfer on freely diffusing molecules: Observation of Forster distance dependence and subpopulations. *Proc. Natl. Acad. Sci. USA* 96: 3670–3675.

Deniz, A. A., T. A. Laurence, G. S. Beligere, M. Dahan, A. B. Martin, D. S. Chemla, P. E. Dawson, P. G. Schultz, and S. Weiss. 2000. Single-molecule protein folding: Diffusion fluorescence resonance energy transfer studies of the denaturation of chymotrypsin inhibitor 2. *Proc. Natl. Acad. Sci. USA* 97: 5179–5184.

Denk, W., W. W. Webb, and A. J. Hudspeth. 1989. Mechanical properties of sensory hair bundles are reflected in their Brownian motion measured with a laser differential interferometer. *Proc. Natl. Acad. Sci. USA* 86: 5371–5375.

Denk, W., J. R. Holt, G. M. Shepherd, and D. P. Corey. 1995. Calcium imaging of single stereocilia in hair cells: Localization of transduction channels at both ends of tip links. *Neuron.* 15: 1311–1321.

Dickson, R. M., A. B. Cubitt, R. Y. Tsien, and W. E. Moerner. 1997. On/off blinking and switching behaviour of single molecules of green fluorescent protein. *Nature* 388: 355–358.

Dillon, P. F., and R. A. Murphy. 1982. High force development and crossbridge attachment in smooth muscle from swine carotid arteries. *Circ. Res.* 50: 799–804.

Ding, R., K. L. McDonald, and J. R. McIntosh. 1993. Three-dimensional reconstruction and analysis of mitotic spindles from the yeast, *Schizosaccharomyces pombe*. *J. Cell Biol.* 120: 141–151.

Dogterom, M., and S. Leibler. 1993. Physical aspects of the growth and regulation of microtubule structures. *Phys. Rev. Lett.* 70: 1347–1350.

Dogterom, M., and B. Yurke. 1997. Measurement of the force-velocity relation for growing microtubules. *Science* 278: 856–860.

Doi, M. 1975. Theory of diffusion-controlled reaction between non-simple molecules. II. *Chem. Phys.* 11: 115–121.

Dominguez, R., Y. Freyzon, K. M. Trybus, and C. Cohen. 1998. Crystal structure of a vertebrate smooth muscle myosin motor domain and its complex with the essential light chain: Visualization of the pre-powerstroke state. *Cell* 94: 559–571.

Drechsel, D. N., and M. W. Kirschner. 1994. The minimum GTP cap required to stabilize microtubules. *Curr. Biol.* 4: 1053–1061.

Drechsel, D. N., A. A. Hyman, M. H. Cobb, and M. W. Kirschner. 1992. Modulation of the dynamic instability of tubulin assembly by the microtubule-associated protein tau. *Mol. Biol. Cell* 3: 1141–1154.

Duke, T., and S. Leibler. 1996. Motor protein mechanics: a stochastic model with minimal mechanochemical coupling. *Biophys. J.* 71: 1235–1247.

Dutcher, S. K., and E. C. Trabuco. 1998. The UNI3 gene is required for assembly of basal bodies of Chlamydomonas and encodes delta-tubulin, a new member of the tubulin superfamily. *Mol. Biol. Cell* 9: 1293–1308.

Ebus, J. P., and G. J. Stienen. 1996. ATPase activity and force production in skinned rat cardiac muscle under isometric and dynamic conditions. *J. Mol. Cell. Cardiol.* 28: 1747–1757.

Eguiluz, V.M., Ospeck, M., Choe, Y., Hudspeth, A.J., and Magnasco, M.O. 2000. Essential nonlinearities in hearing. *Phys. Rev. Lett.* 84: 5232–5235.

Einstein, A. 1956. *Investigations on the Theory of the Brownian Movement.* Translated by A. D. Cowper. Edited by R. Fürth. Dover Publications, New York.

Eisenberg, D. S., and W. Kauzmann. 1969. *The Structure and Properties of Water.* Oxford University Press, New York.

Eisenberg, E., and T. L. Hill. 1978. A cross-bridge model of muscle contraction. *Prog. Biophys. Mol. Biol.* 33: 55–82.

Eisenberg, E., and C. Moos. 1968. The adenosine triphosphatase activity of acto-heavy meromyosin. A kinetic analysis of actin activation. *Biochemistry* 7: 1486–1489.

Eisenberg, E., T. L. Hill, and Y. Chen. 1980. Cross-bridge model of muscle contraction. Quantitative analysis. *Biophys. J.* 29: 195–227.

Elliott, A., and G. Offer. 1978. Shape and flexibility of the myosin molecule. *J. Mol. Biol.* 123: 505–519.

Endow, S. A., and K. W. Waligora. 1998. Determinants of kinesin motor polarity. *Science* 281: 1200–1202.

Endow, S. A., S. Henikoff, and L. Soler-Niedziela. 1990. Mediation of meiotic and early mitotic chromosome segregation in *Drosophila* by a protein related to kinesin. *Nature* 345: 81–83.

Endow, S. A., S. J. Kang, L. L. Satterwhite, M. D. Rose, V. P. Skeen, and E. D. Salmon. 1994. Yeast Kar3 is a minus-end microtubule motor protein that destabilizes microtubules preferentially at the minus ends. *EMBO J.* 13: 2708–2713.

Engelhardt, W. A., and M. N. Ljubimowa. 1939. Myosine and adenosinetriphosphatase. *Nature* 144: 668–668.

Epstein, H. F., R. H. Waterston, and S. Brenner. 1974. A mutant affecting the heavy chain of myosin in *Caenorhabditis elegans. J. Mol. Biol.* 90: 291–300.

Erickson, H. P. 1974. Microtubule surface lattice and subunit structure and observations on reassembly. *J. Cell Biol.* 60: 153–167.

Erickson, H. P. 1989. Co-operativity in protein-protein association. The structure and stability of the actin filament. *J. Mol. Biol.* 206: 465–474.

Erickson, H. P. 1994. Reversible unfolding of fibronectin type III and immunoglobulin domains provides the structural basis for stretch and elasticity of titin and fibronectin. *Proc. Natl. Acad. Sci. USA* 91: 10114–10118.

Erickson, H. P., and E. T. O'Brien. 1992. Microtubule dynamic instability and GTP hydrolysis. *Annu. Rev. Biophys. Biomol. Struct.* 21: 145–166.

Erickson, H. P., and D. Pantaloni. 1981. The role of subunit entropy in cooperative assembly. Nucleation of microtubules and other two-dimensional polymers. *Biophys. J.* 34: 293–309.

Eyring, H., and E. M. Eyring. 1963. *Modern Chemical Kinetics.* Reinhold Publishing Corp., New York.

Fan, J., A. D. Griffiths, A. Lockhart, R. A. Cross, and L. A. Amos. 1996. Microtubule minus ends can be labelled with a phage display antibody specific for alpha-tubulin. *J. Mol. Biol.* 259: 325–330.

Felgner, H., R. Frank, and M. Schliwa. 1996. Flexural rigidity of microtubules measured with the use of optical tweezers. *J. Cell Sci.* 109: 509–516.

Fenn, W. O. 1924. The relation between the work performed and the energy liberated in muscular contraction. *J. Physiol.* 58: 373–395.

Ferenczi, M. A., Y. E. Goldman, and R. M. Simmons. 1984. The dependence of force and shortening velocity on substrate concentration in skinned muscle fibres from *Rana temporaria. J. Physiol. (Lond).* 350: 519–543.

Fersht, A. 1985. *Enzyme Structure and Mechanism.* W. H. Freeman and Company, New York.

Feynman, R. P., R. B. Leighton, and M. L. Sands. 1963. *The Feynman Lectures on Physics.* Addison-Wesley Publishing Co., Reading, MA.

Finer, J. T., R. M. Simmons, and J. A. Spudich. 1994. Single myosin molecule mechanics: Piconewton forces and nanometre steps. *Nature* 368: 113–119.

Fisher, A. J., C. A. Smith, J. B. Thoden, R. Smith, K. Sutoh, H. M. Holden, and I. Rayment. 1995. X-ray structures of the myosin motor domain of *Dictyostelium discoideum* complexed with MgADP, BeFx and MgADP.AlF4. *Biochemistry* 34: 8960–8972.

Fisher, M. E., and A. B. Kolomeisky. 1999. The force exerted by a molecular motor. *Proc. Natl. Acad. Sci. USA* 96: 6597–6602.

Flyvbjerg, H., T. E. Holy, and S. Leibler. 1994. Stochastic dynamics of microtubules: A model for caps and catastrophes. *Phys. Rev. Lett.* 73: 2372–2375.

Flyvbjerg, H., T. E. Holy, and S. Leibler. 1996. Microtubule dynamics: Caps, catastrophes, and coupled hydrolysis. *Phys. Rev. E.* 54: 5538–5560.

Ford, L. E., A. F. Huxley, and R. M. Simmons. 1977. Tension responses to sudden length change in stimulated frog muscle fibres near slack length. *J. Physiol. (Lond).* 269: 441–515.

Ford, L. E., A. F. Huxley, and R. M. Simmons. 1981. The relation between stiffness and filament overlap in stimulated frog muscle fibres. *J. Physiol. (Lond).* 311: 219–249.

Ford, L. E., A. F. Huxley, and R. M. Simmons. 1985. Tension transients during steady shortening of frog muscle fibres. *J. Physiol. (Lond).* 361: 131–150.

Fortune, N. S., M. A. Geeves, and K. W. Ranatunga. 1991. Tension responses to rapid pressure release in glycerinated rabbit muscle fibers. *Proc. Natl. Acad. Sci. USA* 88: 7323–7327.

Frank, J. 1998. The ribosome-structure and functional ligand-binding experiments using cryo-electron microscopy. *J. Struct. Biol.* 124: 142–150.

Fraser, R. D., and T. P. Macrae. 1980. Molecular structure and mechanical properties of keratins. *Symp. Soc. Exp. Biol.* 34: 211–246.

Fujime, S., M. Maruyama, and S. Asakura. 1972. Flexural rigidity of bacterial flagella studied by quasielastic scattering of laser light. *J. Mol. Biol.* 68: 347–359.

Fuller, R. B., and E. J. Applewhite. 1975. *Synergetics: Explorations in the Geometry of Thinking*. Macmillan, New York.

Funatsu, T., Y. Harada, M. Tokunaga, K. Saito, and T. Yanagida. 1995. Imaging of single fluorescent molecules and individual ATP turnovers by single myosin molecules in aqueous solution. *Nature* 374: 555–559.

Fushimi, K., and A. S. Verkman. 1991. Low viscosity in the aqueous domain of cell cytoplasm measured by picosecond polarization microfluorimetry. *J. Cell Biol.* 112: 719–725.

Fygenson, D. K., E. Braun, and A. Libchaber. 1994. The phase diagram of microtubules. *Phys. Rev. E.* 50: 1579–1588.

Fygenson, D. K., J. F. Marko, and A. Libchaber. 1997. Mechanics of microtubule-based membrane extension. *Phys. Rev. Lett.* 79: 4497–4500.

Garcia de la Torre, J. G., and V. A. Bloomfield. 1981. Hydrodynamic properties of complex, rigid, biological macromolecules: theory and applications. *Q. Rev. Biophys.* 14: 81–139.

Gauger, A. K., and L. S. Goldstein. 1993. The *Drosophila* kinesin light chain. Primary structure and interaction with kinesin heavy chain. *J. Biol. Chem.* 268: 13657–13666.

Gee, M. A., J. E. Heuser, and R. B. Vallee. 1997. An extended microtubule-binding structure within the dynein motor domain. *Nature* 390: 636–639.

Geeves, M. A. 1991. The dynamics of actin and myosin association and the crossbridge model of muscle contraction. *Biochem. J.* 274: 1–14.

Geeves, M. A., R. S. Goody, and H. Gutfreund. 1984. Kinetics of acto-S1 interaction as a guide to a model for the crossbridge cycle. *J. Muscle Res. Cell Motil.* 5: 351–361.

Geeves, M. A., and K. C. Holmes. 1999. Structural mechanism of muscle contraction. *Annu. Rev. Biochem.* 68: 687–728.

Geisler, N., J. Schunemann, K. Weber, M. Haner, and U. Aebi. 1998. Assembly and architecture of invertebrate cytoplasmic intermediate filaments reconcile features of vertebrate cytoplasmic and nuclear lamin-type intermediate filaments. *J. Mol. Biol.* 282: 601–617.

Geisterfer-Lowrance, A. A., S. Kass, G. Tanigawa, H. P. Vosberg, W. McKenna, C. E. Seidman, and J. G. Seidman. 1990. A molecular basis for familial hypertrophic cardiomyopathy: A beta cardiac myosin heavy chain gene missense mutation. *Cell* 62: 999–1006.

Gerbal, F., V. Noireaux, C. Sykes, F. Julicher, P. Chaikin, A. Ott, J. Prost, R. M. Golsteyn, E. Friederich, D. Loward, V. Laurent, and Carlier-MF. 1999. On the *Listeria* propulsion mechanism. *Pramana J. Physics.* 53: 155–170.

Gerstein, M., A. M. Lesk, and C. Chothia. 1994. Structural mechanisms for domain movements in proteins. *Biochemistry* 33: 6739–6749.

Gibbons, I. R. 1963. Studies on the protein components of cilia from *Tetrahymena pyriformis*. *Proc. Natl. Acad. Sci. USA*. 50: 1002–1010.

Gibbons, I. R. 1981. Cilia and flagella of eukaryotes. *J. Cell Biol.* 91: 107s–124s.

Gibbons, I. R., and A. J. Rowe. 1965. Dynein: A protein with adenosine triphosphatase activity from cilia. *Science* 149: 424–426.

Gibbons, I. R., B. H. Gibbons, G. Mocz, and D. J. Asai. 1991. Multiple nucleotide-binding sites in the sequence of dynein beta heavy chain. *Nature* 352: 640–643.

Gigant, B., P. A. Curmi, C. Martin-Barbey, E. Charbaut, S. Lachkar, L. Lebeau, S. Siavoshian, A. Sobel, and M. Knossow. 2000. The 4 A X-ray structure of a tubulin:stathmin-like domain complex. *Cell* 102: 809–816.

Gilbert, S. P., M. L. Moyer, and K. A. Johnson. 1998. Alternating site mechanism of the kinesin ATPase. *Biochemistry* 37: 792–799.

Gilson, M. K., T. P. Straatsma, J. A. McCammon, D. R. Ripoll, C. H. Faerman, P. H. Axelsen, I. Silman, and J. L. Sussman. 1994. Open "back door" in a molecular dynamics simulation of acetylcholinesterase. *Science* 263: 1276–1278.

Gittes, F., B. Mickey, J. Nettleton, and J. Howard. 1993. Flexural rigidity of microtubules and actin filaments measured from thermal fluctuations in shape. *J. Cell Biol.* 120: 923–934.

Gittes, F., E. Meyhofer, S. Baek, and J. Howard. 1996. Directional loading of the kinesin motor molecule as it buckles a microtubule. *Biophys. J.* 70: 418–429.

Goel, N. S., and N. Richter-Dyn. 1974. *Stochastic Models in Biology.* Academic Press, New York.

Goffeau, A., B. G. Barrell, H. Bussey, R. W. Davis, B. Dujon, H. Feldmann, F. Galibert, J. D. Hoheisel, C. Jacq, M. Johnston, E. J. Louis, H. W. Mewes, Y. Murakami, P. Philippsen, H. Tettelin, and S. G. Oliver. 1996. Life with 6000 genes. *Science* 274: 546, 563–547.

Gold, T. 1948. Hearing II. The physical basis of the action of the cochlea. *Proc. Roy. Soc. Lond. B.* 135: 492–498.

Goldschmidt-Clermont, P. J., M. I. Furman, D. Wachsstock, D. Safer, V. T. Nachmias, and T. D. Pollard. 1992. The control of actin nucleotide exchange by thymosin beta 4 and profilin. A potential regulatory mechanism for actin polymerization in cells. *Mol. Biol. Cell* 3: 1015–1024.

Goodenough, U. W., and J. E. Heuser. 1982. Substructure of the outer dynein arm. *J. Cell Biol.* 95: 798–815.

Goodenough, U., and J. Heuser. 1984. Structural comparison of purified dynein proteins with in situ dynein arms. *J. Mol. Biol.* 180: 1083–1118.

Goody, R. S., W. Hofmann, and G. H. Mannherz. 1977. The binding constant of ATP to myosin S1 fragment. *Eur. J. Biochem.* 78: 317–324.

Gordon, J. E. 1984. *The New Science of Strong Materials, or, Why You Don't Fall through the Floor.* Princeton University Press, Princeton, NJ.

Govindan, B., R. Bowser, and P. Novick. 1995. The role of Myo2, a yeast class V myosin, in vesicular transport. *J. Cell Biol.* 128: 1055–1068.

Greene, L. E., and E. Eisenberg. 1980. Dissociation of the actin.subfragment 1 complex by adenyl-5'-yl imidodiphosphate, ADP, and PPi. *J. Biol. Chem.* 255: 543–548.

Guharay, F., and F. Sachs. 1984. Stretch-activated single ion channel currents in tissue-cultured embryonic chick skeletal muscle. *J. Physiol. (Lond).* 352: 685–701.

Hackney, D. D. 1988. Kinesin ATPase: Rate-limiting ADP release. *Proc. Natl. Acad. Sci. USA* 85: 6314–6318.

Hackney, D. D. 1994a. Evidence for alternating head catalysis by kinesin during microtubule- stimulated ATP hydrolysis. *Proc. Natl. Acad. Sci. USA* 91: 6865–6869.

Hackney, D. D. 1994b. The rate-limiting step in microtubule-stimulated ATP hydrolysis by dimeric kinesin head domains occurs while bound to the microtubule. *J. Biol. Chem.* 269: 16508–16511.

Hackney, D. D. 1995. Highly processive microtubule-stimulated ATP hydrolysis by dimeric kinesin head domains. *Nature* 377: 448–450.

Hackney, D. D. 1996. The kinetic cycles of myosin, kinesin, and dynein. *Annu. Rev. Physiol.* 58: 731–750.

Hackney, D. D., J. D. Levitt, and J. Suhan. 1992. Kinesin undergoes a 9S to 6S conformational transition. *J. Biol. Chem.* 267: 8696–8701.

Hagerman, P. J. 1988. Flexibility of DNA. *Annu. Rev. Biophys. Biophys. Chem.* 17: 265–286.

Hamasaki, T., M. E. Holwill, K. Barkalow, and P. Satir. 1995. Mechanochemical aspects of axonemal dynein activity studied by in vitro microtubule translocation. *Biophys. J.* 69: 2569–2579.

Hancock, W. O., and J. Howard. 1998. Processivity of the motor protein kinesin requires two heads. *J. Cell Biol.* 140: 1395–1405.

Hancock, W. O., and J. Howard. 1999. Kinesin's processivity results from mechanical and chemical coordination between the ATP hydrolysis cycles of the two motor domains. *Proc. Natl. Acad. Sci. USA* 96: 13147–13152.

Happel, J., and H. Brenner. 1983. *Low Reynolds Number Hydrodynamics, with Special Applications to Particulate Media.* M. Nijhoff, The Hague.

Harada, Y., K. Sakurada, T. Aoki, D. D. Thomas, and T. Yanagida. 1990. Mechanochemical coupling in actomyosin energy transduction studied by in vitro movement assay. *J. Mol. Biol.* 216: 49–68.

Harrison, B. C., S. P. Marchese-Ragona, S. P. Gilbert, N. Cheng, A. C. Steven, and K. A. Johnson. 1993. Decoration of the microtubule surface by one kinesin head per tubulin heterodimer. *Nature* 362: 73–75.

Haugen, P. 1988. The stiffness under isotonic releases during a twitch of a frog muscle fibre. *Adv. Exp. Med. Biol.* 226: 461–471.

Hayden, J. H., S. S. Bowser, and C. L. Rieder. 1990. Kinetochores capture astral microtubules during chromosome attachment to the mitotic spindle: Direct visualization in live newt lung cells. *J. Cell Biol.* 111: 1039–1045.

He, Z. H., R. K. Chillingworth, M. Brune, J. E. Corrie, D. R. Trentham, M. R. Webb, and M. A. Ferenczi. 1997. ATPase kinetics on activation of rabbit and frog permeabilized isometric muscle fibres: a real time phosphate assay. *J. Physiol. (Lond).* 501: 125–148.

Heinemann, S. H., W. Stuhmer, and F. Conti. 1987. Single acetylcholine receptor channel currents recorded at high hydrostatic pressures. *Proc. Natl. Acad. Sci. USA* 84: 3229–3233.

Heins, S., P. C. Wong, S. Muller, K. Goldie, D. W. Cleveland, and U. Aebi. 1993. The rod domain of NF-L determines neurofilament architecture, whereas the end domains specify filament assembly and network formation. *J. Cell Biol.* 123: 1517–1533.

Heinz, W. F., and J. H. Hoh. 1999. Spatially resolved force spectroscopy of biological surfaces using the atomic force microscope. *Trends Biotech.* 4: 143–150.

Henningsen, U., and M. Schliwa. 1997. Reversal in the direction of movement of a molecular motor. *Nature* 389: 93–96.

Hibberd, M. G., J. A. Dantzig, D. R. Trentham, and Y. E. Goldman. 1985. Phosphate release and force generation in skeletal muscle fibers. *Science* 228: 1317–1319.

Higashi-Fujime, S., R. Ishikawa, H. Iwasawa, O. Kagami, E. Kurimoto, K. Kohama, and T. Hozumi. 1995. the fastest actin-based motor protein from the green algae, *Chara*, and its distinct mode of interaction with actin. *FEBS Lett.* 375: 151–154.

Highsmith, S. 1976. Interactions of the actin and nucleotide binding sites on myosin subfragment 1. *J. Biol. Chem.* 251: 6170–6172.

Higuchi, H., and Y. E. Goldman. 1991. Sliding distance between actin and myosin filaments per ATP molecule hydrolysed in skinned muscle fibres. *Nature* 352: 352–354.

Higuchi, H., T. Yanagida, and Y. E. Goldman. 1995. Compliance of thin filaments in skinned fibers of rabbit skeletal muscle. *Biophys. J.* 69: 1000–1010.

Hill, T. L. 1987. *Linear Aggregation Theory in Cell Biology*. Springer-Verlag, New York.

Hill, T. L. 1989. *Free Energy Transduction and Biochemical Cycle Kinetics*. Springer-Verlag, New York.

Hille, B. 1992. *Ionic Channels of Excitable Membranes*. Sinauer Associates, Sunderland, MA.

Hirakawa, E., H. Higuchi, and Y. Y. Toyoshima. 2000. Processive movement of single 22S dynein molecules occurs only at low ATP concentrations. *Proc. Natl. Acad. Sci. USA* 97: 2533–2537.

Hirokawa, N. 1998. Kinesin and dynein superfamily proteins and the mechanism of organelle transport. *Science* 279: 519–526.

Hirokawa, N., K. K. Pfister, H. Yorifuji, M. C. Wagner, S. T. Brady, and G. S. Bloom. 1989. Submolecular domains of bovine brain kinesin identified by electron microscopy and monoclonal antibody decoration. *Cell* 56: 867–878.

Hirose, K., A. Lockhart, R. A. Cross, and L. A. Amos. 1995. Nucleotide-dependent angular change in kinesin motor domain bound to tubulin. *Nature* 376: 277–279.

Hirose, K., A. Lockhart, R. A. Cross, and L. A. Amos. 1996. Three-dimensional cryoelectron microscopy of dimeric kinesin and ncd motor domains on microtubules. *Proc. Natl. Acad. Sci. USA* 93: 9539–9544.

Hirose, K., W. B. Amos, A. Lockhart, R. A. Cross, and L. A. Amos. 1997. Three-dimensional cryoelectron microscopy of 16-protofilament microtubules: Structure, polarity, and interaction with motor proteins. *J. Struct. Biol.* 118: 140–148.

Hirose, K., J. Lowe, M. Alonso, R. A. Cross, and L. A. Amos. 1999. Congruent docking of dimeric kinesin and ncd into three-dimensional electron cryomicroscopy maps of microtubule-motor ADP complexes. *Mol. Biol. Cell* 10: 2063–2074.

Hodgkin, A. L. 1964. *The Conduction of the Nervous Impulse*. Liverpool University Press, Liverpool.

Hohenadl, M., T. Storz, H. Kirpal, K. Kroy, and R. Merkel. 1999. Desmin filaments studied by quasi-elastic light scattering. *Biophys. J.* 77: 2199–2209.

Holmes, K. C. 1997. The swinging lever-arm hypothesis of muscle contraction. *Curr. Biol.* 7:R112–118.

Holmes, K. C. 1998. Muscle contraction. In L. Wolpert (ed.), *The Limits of Reductionism in Biology*, Novartis Foundation Symposium 213: 76–92, Wiley, Chichester.

Holmes, K. C., D. Popp, W. Gebhard, and W. Kabsch. 1990. Atomic model of the actin filament. *Nature* 347: 44–49.

Holy, T. E., and S. Leibler. 1994. Dynamic instability of microtubules as an efficient way to search in space. *Proc. Natl. Acad. Sci. USA* 91: 5682–5685.

Homsher, E., and N. C. Millar. 1990. Caged compounds and striated muscle contraction. *Annu. Rev. Physiol.* 52: 875–896.

Horio, T., and H. Hotani. 1986. Visualization of the dynamic instability of individual microtubules by dark-field microscopy. *Nature* 321: 605–607.

Horton, N., and M. Lewis. 1992. Calculation of the free energy of association for protein complexes. *Protein Sci.* 1: 169–181.

Hoshikawa, H., and R. Kamiya. 1985. Elastic properties of bacterial flagellar filaments. II. Determination of the modulus of rigidity. *Biophys. Chem.* 22: 159–166.

Hotani, H., and H. Miyamoto. 1990. Dynamic features of microtubules as visualized by dark-field microscopy. *Adv. Biophys.* 26: 135–156.

Houdusse, A., V. N. Kalabokis, D. Himmel, A. G. Szent-Györgi, and C. Cohen. 1999. Atomic structure of scallop myosin subfragment S1 complexed with MgADP: a novel conformation of the myosin head. *Cell* 97: 459–470.

Howard, J. 1996. The movement of kinesin along microtubules. *Annu. Rev. Physiol.* 58: 703–729.

Howard, J. 1997. Molecular motors: Structural adaptations to cellular functions. *Nature* 389: 561–567.

Howard, J. 1998. How molecular motors work in muscle. *Nature* 391: 239–240.

Howard, J., and J. F. Ashmore. 1986. Stiffness of sensory hair bundles in the sacculus of the frog. *Hear. Res.* 23: 93–104.

Howard, J., and A. J. Hudspeth. 1987a. Brownian motion of hair bundles from the frog's inner ear. *Biophys. J.* 51: 203a.

Howard, J., and A. J. Hudspeth. 1987b. Mechanical relaxation of the hair bundle mediates adaptation in mechanoelectrical transduction by the bullfrog's saccular hair cell. *Proc. Natl. Acad. Sci. USA* 84: 3064–3068.

Howard, J., and A. J. Hudspeth. 1988. Compliance of the hair bundle associated with gating of mechanoelectrical transduction channels in the bullfrog's saccular hair cell. *Neuron.* 1: 189–199.

Howard, J., and A. A. Hyman. 1993. Preparation of marked microtubules for the assay of the polarity of microtubule-based motors by fluorescence microscopy. *Methods Cell Biol.* 39: 105–113.

Howard, J., and J. A. Spudich. 1996. Is the lever arm of myosin a molecular elastic element? *Proc. Natl. Acad. Sci. USA* 93: 4462–4464.

Howard, J., W. M. Roberts, and A. J. Hudspeth. 1988. Mechanoelectrical transduction by hair cells. *Annu. Rev. Biophys. Biophys. Chem.* 17: 99–124.

Howard, J., A. J. Hudspeth, and R. D. Vale. 1989. Movement of microtubules by single kinesin molecules. *Nature* 342: 154–158.

Howard, J., A. J. Hunt, and S. Baek. 1993. Assay of microtubule movement driven by single kinesin molecules. *Methods Cell Biol.* 39: 137–147.

Hua, W., E. C. Young, M. L. Fleming, and J. Gelles. 1997. Coupling of kinesin steps to ATP hydrolysis. *Nature* 388: 390–393.

Huang, T. G., and D. D. Hackney. 1994. *Drosophila* kinesin minimal motor domain expressed in *Escherichia coli*: Purification and kinetic characterization. *J. Biol. Chem.* 269: 16493–16501.

Hudspeth, A. J., W. M. Roberts, and J. Howard. 1990. Gating compliance, a reduction in hair bundle stiffness associated with the gating of transduction channels in hair cells from the bullfrog's sacculus. In J. P. Wilson and D. T. Kemp (eds.), *Cochlear Mechanisms: Structure, Function, and Models*. Plenum Press, New York, pp. 117–123.

Hunt, A. J., F. Gittes, and J. Howard. 1994. The force exerted by a single kinesin molecule against a viscous load. *Biophys. J.* 67: 766–781.

Hunter, A. W., and L. Wordeman. 2000. How motor proteins influence microtubule polymerization dynamics. *J. Cell Sci.* 113: 4379–4389.

Huxley, A. 1980. *Reflections on Muscle*. Princeton University Press, Princeton, NJ.

Huxley, A. 1998. How molecular motors work in muscle. *Nature* 391: 239–240.

Huxley, A. F. 1957. Muscle structure and theories of contraction. *Prog. Biophys. Biophys. Chem.* 7: 255–318.

Huxley, A. F., and R. Niedergerke. 1954. Structural changes in muscle during contraction. Interference microscopy of living cells. *Nature* 173: 971–976.

Huxley, A. F., and R. M. Simmons. 1971. Proposed mechanism of force generation in striated muscle. *Nature* 233: 533–538.

Huxley, H. E. 1953. Electron microscope studies of the organization of the filaments in striated muscle. *Biochim. Biophys. Acta.* 12: 387–394.

Huxley, H. E. 1957. The double array of filaments in cross-striated muscle. *J. Biophys. Biochem. Cytol.* 3: 631–648.

Huxley, H. E. 1958. The contraction of muscle. *Sci. Am.* 199: 66–82.

Huxley, H. E. 1963. Electron microscopy studies of the structure of natural and synthetic protein filaments from striated muscles. *J. Mol. Biol.* 7: 281–308.

Huxley, H. E. 1996. A personal view of muscle and motility mechanisms. *Annu. Rev. Physiol.* 58: 1–19.

Huxley, H. E., and J. Hanson. 1954. Changes in the cross-striations of muscle during contraction and stretch and their structural interpretion. *Nature* 173: 973–976.

Huxley, H. E., A. Stewart, H. Sosa, and T. Irving. 1994. X-ray diffraction measurements of the extensibility of actin and myosin filaments in contracting muscle. *Biophys. J.* 67: 2411–2421.

Hvidt, S., F. H. Nestler, M. L. Greaser, and J. D. Ferry. 1982. Flexibility of myosin rod determined from dilute solution viscoelastic measurements. *Biochemistry* 21: 4064–4073.

Hyman, A. A., S. Salser, D. N. Drechsel, N. Unwin, and T. J. Mitchison. 1992. Role of GTP hydrolysis in microtubule dynamics: Information from a slowly hydrolyzable analogue, GMPCPP. *Mol. Biol. Cell* 3: 1155–1167.

Inagaki, M., Y. Gonda, S. Ando, S. Kitamura, Y. Nishi, and C. Sato. 1989. Regulation of assembly-disassembly of intermediate filaments in vitro. *Cell. Struct. Funct.* 14: 279–286.

Ingber, D. E. 1997. Tensegrity: The architectural basis of cellular mechanotransduction. *Annu. Rev. Physiol.* 59: 575–599.

Ingber, D. E. 1998. The architecture of life. *Sci. Am.* 278 (Jan.): 48–57.

Isambert, H., P. Venier, A. C. Maggs, A. Fattoum, R. Kassab, D. Pantaloni, and M. F. Carlier. 1995. Flexibility of actin filaments derived from thermal fluctuations. Effect of bound nucleotide, phalloidin, and muscle regulatory proteins. *J. Biol. Chem.* 270: 11437–11444.

Ishijima, A., H. Kojima, T. Funatsu, M. Tokunaga, H. Higuchi, H. Tanaka, and T. Yanagida. 1998. Simultaneous observation of individual ATPase and mechanical events by a single myosin molecule during interaction with actin. *Cell* 92: 161–171.

Israelachvili, J. N. 1992. *Intermolecular and Surface Forces*, 2nd ed. Academic Press, San Diego.

Iwatani, S., A. H. Iwane, H. Higuchi, Y. Ishii, and T. Yanagida. 1999. Mechanical and chemical properties of cysteine-modified kinesin molecules. *Biochemistry* 38: 10318–10323.

Jacobs, R. A., and A. J. Hudspeth. 1990. Ultrastructural correlates of mechanoelectrical transduction in hair cells of the bullfrog's internal ear. *Cold Spring Harb Symp Quant Biol.* 55: 547–561.

Jaeger, J. C., and A. M. Starfield. 1974. *An Introduction to Applied Mathematics*. Clarendon Press, Oxford,.

Jeffrey, D. J., and Y. Onishi. 1981. The slow motion of a cylinder next to a plane wall. *Quant. J. Mech. Appl. Math.* 34: 129–137.

Jencks, W. P. 1981. On the attribution and additivity of binding energies. *Proc. Natl. Acad. Sci. USA.* 78: 4046–4050.

Jewell, B. R., and D. R. Wilkie. 1958. An analysis of the mechanical components in frog's striated muscle. *J. Physiol. Lond.* 143: 515–540.

Johnson, K. A. 1985. Pathway of the microtubule-dynein ATPase and the structure of dynein: a comparison with actomyosin. *Annu. Rev. Biophys. Biophys. Chem.* 14: 161–188.

Jontes, J. D., E. M. Wilson-Kubalek, and R. A. Milligan. 1995. A 32 degree tail swing in brush border myosin I on ADP release. *Nature* 378: 751–753.

Jung, G., E. D. Korn, and J. A. d. Hammer. 1987. The heavy chain of *Acanthamoeba* myosin IB is a fusion of myosin-like and non-myosin-like sequences. *Proc. Natl. Acad. Sci. USA* 84: 6720–6724.

Junge, D. 1992. *Nerve and Muscle Excitation*, 3rd ed. Sinauer Associates, Sunderland, MA.

Kabsch, W., H. G. Mannherz, D. Suck, E. F. Pai, and K. C. Holmes. 1990. Atomic structure of the actin:DNase I complex. *Nature* 347: 37–44.

Kasai, M., and F. Oosawa. 1969. Behavior of divalent cations and nucleotides bound to F-actin. *Biochim. Biophys. Acta.* 172: 300–310.

Katz, B. 1966. *Nerve, Muscle, and Synapse*. McGraw-Hill, New York.

Kawaguchi, K., and S. Ishiwata. 2000. Temperature dependence of force, velocity, and processivity of single kinesin molecules. *Biochem. Biophys. Res. Commun.* 272: 895–899.

Kawaguchi, K., and S. Ishiwata. In press. Bimodal nucleotide-dependent binding of single kinesin molecules.

Kaye, G. W. C., and T. H. Laby. 1986. *Tables of Physical and Chemical Constants and Some Mathematical Functions*. Longman, New York.

Kellermayer, M. S., S. B. Smith, H. L. Granzier, and C. Bustamante. 1997. Folding-unfolding transitions in single titin molecules characterized with laser tweezers. *Science* 276: 1112–1116.

Kikkawa, M., T. Ishikawa, T. Nakata, T. Wakabayashi, and N. Hirokawa. 1994. Direct visualization of the microtubule lattice seam both in vitro and in vivo. *J. Cell Biol.* 127: 1965–1971.

Kikkawa, M., Y. Okada, and N. Hirokawa. 2000. A resolution model of the monomeric kinesin motor, KIF1A. *Cell* 100: 241–252.

King, S. J., and T. A. Schroer. 2000. Dynactin increases the processivity of the cytoplasmic dynein motor. *Nat. Cell Biol.* 2: 20–24.

Kinosian, H. J., L. A. Selden, J. E. Estes, and L. C. Gershman. 1993. Nucleotide binding to actin. Cation dependence of nucleotide dissociation and exchange rates. *J. Biol. Chem.* 268: 8683–8691.

Kirschner, M., and T. Mitchison. 1986. Beyond self-assembly: from microtubules to morphogenesis. *Cell* 45: 329–342.

Kitamura, K., M. Tokunaga, A. H. Iwane, and T. Yanagida. 1999. A single myosin head moves along an actin filament with regular steps of 5.3 nanometres. *Nature* 397: 129–134.

Kittel, C. 1996. *Introduction to Solid State Physics*. Wiley, New York.

Kocks, C., E. Gouin, M. Tabouret, P. Berche, H. Ohayon, and P. Cossart. 1992. *L. monocytogenes*-induced actin assembly requires the actA gene product, a surface protein. *Cell* 68: 521–531.

Kojima, H., A. Ishijima, and T. Yanagida. 1994. Direct measurement of stiffness of single actin filaments with and without tropomyosin by in vitro nanomanipulation. *Proc. Natl. Acad. Sci. USA* 91: 12962–12966.

Kojima, H., E. Muto, H. Higuchi, and T. Yanagida. 1997. Mechanics of single kinesin molecules measured by optical trapping nanometry. *Biophys. J.* 73: 2012–2022.

Kozielski, F., S. Sack, A. Marx, M. Thormahlen, E. Schonbrunn, V. Biou, A. Thompson, E. M. Mandelkow, and E. Mandelkow. 1997. The crystal structure of dimeric kinesin and implications for microtubule-dependent motility. *Cell* 91: 985–994.

Kozminski, K. G., K. A. Johnson, P. Forscher, and J. L. Rosenbaum. 1993. A motility in the eukaryotic flagellum unrelated to flagellar beating. *Proc. Natl. Acad. Sci. USA* 90: 5519–5523.

Kozminski, K. G., P. L. Beech, and J. L. Rosenbaum. 1995. The *Chlamydomonas* kinesin-like protein FLA10 is involved in motility associated with the flagellar membrane. *J. Cell Biol.* 131: 1517–1527.

Kramers, H. A. 1940. Brownian motion in a field of force and the diffusion model of chemical reactions. *Physica.* 7: 284–304.

Krammer, A., H. Lu, B. Isralewitz, K. Schulten, and V. Vogel. 1999. Forced unfolding of the fibronectin type III module reveals a tensile molecular recognition switch. *Proc. Natl. Acad. Sci. USA* 96: 1351–1356.

Kreis, T., and R. Vale. 1999. *Guidebook to the Cytoskeletal and Motor Proteins.* Oxford University Press, New York.

Kreis, T. E., B. Geiger, and J. Schlessinger. 1982. Mobility of microinjected rhodamine actin within living chicken gizzard cells determined by fluorescence photobleaching recovery. *Cell* 29: 835–845.

Kühne, W., Wagenschieber, H. Kühne, C. A. Kofoid, and Lees Museum. 1864. *Untersuchungen über das Protoplasma und die Contractilität.* W. Engelmann, Leipzig.

Kull, F. J., E. P. Sablin, R. Lau, R. J. Fletterick, and R. D. Vale. 1996. Crystal structure of the kinesin motor domain reveals a structural similarity to myosin. *Nature* 380: 550–555.

Kull, F. J., R. D. Vale, and R. J. Fletterick. 1998. The case for a common ancestor: kinesin and myosin motor proteins and G proteins. *J. Muscle Res. Cell Motil.* 19: 877–886.

Kurimoto, E., and R. Kamiya. 1991. Microtubule sliding in flagellar axonemes of *Chlamydomonas* mutants missing inner- or outer-arm dynein: Velocity measurements on new types of mutants by an improved method. *Cell Motil Cytoskeleton.* 19: 275–281.

Kurz, J. C., and R. C. Williams, Jr. 1995. Microtubule-associated proteins and the flexibility of microtubules. *Biochemistry* 34: 13374–13380.

Kushmerick, M. J., and R. E. Davies. 1969. The chemical energetics of muscle contraction. II. The chemistry, efficiency and power of maximally working sartorius muscles. Appendix. Free energy and enthalpy of atp hydrolysis in the sarcoplasm. *Proc. R. Soc. Lond. B.* 174: 315–353.

Landau, L. D., and E. M. Lifshits. 1970. *Theory of Elasticity.* Pergamon Press, New York.

Landau, L. D., and E. M. Lifshits. 1976. *Mechanics.* Pergamon Press, New York.

Landau, L. D., and E. M. Lifshits. 1987. *Fluid Mechanics.* Pergamon Press, New York.

Landau, L. D., E. M. Lifshits, and L. P. Pitaevskii. 1980. *Statistical Physics.* Pergamon Press, New York.

Langevin, P. 1908. Sur la theorie du movement Brownien. *C. R. Acad. Sci.* 146: 530–532.

Lasek, R. J., and S. T. Brady. 1985. Attachment of transported vesicles to microtubules in axoplasm is facilitated by AMP-PNP. *Nature* 316: 645–647.

Lecar, H., and C. E. Morris. 1993. Biophysics of mechanotransduction. *In* G. M. Rubanyi (ed.), *Mechanoreception by the Vascular Wall.* Futura Publishing Co., Mount Kisco, New York, pp. 1–11.

Leibler, S., and D. A. Huse. 1993. Porters versus rowers: a unified stochastic model of motor proteins. *J. Cell Biol.* 121: 1357–1368.

Levitt, M. 1974. Energy refinement of hen egg-white lysozyme. *J. Mol. Biol.* 82: 393–420.

Lewis, E. R., E. L. Leverenz, and W. S. Bialek. 1985. *The Vertebrate Inner Ear.* CRC Press, Boca Raton, FL.

Lipmann, F. 1941. Metabolic generation and utilization of phosphate bond energy. *Adv. Enz.* 1: 99–162.

Lighthill, M. J. 1975. *Mathematical Biofluiddynamics.* Society for Industrial and Applied Mathematics, Philadelphia.

Littlefield, R., and V. M. Fowler. 1998. Defining actin filament length in striated muscle: Rulers and caps or dynamic stability? *Annu. Rev. Cell. Dev. Biol.* 14: 487–525.

Lockhart, A., and R. A. Cross. 1994. Origins of reversed directionality in the ncd molecular motor. *EMBO J.* 13: 751–757.

Lockhart, A., and R. A. Cross. 1996. Kinetics and motility of the Eg5 microtubule motor. *Biochemistry* 35: 2365–2373.

Lockhart, A., I. M. Crevel, and R. A. Cross. 1995. Kinesin and ncd bind through a single head to microtubules and compete for a shared MT binding site. *J. Mol. Biol.* 249: 763–771.

Lodish, H. F., A. Berk, S. L. Zipursky, P. Matsudaira, D. Baltimore, and J. E. Darnell. 2000. *Molecular Cell Biology,* 4th ed. W. H. Freeman and Company, New York.

Lohmann, K. 1929. Ueber die Pyrophatfraktion im Muskel. *Naturwissensch.* 17: 624–625.

Loisel, T. P., R. Boujemaa, D. Pantaloni, and M. F. Carlier. 1999. Reconstitution of actin-based motility of *Listeria* and *Shigella* using pure proteins. *Nature* 401: 613–616.

Lombardi, V., G. Piazzesi, and M. Linari. 1992. Rapid regeneration of the actin-myosin power stroke in contracting muscle. *Nature* 355: 638–641.

Lorenz, M., D. Popp, and K. C. Holmes. 1993. Refinement of the F-actin model against X-ray fiber diffraction data by the use of a directed mutation algorithm. *J. Mol. Biol.* 234: 826–836.

Lowe, J., and L. A. Amos. 1998. Crystal structure of the bacterial cell-division protein FtsZ. *Nature* 391: 203–206.

Lowey, S., H. S. Slayter, A. G. Weeds, and H. Baker. 1969. Substructure of the myosin molecule. I. Subfragments of myosin by enzymic degradation. *J. Mol. Biol.* 42: 1–29.

Luby-Phelps, K., P. E. Castle, D. L. Taylor, and F. Lanni. 1987. Hindered diffusion of inert tracer particles in the cytoplasm of mouse 3T3 cells. *Proc. Natl. Acad. Sci. USA* 84: 4910–4913.

Lupas, A. 1996. Coiled coils: New structures and new functions. *Trends Biochem. Sci.* 21: 375–382.

Lymn, R. W., and E. W. Taylor. 1971. Mechanism of adenosine triphosphate hydrolysis by actomyosin. *Biochemistry* 10: 4617–4624.

Ma, Y. Z., and E. W. Taylor. 1994. Kinetic mechanism of myofibril ATPase. *Biophys. J.* 66: 1542–1553.

Ma, Y. Z., and E. W. Taylor. 1995a. Kinetic mechanism of kinesin motor domain. *Biochemistry* 34: 13233–13241.

Ma, Y. Z., and E. W. Taylor. 1995b. Mechanism of microtubule kinesin ATPase. *Biochemistry* 34: 13242–13251.

Ma, Y. Z., and E. W. Taylor. 1997. Interacting head mechanism of microtubule-kinesin ATPase. *J. Biol. Chem.* 272: 724–730.

Machesky, L. M., S. J. Atkinson, C. Ampe, J. Vandekerckhove, and T. D. Pollard. 1994. Purification of a cortical complex containing two unconventional actins from *Acanthamoeba* by affinity chromatography on profilin-agarose. *J. Cell Biol.* 127: 107–115.

Machesky, L. M., R. D. Mullins, H. N. Higgs, D. A. Kaiser, L. Blanchoin, R. C. May, M. E. Hall, and T. D. Pollard. 1999. Scar, a WASp-related protein, activates nucleation of actin filaments by the Arp2/3 complex. *Proc. Natl. Acad. Sci. USA* 96: 3739–3744.

Magnasco, M. O. 1994. Molecular combustion motors. *Phys. Rev. Lett.* 72: 2656–2659.

Mandelkow, E. M., R. Schultheiss, R. Rapp, M. Muller, and E. Mandelkow. 1986. On the surface lattice of microtubules: helix starts, protofilament number, seam, and handedness. *J. Cell Biol.* 102: 1067–1073.

Mandelkow, E. M., E. Mandelkow, and R. A. Milligan. 1991. Microtubule dynamics and microtubule caps: a time-resolved cryo-electron microscopy study. *J. Cell Biol.* 114: 977–991.

Mandelkow, E., Y.-H. Song, and E.-M. Mandeljkow. 1995. The microtubule lattice—dynamic instability of concepts. *Trends Cell Biol.* 5: 262–266.

Manning, B. D., and M. Snyder. 2000. Drivers and passengers wanted! The role of kinesin-associated proteins. *Trends Cell Biol.* 10: 281–289.

Marcus, R. A. 1996. Electron transfer in chemistry. Theory and experiment. *In* D. S. Bendall (ed.), *Protein Electron Transfer*. BIOS Scientific Publishers, Oxford, pp. 249–272.

Margolis, R. L., and L. Wilson. 1978. Opposite end assembly and disassembly of microtubules at steady state in vitro. *Cell* 13: 1–8.

Margossian, S. S., and S. Lowey. 1978. Interaction of myosin subfragments with F-actin. *Biochemistry* 17: 5431–5439.

Markin, V. S., and A. J. Hudspeth. 1995. Gating-spring models of mechanoelectrical transduction by hair cells of the internal ear. *Annu. Rev. Biophys. Biomol. Struct.* 24: 59–83.

Marszalek, P. E., H. Lu, H. Li, M. Carrion-Vazquez, A. F. Oberhauser, K. Schulten, and J. M. Fernandez. 1999. Mechanical unfolding intermediates in titin modules. *Nature* 402: 100–103.

Martonosi, A., M. A. Gouvea, and J. Gergely. 1960. Studies on actin. *J. Biol. Chem.* 235: 1700–1703.

McCammon, J. A., and S. C. Harvey. 1987. *Dynamics of Proteins and Nucleic Acids*. Cambridge University Press, New York.

McCammon, J. A., B. R. Gelin, M. Karplus, and P. G. Wolynes. 1976. The hinge-bending mode in lysozyme. *Nature* 262: 325–326.

McCammon, J. A., B. R. Gelin, and M. Karplus. 1977. Dynamics of folded proteins. *Nature* 267: 585–590.

McDonald, H. B., R. J. Stewart, and L. S. Goldstein. 1990. The kinesin-like ncd protein of *Drosophila* is a minus end-directed microtubule motor. *Cell* 63: 1159–1165.

McLachlan, A. D., and J. Karn. 1983. Periodic features in the amino acid sequence of nematode myosin rod. *J. Mol. Biol.* 164: 605–626.

McNally, F. J., and R. D. Vale. 1993. Identification of katanin, an ATPase that severs and disassembles stable microtubules. *Cell* 75: 419–429.

Mehta, A. D., J. T. Finer, and J. A. Spudich. 1997. Detection of single-molecule interactions using correlated thermal diffusion. *Proc. Natl. Acad. Sci. USA* 94: 7927–7931.

Mehta, A. D., R. S. Rock, M. Rief, J. A. Spudich, M. S. Mooseker, and R. E. Cheney. 1999. Myosin-V is a processive actin-based motor. *Nature* 400: 590–593.

Mendelson, R. A., M. F. Morales, and J. Botts. 1973. Segmental flexibility of the S-1 moiety of myosin. *Biochemistry* 12: 2250–2255.

Mercer, J. A., P. K. Seperack, M. C. Strobel, N. G. Copeland, and N. A. Jenkins. 1991. Novel myosin heavy chain encoded by murine dilute coat colour locus. *Nature* 349: 709–713.

Meyhöfer, E., and J. Howard. 1995. The force generated by a single kinesin molecule against an elastic load. *Proc. Natl. Acad. Sci. USA* 92: 574–578.

Mickey, B., and J. Howard. 1995. Rigidity of microtubules is increased by stabilizing agents. *J. Cell Biol.* 130: 909–917.

Milisav, I. 1998. Dynein and dynein-related genes. *Cell Motil Cytoskeleton.* 39: 261–272.

Millar, N. C., and M. A. Geeves. 1983. The limiting rate of the ATP-mediated dissociation of actin from rabbit skeletal muscle myosin subfragment 1. *FEBS Lett.* 160: 141–148.

Miller, R. H., and R. J. Lasek. 1985. Cross-bridges mediate anterograde and retrograde vesicle transport along microtubules in squid axoplasm. *J. Cell Biol.* 101: 2181–2193.

Mitchison, T., and M. Kirschner. 1984a. Dynamic instability of microtubule growth. *Nature* 312: 237–242.

Mitchison, T., and M. Kirschner. 1984b. Microtubule assembly nucleated by isolated centrosomes. *Nature* 312: 232–237.

Mitchison, T. J. 1989. Polewards microtubule flux in the mitotic spindle: Evidence from photoactivation of fluorescence. *J. Cell Biol.* 109: 637–652.

Mitchison, T. J. 1993. Localization of an exchangeable GTP binding site at the plus end of microtubules. *Science* 261: 1044–1047.

Mockrin, S. C., and E. D. Korn. 1980. *Acanthamoeba* profilin interacts with G-actin to increase the rate of exchange of actin-bound adenosine 5′-triphosphate. *Biochemistry* 19: 5359–5362.

Moerner, W. E., and L. Kador. 1989. Optical detection and spectroscopy of single molecules in a solid. *Phys. Rev. Lett.* 62: 2535–2538.

Mogilner, A., and G. Oster. 1996. Cell motility driven by actin polymerization. *Biophys. J.* 71: 3030–3045.

Molloy, J. E., J. E. Burns, J. Kendrick-Jones, R. T. Tregear, and D. C. White. 1995a. Movement and force produced by a single myosin head. *Nature* 378: 209–212.

Molloy, J. E., J. E. Burns, J. C. Sparrow, R. T. Tregear, J. Kendrick-Jones, and D. C. S. White. 1995b. Single-molecule mechanics of heavy meromyosin and S1 interacting with rabbit or *Drosophila* actins using optical tweezers. *Biophys. J.* 68: 298s–305s.

Moore, P. B., H. E. Huxley, and D. J. DeRosier. 1970. Three-dimensional reconstruction of F-actin, thin filaments and decorated thin filaments. *J. Mol. Biol.* 50: 279–295.

Moore, W. J. 1972. *Physical Chemistry*. Prentice-Hall, Englewood Cliffs, NJ.

Morimatsu, M., A. Nakamura, H. Sumiyoshi, N. Sakaba, H. Taniguchi, K. Kohama, and S. Higashi-Fujime. 2000. The molecular structure of the fastest myosin from green algae, *Chara*. *Biochem. Biophys. Res. Commun.* 270: 147–152.

Murray, A. W., and T. Hunt. 1993. *The Cell Cycle: An Introduction*. W. H. Freeman and Company, New York.

Nagashima, H., and S. Asakura. 1980. Dark-field light microscopic study of the flexibility of F-actin complexes. *J. Mol. Biol.* 136: 169–182.

Nakata, T., S. Terada, and N. Hirokawa. 1998. Visualization of the dynamics of synaptic vesicle and plasma membrane proteins in living axons. *J. Cell Biol.* 140: 659–674.

Needham, D. M. 1971. *Machina Carnis: The Biochemistry of Muscular Contraction in Its Historical Development*. Cambridge University Press, Cambridge.

Neher, E., and B. Sakmann. 1976. Single-channel currents recorded from membrane of denervated frog muscle fibres. *Nature* 260: 799–802.

Neuwald, A. F., L. Aravind, J. L. Spouge, and E. V. Koonin. 1999. AAA+: A class of chaperone-like ATPases associated with the assembly, operation, and disassembly of protein complexes. *Genome Res.* 9: 27–43.

Nicklas, R. B. 1983. Measurements of the force produced by the mitotic spindle in anaphase. *J. Cell Biol.* 97: 542–548.

Nicklas, R. B. 1988. The forces that move chromosomes in mitosis. *Annu. Rev. Biophys. Biophys. Chem.* 17: 431–449.

Nogales, E., S. G. Wolf, and K. H. Downing. 1998. Structure of the alpha beta tubulin dimer by electron crystallography. *Nature* 391: 199–203.

Nogales, E., M. Whittaker, R. A. Milligan, and K. H. Downing. 1999. High-resolution model of the microtubule. *Cell* 96: 79–88.

Noji, H., R. Yasuda, M. Yoshida, and K. Kinosita, Jr. 1997. Direct observation of the rotation of F1-ATPase. *Nature* 386: 299–302.

Nonaka, S., Y. Tanaka, Y. Okada, S. Takeda, A. Harada, Y. Kanai, M. Kido, and N. Hirokawa. 1998. Randomization of left-right asymmetry due to loss of nodal cilia generating leftward flow of extraembryonic fluid in mice lacking KIF3B motor protein. *Cell* 95: 829–837.

Northrup, S. H., and H. P. Erickson. 1992. Kinetics of protein-protein association explained by Brownian dynamics computer simulation. *Proc. Natl. Acad. Sci. USA* 89: 3338–3342.

O'Shea, E. K., J. D. Klemm, P. S. Kim, and T. Alber. 1991. X-ray structure of the GCN4 leucine zipper, a two-stranded, parallel coiled coil. *Science* 254: 539–544.

Oakley, B. R., C. E. Oakley, Y. Yoon, and M. K. Jung. 1990. Gamma-tubulin is a component of the spindle pole body that is essential for microtubule function in *Aspergillus nidulans*. *Cell* 61: 1289–1301.

Okada, Y., and N. Hirokawa. 1999. A processive single-headed motor: Kinesin superfamily protein KIF1A. *Science* 283: 1152–1157.

Okada, Y., and N. Hirokawa. 2000. Mechanism of the single-headed processivity: diffusional anchoring between the K-loop of kinesin and the C terminus of tubulin. *Proc. Natl. Acad. Sci. USA* 97: 640–645.

Okada, Y., H. Yamazaki, Y. Sekine-Aizawa, and N. Hirokawa. 1995. The neuron-specific kinesin superfamily protein KIF1A is a unique monomeric motor for anterograde axonal transport of synaptic vesicle precursors. *Cell* 81: 769–780.

Okuno, M. 1980. Inhibition and relaxation of sea urchin sperm flagella by vanadate. *J. Cell Biol.* 85: 712–725.

Okuno, M., and Y. Hiramoto. 1979. Direct measurements of the stiffness of echinoderm sperm flagella. *J. exp. Biol.* 79: 235–243.

Ostap, E. M., and T. D. Pollard. 1996. Biochemical kinetic characterization of the *Acanthamoeba* myosin-I ATPase. *J. Cell Biol.* 132: 1053–1060.

Oster, G. F., and A. S. Perelson. 1987. The physics of cell motility. *J. Cell Sci. Suppl.* 8: 35–54.

Papoulis, A. 1991. *Probability, Random Variables, and Stochastic Processes*. McGraw-Hill, New York.

Parry, D. A. D., and P. M. Steinert. 1995. *Intermediate Filament Structure*. Springer-Verlag, New York.

Paschal, B. M., H. S. Shpetner, and R. B. Vallee. 1987. MAP 1C is a microtubule-activated ATPase which translocates microtubules in vitro and has dynein-like properties. *J. Cell Biol.* 105: 1273–1282.

Pasternak, C., J. A. Spudich, and E. L. Elson. 1989. Capping of surface receptors and concomitant cortical tension are generated by conventional myosin. *Nature* 341: 549–551.

Pate, E., and R. Cooke. 1989. Addition of phosphate to active muscle fibers probes actomyosin states within the powerstroke. *Pflugers Arch.* 414: 73–81.

Pate, E., H. White, and R. Cooke. 1993. Determination of the myosin step size from mechanical and kinetic data. *Proc. Natl. Acad. Sci. USA* 90: 2451–2455.

Pate, E., G. J. Wilson, M. Bhimani, and R. Cooke. 1994. Temperature dependence of the inhibitory effects of orthovanadate on shortening velocity in fast skeletal muscle. *Biophys. J.* 66: 1554–1562.

Pauling, L. 1970. *General Chemistry*. W. H. Freeman and Company, San Francisco.

Pauling, L., and R. B. Corey. 1953. Compound helical configurations of polypeptide chains: Structure of proteins of the α-keratin type. *Nature* 171: 59–61.

Penrose, R. 1994. *Shadows of the Mind: A Search for the Missing Science of Consciousness*. Oxford University Press, New York.

Perrin, F. 1934. Mouvement Brownien d'un ellipsoïd (I). Dispersion diélectrique pour des molécules elllipsoïdales. *Le Journal de Physique et le Radium*, Série 7. 5: 497–511.

Perrin, F. 1936. Mouvement Brownien d'un ellipsoïd (II). Rotation libre et dépolarisation des fluorescences. Translation et diffusion de molécules ellipsoïdales. *Le Journal de Physique et le Radium*, Série 7. 7: 1–11.

Peskin, C. S., G. M. Odell, and G. F. Oster. 1993. Cellular motions and thermal fluctuations: The Brownian ratchet. *Biophys. J.* 65: 316–324.

Peskin, C. S., and G. F. Oster. 1995. Coordinated hydrolysis explains the mechanical behavior of kinesin. *Biophys. J.* 68: 202S–210S.

Peteanu, L. A., R. W. Schoenlein, Q. Wang, R. A. Mathies, and C. V. Shank. 1993. The first step in vision occurs in femtoseconds: Complete blue and red spectral studies. *Proc. Natl. Acad. Sci. USA* 90: 11762–11766.

Phillips, G. N., Jr. 1992. What is the pitch of the alpha-helical coiled coil? *Proteins*. 14: 425–429.

Phillips, G. N., Jr., and S. Chacko. 1996. Mechanical properties of tropomyosin and implications for muscle regulation. *Biopolymers*. 38: 89–95.

Pickles, J. O. 1988. *An Introduction to the Physiology of Hearing*. Academic Press, San Diego.

Pierce, D. W., N. Hom-Booher, and R. D. Vale. 1997. Imaging individual green fluorescent proteins. *Nature* 388: 338.

Pollard, T. D. 1986. Rate constants for the reactions of ATP- and ADP-actin with the ends of actin filaments. *J. Cell Biol.* 103: 2747–2754.

Pollard, T. D., and E. D. Korn. 1973. *Acanthamoeba* myosin. I. Isolation from *Acanthamoeba castellanii* of an enzyme similar to muscle myosin. *J. Biol. Chem.* 248: 4682–4690.

Pollard, T. D., and A. G. Weeds. 1984. The rate constant for ATP hydrolysis by polymerized actin. *FEBS Lett.* 170: 94–98.

Pollard, T. D., L. Blanchoin, and R. D. Mullins. 2000. Molecular mechanisms controlling actin filament dynamics in nonmuscle cells. *Annu. Rev. Biophys. Biomol. Struct.* 29: 545–576.

Press, W. H. 1997. *Numerical Recipes in C: The Art of Scientific Computing*. Cambridge University Press, New York.

Quinlan, R., C. Hutchison, and B. Lane. 1995. Intermediate filament proteins. *Protein Profile.* 2: 795–952.

Raff, E. C. 1994. The role of multiple tubulin isoforms in cellular microtubule function. *In* J. S. Hyams and C. W. Lloyd (eds.), *Microtubules*, Wiley-Liss, New York, pp. 85–109.

Ray, S., E. Meyhofer, R. A. Milligan, and J. Howard. 1993. Kinesin follows the microtubule's protofilament axis. *J. Cell Biol.* 121: 1083–1093.

Rayment, I. 1996. Kinesin and myosin: molecular motors with similar engines. *Structure* 4: 501–504.

Rayment, I., H. M. Holden, M. Whittaker, C. B. Yohn, M. Lorenz, K. C. Holmes, and R. A. Milligan. 1993a. Structure of the actin-myosin complex and its implications for muscle contraction. *Science* 261: 58–65.

Rayment, I., W. R. Rypniewski, K. Schmidt-Base, R. Smith, D. R. Tomchick, M. M. Benning, D. A. Winkelmann, G. Wesenberg, and H. M. Holden. 1993b. Three-dimensional structure of myosin subfragment-1: A molecular motor. *Science* 261: 50–58.

Resnick, R., D. Halliday, and K. S. Krane. 1992. *Physics*. Wiley, New York.

Rice, S., A. W. Lin, D. Safer, C. L. Hart, N. Naber, B. O. Carragher, S. M. Cain, E. Pechatnikova, E. M. Wilson-Kubalek, M. Whittaker, E. Pate, R. Cooke, E. W. Taylor, R. A. Milligan, and R. D. Vale. 1999. A structural change in the kinesin motor protein that drives motility. *Nature* 402: 778–784.

Rice, S. O. 1954. Mathematical analysis of random noise. *In* N. Wax (ed.), *Selected Papers on Noise and Stochastic Processes*, Dover, New York, pp. 133–294.

Rickwood, D. 1984. *Centrifugation: A Practical Approach*. IRL Press, Washington, D.C.

Rieder, C. L., E. A. Davison, L. C. Jensen, L. Cassimeris, and E. D. Salmon. 1986. Oscillatory movements of monooriented chromosomes and their position relative to the spindle pole result from the ejection properties of the aster and half-spindle. *J. Cell Biol.* 103: 581–591.

Rief, M., M. Gautel, F. Oesterhelt, J. M. Fernandez, and H. E. Gaub. 1997. Reversible unfolding of individual titin immunoglobulin domains by AFM. *Science* 276: 1109–1112.

Rief, M., R. S. Rock, A. D. Mehta, M. S. Mooseker, R. E. Cheney, and J. A. Spudich. 2000. Myosin-V stepping kinetics: A molecular model for processivity. *Proc. Natl. Acad. Sci. USA* 97: 9482–9486.

Rivolta, M. N., R. Urrutia, and B. Kachar. 1995. A soluble motor from the alga *Nitella* supports fast movement of actin filaments in vitro. *Biochim. Biophys. Acta.* 1232: 1–4.

Rodionov, V. I., and G. G. Borisy. 1997. Microtubule treadmilling in vivo. *Science* 275: 215–218.

Rosenbaum, J. L., and F. M. Child. 1967. Flagellar regeneration in protozoan flagellates. *J. Cell Biol.* 34: 345–364.

Rosenfeld, S. S., B. Rener, J. J. Correia, M. S. Mayo, and H. C. Cheung. 1996. Equilibrium studies of kinesin-nucleotide intermediates. *J. Biol. Chem.* 271: 9473–9482.

Rousselet, J., L. Salome, A. Ajdari, and J. Prost. 1994. Directional motion of brownian particles induced by a periodic asymmetric potential. *Nature* 370: 446–448.

Ruff, C., M. Furch, B. Brenner, D. J. Manstein, and E. Meyhöfer. In press. Single-molecule tracking of myosin with genetically engineered amplifier domains. *Nat. Struct. Biol.*

Ruggero, M. A., N. C. Rich, A. Recio, S. S. Narayan, and L. Robles. 1997. Basilar-membrane responses to tones at the base of the chinchilla cochlea. *J. Acoust. Soc. Am.* 101: 2151–2163.

Sablin, E. P., R. B. Case, S. C. Dai, C. L. Hart, A. Ruby, R. D. Vale, and R. J. Fletterick. 1998. Direction determination in the minus-end-directed kinesin motor ncd. *Nature* 395: 813–816.

Sachs, F., and H. Lecar. 1991. Stochastic models for mechanical transduction. *Biophys. J.* 59: 1143–1145.

Safer, D., M. Elzinga, and V. T. Nachmias. 1991. Thymosin beta 4 and Fx, an actin-sequestering peptide, are indistinguishable. *J. Biol. Chem.* 266: 4029–4032.

Saito, K., T. Aoki, and T. Yanagida. 1994. Movement of single myosin filaments and myosin step size on an actin filament suspended in solution by a laser trap. *Biophys. J.* 66: 769–777.

Sakakibara, H., H. Kojima, Y. Sakai, E. Katayama, and K. Oiwa. 1999. Inner-arm dynein c of *Chlamydomonas* flagella is a single-headed processive motor. *Nature* 400: 586–590.

Sakamoto, T., I. Amitani, E. Yokota, and T. Ando. 2000. Direct observation of processive movement by individual myosin V molecules. *Biochem. Biophys. Res. Commun.* 272: 586–590.

Sakmann, B., and E. Neher. 1995. *Single-Channel Recording*. Plenum Press, New York.

Sako, Y., S. Minoghchi, and T. Yanagida. 2000. Single-molecule imaging of EGFR signalling on the surface of living cells. *Nat. Cell Biol.* 2: 168–172.

Sale, W. S., and P. Satir. 1977. Direction of active sliding of microtubules in *Tetrahymena* cilia. *Proc. Natl. Acad. Sci. USA* 74: 2045–2049.

Sale, W. S., L. A. Fox, and E. F. Smith. 1993. Assays of axonemal dynein-driven motility. *Methods Cell Biol.* 39: 89–104.

Sammak, P. J., and G. G. Borisy. 1988. Direct observation of microtubule dynamics in living cells. *Nature* 332: 724–726.

Samso, M., M. Radermacher, J. Frank, and M. P. Koonce. 1998. Structural characterization of a dynein motor domain. *J. Mol. Biol.* 276: 927–937.

Sase, I., H. Miyata, S. Ishiwata, and K. Kinosita, Jr. 1997. Axial rotation of sliding actin filaments revealed by single-fluorophore imaging. *Proc. Natl. Acad. Sci. USA* 94: 5646–5650.

Saunders, W. S., and M. A. Hoyt. 1992. Kinesin-related proteins required for structural integrity of the mitotic spindle. *Cell* 70: 451–458.

Savage, C., M. Hamelin, J. G. Culotti, A. Coulson, D. G. Albertson, and M. Chalfie. 1989. mec-7 is a beta-tubulin gene required for the production of 15- protofilament microtubules in *Caenorhabditis elegans*. *Genes Dev.* 3: 870–881.

Saxton, W. M., D. L. Stemple, R. J. Leslie, E. D. Salmon, M. Zavortink, and J. R. McIntosh. 1984. Tubulin dynamics in cultured mammalian cells. *J. Cell Biol.* 99: 2175–2186.

Schafer, D. A., S. R. Gill, J. A. Cooper, J. E. Heuser, and T. A. Schroer. 1994. Ultrastructural analysis of the dynactin complex: an actin-related protein is a component of a filament that resembles F-actin. *J. Cell Biol.* 126: 403–412.

Schief, W. R., and J. Howard. 2001. Conformational changes during kinesin motility. *Curr. Opin. Cell Biol.* 13: 19–28.

Schliwa, M. 1986. *The Cytoskeleton : An Introductory Survey*. Springer-Verlag, New York.

Schnitzer, M. J., K. Visscher, and S. M. Block. 2000. Force production by single kinesin motors. *Nat. Cell Biol.* 2: 718–723.

Schoenberg, M. 1985. Equilibrium muscle cross-bridge behavior. Theoretical considerations. *Biophys. J.* 48: 467–475.

Schoenberg, M., and E. Eisenberg. 1985. Muscle cross-bridge kinetics in rigor and in the presence of ATP analogues. *Biophys. J.* 48: 863–871.

Scholey, J. M. 1993. *Motility Assays for Motor Proteins*. Academic Press, San Diego.

Scholey, J. M., J. Heuser, J. T. Yang, and L. S. Goldstein. 1989. Identification of globular mechanochemical heads of kinesin. *Nature* 338: 355–357.

Schroeder, T. E. 1969. The role of the "contractile ring" filaments in dividing *Arbacia* egg. *Biol. Bull.* 137: 413–414.

Schurr, J. M. 1970a. The role of diffusion in bimolecular solution kinetics. *Biophys. J.* 10: 700–716.

Schurr, J. M. 1970b. The role of diffusion in enzyme kinetics. *Biophys. J.* 10: 717–727.

Schutz, G. J., G. Kada, V. P. Pastushenko, and H. Schindler. 2000. Properties of lipid microdomains in a muscle cell membrane visualized by single molecule microscopy. *EMBO J.* 19: 892–901.

Seksek, O., J. Biwersi, and A. S. Verkman. 1997. Translational diffusion of macromolecule-sized solutes in cytoplasm and nucleus. *J. Cell Biol.* 138: 131–142.

Sellers, J. R., and H. V. Goodson. 1995. Motor proteins 2: myosin. *Protein Profile.* 2: 1323–1423.

Sheetz, M. P., and J. A. Spudich. 1983. Movement of myosin-coated fluorescent beads on actin cables in vitro. *Nature* 303: 31–35.

Sheterline, P., J. Clayton, and J. Sparrow. 1995. Actin. *Protein Profile.* 2: 1–103.

Shingyoji, C., H. Higuchi, M. Yoshimura, E. Katayama, and T. Yanagida. 1998. Dynein arms are oscillating force generators. *Nature* 393: 711–714.

Shpetner, H. S., B. M. Paschal, and R. B. Vallee. 1988. Characterization of the microtubule-activated ATPase of brain cytoplasmic dynein (MAP 1C). *J. Cell Biol.* 107: 1001–1009.

Siemankowski, R. F., M. O. Wiseman, and H. D. White. 1985. ADP dissociation from actomyosin subfragment 1 is sufficiently slow to limit the unloaded shortening velocity in vertebrate muscle. *Proc. Natl. Acad. Sci. USA* 82: 658–662.

Sigworth, F. J. 1995. Electronic design of the patch clamp. *In* B. Sakmann and E. Neher (eds.), *Single-Channel Recording*, Plenum Press, New York, pp. 95–127.

Sleep, J. A., D. D. Hackney, and P. D. Boyer. 1978. Characterization of phosphate oxygen exchange reactions catalyzed by myosin through measurement of the distribution of 18-O-labeled species. *J. Biol. Chem.* 253: 5235–5238.

Small, J. V. 1988. The actin cytoskeleton. *Electron Microsc. Rev.* 1: 155–174.

Smith, C. A., and I. Rayment. 1996. Active site comparisons highlight structural similarities between myosin and other P-loop proteins. *Biophys. J.* 70: 1590–1602.

Smith, S. B., L. Finzi, and C. Bustamante. 1992. Direct mechanical measurements of the elasticity of single DNA molecules by using magnetic beads. *Science* 258: 1122–1126.

Soltys, B. J., and G. G. Borisy. 1985. Polymerization of tubulin in vivo: Direct evidence for assembly onto microtubule ends and from centrosomes. *J. Cell Biol.* 100: 1682–1689.

Soong, R. K., G. D. Bachand, H. P. Neves, A. G. Olkhovets, H. G. Craighead, and C. D. Montemagno. 2000. Powering an inorganic nanodevice with a biomolecular motor. *Science* 290: 1555–1558.

Spencer, M. 1982. *Fundamentals of Light Microscopy*. Cambridge University Press, New York.

Spudich, J. A., S. J. Kron, and M. P. Sheetz. 1985. Movement of myosin-coated beads on oriented filaments reconstituted from purified actin. *Nature* 315: 584–586.

Squire, J. 1981. *The Structural Basis of Muscular Contraction*. Plenum Press, New York.

Stein, L. A., R. P. Schwarz, Jr., P. B. Chock, and E. Eisenberg. 1979. Mechanism of actomyosin adenosine triphosphatase. Evidence that adenosine 5′-triphosphate hydrolysis can occur without dissociation of the actomyosin complex. *Biochemistry* 18: 3895–3909.

Steuer, E. R., L. Wordeman, T. A. Schroer, and M. P. Sheetz. 1990. Localization of cytoplasmic dynein to mitotic spindles and kinetochores. *Nature* 345: 266–268.

Stewart, R. J., J. P. Thaler, and L. S. Goldstein. 1993. Direction of microtubule movement is an intrinsic property of the motor domains of kinesin heavy chain and *Drosophila* ncd protein. *Proc. Natl. Acad. Sci. USA* 90: 5209–5213.

Stock, M. F., J. Guerrero, B. Cobb, C. T. Eggers, T. G. Huang, X. Li, and D. D. Hackney. 1999. Formation of the compact conformer of kinesin requires a COOH-terminal heavy chain domain and inhibits microtubule-stimulated ATPase activity. *J. Biol. Chem.* 274: 14617–14623.

Straub, F. B. 1941–1942. Actin. *Stud. Inst. Med. Chem. Univ. Szeged.* 2: 3–15.

Straub, F. B., and G. Feuer. 1950. Adenosinetriphosphate. The functional group of actin. *Biochim. Biophys. Acta.* 4: 455–470.

Stryer, L. 1995. *Biochemistry*, 4th ed. W. H. Freeman and Company, New York.

Stryer, L., and R. P. Haugland. 1967. Energy transfer: a spectroscopic ruler. *Proc. Natl. Acad. Sci. USA* 58: 719–726.

Suezaki, Y., and N. Go. 1976. Fluctuations and mechanical strength of alpha-helices of polyglycine and poly(L-alanine). *Biopolymers.* 15: 2137–2153.

Supp, D. M., D. P. Witte, S. S. Potter, and M. Brueckner. 1997. Mutation of an axonemal dynein affects left-right asymmetry in inversus viscerum mice. *Nature* 389: 963–966.

Suzuki, N., H. Miyata, S. Ishiwata, and K. Kinosita, Jr. 1996. Preparation of bead-tailed actin filaments: estimation of the torque produced by the sliding force in an in vitro motility assay. *Biophys. J.* 70: 401–408.

Svitkina, T. M., and G. G. Borisy. 1999. Arp2/3 complex and actin depolymerizing factor/cofilin in dendritic organization and treadmilling of actin filament array in lamellipodia. *J. Cell Biol.* 145: 1009–1026.

Svoboda, K., and S. M. Block. 1994a. Biological applications of optical forces. *Annu. Rev. Biophys. Biomol. Struct.* 23: 247–285.

Svoboda, K., and S. M. Block. 1994b. Force and velocity measured for single kinesin molecules. *Cell* 77: 773–784.

Svoboda, K., C. F. Schmidt, B. J. Schnapp, and S. M. Block. 1993. Direct observation of kinesin stepping by optical trapping interferometry. *Nature* 365: 721–727.

Szent-Györgyi, A. 1941. Contraction of myosin threads. *Stud. Inst. Med. Chem. Univ. Szeged.* 1: 17–26.

Szent-Györgyi, A. 1941–1942. Discussion. *Stud. Inst. Med. Chem. Univ. Szeged.* 1: 67–71.

Szent-Györgyi, A. 1953. *Arch. Biochem. Biophys.* 42: 305–320.

Tamm, S. L. 1973. Mechanisms of ciliary co-ordination in ctenophores. *J. Exp. Biol.* 59: 231–245.

Tawada, K., and K. Sekimoto. 1991. Protein friction exerted by motor enzymes through a weak-binding interaction. *J. Theor. Biol.* 150: 193–200.

Taylor, E. W. 1979. Mechanism of actomyosin ATPase and the problem of muscle contraction. *CRC Crit. Rev. Biochem.* 6: 103–164.

Tennent, R. M. 1971. *Science Data Book*. Oliver & Boyd, Edinburgh.

Tesi, C., T. Barman, and F. Travers. 1990. Is a four-state model sufficient to describe actomyosin ATPase?. *FEBS Lett.* 260: 229–232.

Theriot, J. A., and T. J. Mitchison. 1991. Actin microfilament dynamics in locomoting cells. *Nature* 352: 126–131.

Theriot, J. A., T. J. Mitchison, L. G. Tilney, and D. A. Portnoy. 1992. The rate of actin-based motility of intracellular Listeria monocytogenes equals the rate of actin polymerization. *Nature* 357: 257–260.

Tilney, L. G., and S. Inoue. 1982. Acrosomal reaction of Thyone sperm. II. The kinetics and possible mechanism of acrosomal process elongation. *J. Cell Biol.* 93: 820–827.

Tilney, L. G., and D. A. Portnoy. 1989. Actin filaments and the growth, movement, and spread of the intracellular bacterial parasite, Listeria monocytogenes. *J. Cell Biol.* 109: 1597–1608.

Tilney, L. G., and M. S. Tilney. 1988. The actin filament content of hair cells of the bird cochlea is nearly constant even though the length, width, and number of stereocilia vary depending on the hair cell location. *J. Cell Biol.* 107: 2563–2574.

Tilney, L. G., J. Bryan, D. J. Bush, K. Fujiwara, M. S. Mooseker, D. B. Murphy, and D. H. Snyder. 1973. Microtubules: evidence for 13 protofilaments. *J. Cell Biol.* 59: 267–275.

Tilney, L. G., D. J. Derosier, and M. J. Mulroy. 1980. The organization of actin filaments in the stereocilia of cochlear hair cells. *J. Cell Biol.* 86: 244–259.

Tilney, L. G., D. J. DeRosier, and M. S. Tilney. 1992. How Listeria exploits host cell actin to form its own cytoskeleton. I. Formation of a tail and how that tail might be involved in movement. *J. Cell Biol.* 118: 71–81.

Tirado, M. M. and Garcia de la Torre, J. 1979. Translational frictional coefficients of rigid, symmetric top macromolecules. Application to circular cylinders. *J. Chem. Phys.* 71: 2581–2587.

Tirado, M. M. and Garcia de la Torre, J. 1980. Rotational dynamics of rigid, symmetric top macromolecules. Application to circular cylinders. *J. Chem. Phys.* 73: 1986–1993.

Toyoshima, Y. Y., S. J. Kron, E. M. McNally, K. R. Niebling, C. Toyoshima, and J. A. Spudich. 1987. Myosin subfragment-1 is sufficient to move actin filaments in vitro. *Nature* 328: 536–539.

Toyoshima, Y. Y., S. J. Kron, and J. A. Spudich. 1990. The myosin step size: measurement of the unit displacement per ATP hydrolyzed in an in vitro assay. *Proc. Natl. Acad. Sci. USA* 87: 7130–7134.

Tskhovrebova, L., J. Trinick, J. A. Sleep, and R. M. Simmons. 1997. Elasticity and unfolding of single molecules of the giant muscle protein titin. *Nature* 387: 308–312.

Tsuda, Y., H. Yasutake, A. Ishijima, and T. Yanagida. 1996. Torsional rigidity of single actin filaments and actin-actin bond breaking force under torsion measured directly by in vitro micromanipulation. *Proc. Natl. Acad. Sci. USA* 93: 12937–12942.

Tung, C. S., S. C. Harvey, and J. A. McCammon. 1984. Large-amplitude bending motions in phenylalanine transfer RNA. *Biopolymers*. 23: 2173–2193.

Tyska, M. J., D. E. Dupuis, W. H. Guilford, J. B. Patlak, G. S. Waller, K. M. Trybus, D. M. Warshaw, and S. Lowey. 1999. Two heads of myosin are better than one for generating force and motion. *Proc. Natl. Acad. Sci. USA* 96: 4402–4407.

Uyeda, T. Q., S. J. Kron, and J. A. Spudich. 1990. Myosin step size. Estimation from slow sliding movement of actin over low densities of heavy meromyosin. *J. Mol. Biol.* 214: 699–710.

Valberg, P. A., and H. A. Feldman. 1987. Magnetic particle motions within living cells. Measurement of cytoplasmic viscosity and motile activity. *Biophys. J.* 52: 551–561.

Vale, R. D. 1996. Switches, latches, and amplifiers: common themes of G proteins and molecular motors. *J. Cell Biol.* 135: 291–302.

Vale, R. D., and R. A. Milligan. 2000. The way things move: Looking under the hood of molecular motor proteins. *Science* 288: 88–95.

Vale, R. D., and F. Oosawa. 1990. Protein motors and Maxwell's demons: Does mechanochemical transduction involve a thermal ratchet? *Adv. Biophys.* 26: 97–134.

Vale, R. D., and Y. Y. Toyoshima. 1988. Rotation and translocation of microtubules in vitro induced by dyneins from Tetrahymena cilia. *Cell* 52: 459–469.

Vale, R. D., T. S. Reese, and M. P. Sheetz. 1985. Identification of a novel force-generating protein, kinesin, involved in microtubule-based motility. *Cell* 42: 39–50.

Vale, R. D., T. Funatsu, D. W. Pierce, L. Romberg, Y. Harada, and T. Yanagida. 1996. Direct observation of single kinesin molecules moving along microtubules. *Nature* 380: 451–453.

van Netten, S. M. 1991. Hydrodynamics of the excitation of the cupula in the fish canal lateral line. *J. Acoust. Soc. Am.* 89: 310–319.

Vassar, R., P. A. Coulombe, L. Degenstein, K. Albers, and E. Fuchs. 1991. Mutant keratin expression in transgenic mice causes marked abnormalities resembling a human genetic skin disease. *Cell* 64: 365–380.

Vater, M., and W. Siefer. 1995. The cochlea of *Tadarida brasiliensis*: Specialized functional organization in a generalized bat. *Hear. Res.* 91: 178–195.

Veigel, C., M. L. Bartoo, D. C. White, J. C. Sparrow, and J. E. Molloy. 1998. The stiffness of rabbit skeletal actomyosin cross-bridges determined with an optical tweezers transducer. *Biophys. J.* 75: 1424–1438.

Veigel, C., L. M. Coluccio, J. D. Jontes, J. C. Sparrow, R. A. Milligan, and J. E. Molloy. 1999. The motor protein myosin-I produces its working stroke in two steps. *Nature* 398: 530–533.

Verde, F., M. Dogterom, E. Stelzer, E. Karsenti, and S. Leibler. 1992. Control of microtubule dynamics and length by cyclin A- and cyclin B- dependent kinases in *Xenopus* egg extracts. *J. Cell Biol.* 118: 1097–1108.

Visscher, K., M. J. Schnitzer, and S. M. Block. 1999. Single kinesin molecules studied with a molecular force clamp. *Nature* 400: 184–189.

Voter, W. A., E. T. O'Brien, and H. P. Erickson. 1991. Dilution-induced disassembly of microtubules: relation to dynamic instability and the GTP cap. *Cell Motil Cytoskeleton.* 18: 55–62.

Wade, R. H., and D. Chretien. 1993. Cryoelectron microscopy of microtubules. *J. Struct. Biol.* 110: 1–27.

Wainwright, S. A., W. D. Biggs, J. D. Currey, and J. M. Gosline. 1976. *Mechanical Design in Organisms*. Princeton University Press, Princeton, NJ.

Wakabayashi, K., Y. Sugimoto, H. Tanaka, Y. Ueno, Y. Takezawa, and Y. Amemiya. 1994. X-ray diffraction evidence for the extensibility of actin and myosin filaments during muscle contraction. *Biophys. J.* 67: 2422–2435.

Walczak, C. E., T. J. Mitchison, and A. Desai. 1996. XKCM1: A *Xenopus* kinesin-related protein that regulates microtubule dynamics during mitotic spindle assembly. *Cell* 84: 37–47.

Walker, J. E., M. Saraste, M. J. Runswick, and N. J. Gay. 1982. Distantly related sequences in the alpha- and beta-subunits of ATP synthase, myosin, kinases and other ATP-requiring enzymes and a common nucleotide binding fold. *EMBO J.* 1: 945–951.

Walker, M. L., S. A. Burgess, J. R. Sellers, F. Wang, J. A. Hammer, J. Trinick, and P. J. Knight. 2000. Two-headed binding of a processive myosin to F-actin. *Nature* 405: 804–807.

Walker, R. A., E. T. O'Brien, N. K. Pryer, M. F. Soboeiro, W. A. Voter, H. P. Erickson, and E. D. Salmon. 1988. Dynamic instability of individual microtubules analyzed by video light microscopy: rate constants and transition frequencies. *J. Cell Biol.* 107: 1437–1448.

Walker, R. A., E. D. Salmon, and S. A. Endow. 1990. The *Drosophila* claret segregation protein is a minus-end directed motor molecule. *Nature* 347: 780–782.

Walker, R. A., N. K. Pryer, and E. D. Salmon. 1991. Dilution of individual microtubules observed in real time in vitro: Evidence that cap size is small and independent of elongation rate. *J. Cell Biol.* 114: 73–81.

Walther, Z., M. Vashishtha, and J. L. Hall. 1994. The *Chlamydomonas* FLA10 gene encodes a novel kinesin-homologous protein. *J. Cell Biol.* 126: 175–188.

Wang, M. D., M. J. Schnitzer, H. Yin, R. Landick, J. Gelles, and S. M. Block. 1998. Force and velocity measured for single molecules of RNA polymerase. *Science* 282: 902–907.

Wang, Y. L. 1985. Exchange of actin subunits at the leading edge of living fibroblasts: possible role of treadmilling. *J. Cell Biol.* 101: 597–602.

Warshaw, D. M., J. M. Desrosiers, S. S. Work, and K. M. Trybus. 1990. Smooth muscle myosin cross-bridge interactions modulate actin filament sliding velocity in vitro. *J. Cell Biol.* 111: 453–463.

Webb, M. R., G. G. McDonald, and D. R. Trentham. 1978. Kinetics of oxygen-18 exchange between inorganic phosphate and water catalyzed by myosin subfragment 1, using the 18O shift in 31P NMR. *J. Biol. Chem.* 253: 2908–2911.

Weber, I. T., and T. A. Steitz. 1987. Structure of a complex of catabolite gene activator protein and cyclic AMP refined at 2.5 A resolution. *J. Mol. Biol.* 198: 311–326.

Weber, K. 1999. Evolutionary aspects of IF proteins. In T. Kreis and R. Vale (eds.), *Guidebook to the Cytoskeletal and Motor Proteins*, Oxford University Press, Oxford, pp. 291–293.

Wegner, A. 1976. Head to tail polymerization of actin. *J. Mol. Biol.* 108: 139–150.

Wegner, A., and G. Isenberg. 1983. 12-fold difference between the critical monomer concentrations of the two ends of actin filaments in physiological salt conditions. *Proc. Natl. Acad. Sci. USA* 80: 4922–4925.

Weil, D., S. Blanchard, J. Kaplan, P. Guilford, F. Gibson, J. Walsh, P. Mburu, A. Varela, J. Levilliers, M. D. Weston, and et al. 1995. Defective myosin VIIA gene responsible for Usher syndrome type 1B. *Nature* 374: 60–61.

Weisenberg, R. C., W. J. Deery, and P. J. Dickinson. 1976. Tubulin-nucleotide interactions during the polymerization and depolymerization of microtubules. *Biochemistry* 15: 4248–4254.

Welch, M. D., A. Iwamatsu, and T. J. Mitchison. 1997. Actin polymerization is induced by Arp2/3 protein complex at the surface of *Listeria monocytogenes*. *Nature* 385: 265–269.

Welch, M. D., J. Rosenblatt, J. Skoble, D. A. Portnoy, and T. J. Mitchison. 1998. Interaction of human Arp2/3 complex and the *Listeria monocytogenes* ActA protein in actin filament nucleation. *Science* 281: 105–108.

Wells, A. L., A. W. Lin, L. Q. Chen, D. Safer, S. M. Cain, T. Hasson, B. O. Carragher, R. A. Milligan, and H. L. Sweeney. 1999. Myosin VI is an actin-based motor that moves backwards. *Nature* 401: 505–508.

Wessells, N. K., B. S. Spooner, J. F. Ash, M. O. Bradley, M. A. Luduena, E. L. Taylor, J. T. Wrenn, and K. Yamaa. 1971. Microfilaments in cellular and developmental processes. *Science* 171: 135–143.

White, J. G., and G. H. Rao. 1998. Microtubule coils versus the surface membrane cytoskeleton in maintenance and restoration of platelet discoid shape. *Am. J. Pathol.* 152: 597–609.

Whittaker, M., E. M. Wilson-Kubalek, J. E. Smith, L. Faust, R. A. Milligan, and H. L. Sweeney. 1995. A 35-A movement of smooth muscle myosin on ADP release. *Nature* 378: 748–751.

Woledge, R. C., N. A. Curtin, and E. Homsher. 1985. *Energetic Aspects of Muscle Contraction*. Academic Press, Orlando, FL.

Wong, E. W., P. E. Sheehan, and C. Lieber, M. 1997. Nanobeam mechanics: Elasticity, strength, and toughness of nanorods and nanotubes. *Science* 277: 1971–1975.

Wordeman, L., and T. J. Mitchison. 1995. Identification and partial characterization of mitotic centromere- associated kinesin, a kinesin-related protein that associates with centromeres during mitosis. *J. Cell Biol.* 128: 95–104.

Wright, A. 1984. Dimensions of the cochlear stereocilia in man and the guinea pig. *Hear. Res.* 13: 89–98.

Wuite, G. J., S. B. Smith, M. Young, D. Keller, and C. Bustamante. 2000. Single-molecule studies of the effect of template tension on T7 DNA polymerase activity. *Nature* 404: 103–106.

Yanagida, T., M. Nakase, K. Nishiyama, and F. Oosawa. 1984. Direct observation of motion of single F-actin filaments in the presence of myosin. *Nature* 307: 58–60.

Yanagida, T., T. Arata, and F. Oosawa. 1985. Sliding distance of actin filament induced by a myosin crossbridge during one ATP hydrolysis cycle. *Nature* 316: 366–369.

Yang, J. T., R. A. Laymon, and L. S. Goldstein. 1989. A three-domain structure of kinesin heavy chain revealed by DNA sequence and microtubule binding analyses. *Cell* 56: 879–889.

Yang, J. T., W. M. Saxton, R. J. Stewart, E. C. Raff, and L. S. Goldstein. 1990. Evidence that the head of kinesin is sufficient for force generation and motility in vitro. *Science* 249: 42–47.

Yasuda, R., H. Miyata, and K. Kinosita, Jr. 1996. Direct measurement of the torsional rigidity of single actin filaments. *J. Mol. Biol.* 263: 227–236.

Yasuda, R., H. Noji, K. Kinosita, Jr., and M. Yoshida. 1998. F1-ATPase is a highly efficient molecular motor that rotates with discrete 120 degree steps. *Cell* 93: 1117–1124.

Yeh, E., R. V. Skibbens, J. W. Cheng, E. D. Salmon, and K. Bloom. 1995. Spindle dynamics and cell cycle regulation of dynein in the budding yeast, *Saccharomyces cerevisiae*. *J. Cell Biol.* 130: 687–700.

Yen, T. J., G. Li, B. T. Schaar, I. Szilak, and D. W. Cleveland. 1992. CENP-E is a putative kinetochore motor that accumulates just before mitosis. *Nature* 359: 536–539.

Yguerabide, J., H. F. Epstein, and L. Stryer. 1970. Segmental flexibility in an antibody molecule. *J. Mol. Biol.* 51: 573–590.

Yin, H., M. D. Wang, K. Svoboda, R. Landick, S. M. Block, and J. Gelles. 1995. Transcription against an applied force. *Science* 270: 1653–1657.

Yin, H. L., and T. P. Stossel. 1979. Control of cytoplasmic actin gel-sol transformation by gelsolin, a calcium-dependent regulatory protein. *Nature* 281: 583–586.

Yokota, E., and I. Mabuchi. 1994. C/A dynein isolated from sea urchin sperm flagellar axonemes. Enzymatic properties and interaction with microtubules. *J. Cell Sci.* 107: 353–361.

Yonekawa, Y., A. Harada, Y. Okada, T. Funakoshi, Y. Kanai, Y. Takei, S. Terada, T. Noda, and N. Hirokawa. 1998. Defect in synaptic vesicle precursor transport and neuronal cell death in KIF1A motor protein-deficient mice. *J. Cell Biol.* 141: 431–441.

Zaner, K. S., and P. A. Valberg. 1989. Viscoelasticity of F-actin measured with magnetic microparticles. *J. Cell Biol.* 109: 2233–2243.

Zeeberg, B., and M. Caplow. 1979. Determination of free and bound microtubular protein and guanine nucleotide under equilibrium conditions. *Biochemistry* 18: 3880–3886.

Zhai, Y., and G. G. Borisy. 1994. Quantitative determination of the proportion of microtubule polymer present during the mitosis-interphase transition. *J. Cell Sci.* 107: 881–890.

Zheng, Y., M. L. Wong, B. Alberts, and T. Mitchison. 1995. Nucleation of microtubule assembly by a gamma-tubulin-containing ring complex. *Nature* 378: 578–583.

Zhou, H. X., and Y. D. Chen. 1996. Chemically driven motility of Brownian particles. *Phys. Rev. Lett.* 77: 194–197.

Zhuang, X., L. E. Bartley, H. P. Babcock, R. Russell, T. Ha, D. Herschlag, and S. Chu. 2000. A single-molecule study of RNA catalysis and folding. *Science* 288: 2048–2051.

Zimmerberg, J., F. Bezanilla, and V. A. Parsegian. 1990. Solute inaccessible aqueous volume changes during opening of the potassium channel of the squid giant axon. *Biophys. J.* 57: 1049–1064.

Zimmerberg, J., and V. A. Parsegian. 1986. Polymer inaccessible volume changes during opening and closing of a voltage-dependent ionic channel. *Nature* 323: 36–39.

Zot, H. G., S. K. Doberstein, and T. D. Pollard. 1992. Myosin-I moves actin filaments on a phospholipid substrate: Implications for membrane targeting. *J. Cell Biol.* 116: 367–376.

Index

A

Acanthamoeba, 24, 124
Acetone, viscosity, 38
Acetylcholine, binding of, 82
Acrosomal reaction, 328–329
Actin
 bending stiffness, 138–140
 family of proteins, 124
 force generation in vitro, 169
 longitudinal stiffness, 136–138
 M·D·P binding, 237–238
 monomer structures, 121–122
 motility of viruses and bacteria, 166–168
 myosin and, 90, 257–258
 myosin hydrolysis of ATP, 232–238
 nucleotide hydrolysis, 191–193
 polymerization, 3, 93, 166–167
 polymerization-driven motility, 174–175, 176
 properties, 121
 structures, 119, 121
 treadmilling, 184
 Young's modulus, 137
Actin-binding proteins (ABPs), 44, 190
Actin filaments
 buckling, 170–171
 cell locomotion, 4
 crosslinking and bending, 140
 diameter, 119
 rigidity in vitro, 142–143
 severing by gelsolin, 152
 structure, 125–127
 thermal fluctuations, 143
Actin-related proteins (Arps), 124, 126, 158, 166–167
Activated states, 84–85, 89–91
Active sites, conservation of, 205
Adenosine triphosphate. See ATP
Alanine, density, 30
Amos and Klug model, 129
Anisotropy, of materials, 33
Annealing/breakage reaction, 324–325
Arc length, definition, 102
Arrhenius equation, 85
Association rate constants, 92–93, 307–309
Atomic force microscope (AFM), 5, 91, 248–249, 251

ATP (adenosine triphosphate)
 fluorescence labeling, 251
 function of, 229–238
 hydrolysis, 229–244
 hydrolysis by kinesin, 238–242
 hydrolysis by myosin, 232–238
 in-line hydrolysis mechanism, 231
 motor domain distance per hydrolysis cycle, 245
 and rigor, 140–141
 steps per molecule, 279
 structure of, 230
ATP synthase, 116, 261
ATPases
 assays, 222–225
 dynein motor domain, 206
 and filament concentration, 242
 kinesin and myosin cycles, 242–244
 microtubule-stimulated, 192
 rates and speeds, 276
 rates per molecule, 264
Autocorrelation functions, 64, 298–299, 300–301, 302–303
Avogadro's number, 55

B

Bacteria. See also Specific bacteria
 acceleration, 38
 chemotaxis, 115–116
 flagellar rigidity in vitro, 146
 inertia of, 18
 motility of, 166
 motors, 16
 Reynolds numbers, 39
Bats, 322
Bead assays
 crossbridge forces, 259
 kinesin movement, 252
 motility, 216–219
 myosin steps, 256
 single molecules, 247
Beam equation, 100, 102–105
Beams, cantilevered, 103–104
Bending
 axes of, 312–313
 dynamics, 107–110
 energy of, 311
 hydrodynamic modes, 314–315
 slender rods, 100
Bending moment, definition, 100

Bent rods
 drag force on, 108
 energy in, 312
 energy of bending, 311
 geometry of, 102
Bimolecular reactions, 92–94
Binding energies, and entropy, 159–161
Biomolecules and gravity, 9
Boltzmann's law, 50–55, 76–77, 169, 292–294
Brownian motion, 49, 55, 56
Brownian ratchet models, 172–174, 325–328
Buckling
 actin filaments, 170–171
 compressive forces, 104–105
 dynamics, 107–110
 energy of, 313–314
 slender rods, 316
Budding yeast. See Saccharomyces cerevisiae

C

Caenorhabditis elegans, 123, 204
Cantilever force transducer, 247
Cantilever springs, 32, 315–316
Cantilevered beams, 103–104, 321
Catastrophe rate, 186–187
Cell cycles, 167–168, 185, 204
Cell motion, 4, 44–45
Cellular transport, 60
Centrifugal forces, 11, 13
Centrifugation, analytical, 13, 52–53
Chains, freely jointed, 112–114, 320
Chemical bonds. See also van der Waals forces
 energy of, 25
 strains, 20
 vibration of, 21
 Young's modulus, 35
Chemical equilibrium, 76–80
Chemical-force equation, 170
Chemical forces, 75–98
Chemical kinetics, 306–307
Chemical rate constants, 89–92
Chemical reactions, rate theories, 83–88
Chemotaxis, bacterial, 115–116
Chromosomes, density, 30
Cilia, 199, 204

362 INDEX

Coiled coils. *See also* leucine zippers, 130
  description, 119
  dimer structure, 131
  flexural rigidity, 104
  myosin subfragment 1, 200–201
  properties of, 121
  rigidity of in vitro, 145
  structure, 130–131
Coiled springs, stiffness, 33
Collision rate constants, 307–308
Collision rates, orientations and, 308
Collisional forces, single-molecule level, 11, 12
Compliance, 27, 82
Compressive forces, 104–105, 189, 312
Concentrations, 56, 57
Conformational changes, 83–88
Conformational states, 76, 77, 303–306
Conservation of energy. *See* Thermodynamics, first law of
Contracting-spring model, 271
Contraction. *See* Muscle contractions
Converter domains, 207–208, 218, 266
Correlation times, 63–66
Coulomb's law, equation, 27
Covalent forces, 11, 21, 25
Creeping flow, definition, 38
Critical concentration, 158
Critical force. *See* Euler force
Crosslinking, 39–40, 140
Cross-section, second moments, 101
Crossbridges
  cycles, 221–226
  definition, 199–200
  drag-stroke distance, 266
  drag time, 266
  electron microscopy, 199
  model for kinesin, 280–282
  model for muscle contraction, 269–273
  model for myosin, 333–334
  motor proteins, 198–202
  single-motor techniques, 259
  stiffness, 263–264
  thick filaments, 216
  time-average force, 263
Curvature, of bends, 100
Cyclic reactions, energy transduction, 94–97
Cytoskeletal filaments, 120
  force generation, 165–178
  polymerization, 151–163
  rigidity in vivo, 136–142
  stiffness, 99
  structures of, 119–134
Cytoskeletal protein families, 123–125
Cytoskeleton
  description, 117
  force generation, 3
  mechanics of, 135–149
  motions of, 44–45

D

Damping, 37–39, 40, 287, 291, 322–323
Dashpot, 14–16
Dashpot and mass model, 16–18
Dashpot and spring model, 19
Degrees of freedom, 292–293
Densities, of proteins, 30
Depolymerization, 166–169, 188–191
Desmins, 125, 145–146
Detailed Balance, Principle of, 95, 293
Dictyostelium, 124, 208
Dielectric constant, 27
Diffusion
  association rate constants limited by, 307–309
  and cellular transport, 60–61
  definition, 57
  and drag coefficients, 295
  free proteins, 65
  molecular, 60
  from point source, 60–61
  random walk, 55–58
  reaction speed, 173
  single-molecule level, 12
  tethered proteins, 66
  and thermal forces, 49–73
Diffusion coefficients, 57, 2329
Diffusion equation, 57–58, 59–63, 294–295
Diffusion-limited collisions, 92–93
Diffusion-limited rates, 61, 173, 308
Displacement, and structural changes, 78
Dissociation constants, definition, 92, 152–154
Dissociation rate constants, 92
Distance per ATP, definition, 245
DNA, 147
DNA polymerase, forces generated, 261
Docking, motors to filaments, 210–212
Domain organization, motor proteins, 198–202
Drag coefficients
  bending filaments, 314
  cylinder at a plane surface, 106
  dashpots, 15
  and diffusion, 295
  and gliding assays, 107
  single-molecule level, 12
  and speed, 40
  sperm, 107
  unbounded solutions, 106
Drag forces, 38, 41, 290–291
Drag-stroke distance, definition, 266
Drag time, 266
Duty ratios
  concept of, 213
  definition, 222
  description, 264
  and hydrolysis cycle, 221–226
  muscles and, 277–279, 283
  and sliding speed, 273
  structural basis for, 261–262
Dynamic instability, 185–186, 187–188, 329–330
Dyneins
  crossbridges, 192, 199
  duty ratios, 222
  identification, 1
  mechanical lever, 212
  microtubule-binding stalk, 209
  motor domain, 206
  outer-arm, 261, 330
  rotating crossbridge model, 197
  speeds, 213–216

E

E sites, definition, 122
Einstein, Albert, 55
Einstein polymers, 152–153
Einstein relation, 58–59, 295
Elastic forces, 11, 12, 13
Elastic modulus, definition, 31
Elasticity
  deformation, 37, 47
  freely jointed chains, 112–114
  molecular basis of, 34–37
  of proteins, 30–34
Electric constant, 27
Electrical circuits, thermal noise, 70
Electrochemical reactions, in solution, 87–88
Electrostatic bonds, and rigidity, 34
Electrostatic forces, single-molecule level, 11, 13
Elongation
  and bending moment, 100
  proteins and filaments, 291
End-to-end length
  Freely jointed chains, 320
  slender rods, 317–318
Energy. *See also* Free energies
  of chemical bonds, 25
  definition, 222

INDEX

equipartition of, 53–55
and force, 24–25, 90
law of conservation of, 25–26
storage in conformational changes, 26
Enthalpy, 77
Entropic springs, as models, 112
Entropy, 39, 77, 159–161, 303
Enzymes, Michaelis–Menten equation, 93–94
Equilibrium, 51, 95
Equilibrium constant, 77
Equilibrium force, 96, 169, 170
Equilibrium polymers, passive polymerization, 152–154
Equipartition of Energy, Principle of, 54–55, 111, 318
Euler force, 104–105
Eyring rate theory, 85–85, 91, 177

F

Families, of proteins, 124
Fenn effect, 272
Fibroblasts, 30, 167
Fick's law, 57, 294–295, 307–308
Filament assays, 247
Filaments
  ATPase and, 242
  bending, 314
  dimensionality, 161–162
  docking of motors to, 210–212
  elongated proteins and, 291
  length, 132–133
  multistranded, 155–156, 156–158
  polarity, 133–134
  rigidity in vitro, 142–147
  shape of, 318
  single-stranded, 154–155
  sliding, 330–331
  straightness, 133
  strength, 133
  thermal bending, 110–115
First-passage times, 61–63, 296–298. See also Kramers time
Flagella, bacterial, 146
Flexibility
  in converter domains, 266
  and entropy, 39
  light-chain domains, 266
Flexural rigidity
  bacterial flagellum, 146
  definition, 100
  of microtubules, 144
  and persistence lengths, 111
  sperm and microtubules, 141
  and thermal fluctuations, 142–143

Fluctuation–dissipation theorem, 71, 303
Fluorescence, single-molecule techniques, 251
Fluorescence resonance energy transfer (FRET), 251
Fluorophores, 251
Flux
  and concentration change, 57
  definition, 94
  at equilibrium, 95
Fokker–Planck equation, 58, 59, 295
Force–extension curves, DNA, 147
Forces
  conformational changes, 245–262
  definition, 10
  dependence on speed, 274–275
  generation, 165–178
  and potential energy, 24–25
  single-motor techniques, 247–250
  work, energy and, 24–25
Force–velocity curves, 97, 273–276, 280–282
Fourier analysis, 66–69, 68, 298–304
Fourier series, 67
Franck–Condon principle, 87
Free energies
  conformational states, 303–306
  in cyclic reactions, 94–97
  and dissociation constants, 153
  and hydrolysis reactions, 236
  of reactions, 310
  of a spring, 305–306
  standard, 96
  structural state, 76
Freely jointed chains, 112–114, 320
Frequency factor, definition, 85
Function, and adaptation, 227

G

G-proteins, 42, 205–209
Gibbs free energy, 77
Glial fibrillary acidic proteins, 125
Gliding assays, motility, 217–219
Globular proteins, 12, 14, 18, 42
Glutamic acid, density, 30
Goldman-Hodgkin-Katz (GHK) current equation, 174
Gravitational forces, single-molecule level, 11, 13
GTP-cap model, 187–188
GTP (guanosine triphosphate), hydrolysis, 185, 191–192

H

Hair bundles, damping, 322–323
Hair cells, 45, 80–81, 138–140
Hand-over-hand model, 240–241, 281, 335
Harmonic motion, 21, 22
Helical rods, second moment of, 313
Helical structures, 3-start, 127
Helically symmetrical, definition, 130
α-Helices, hydrophobicity, 130–131
Helmholtz free energy, 77
Hemiparabolic energy wells, 63
Hemoglobin, density, 30
Hipposideros bicolor (bats), 322
HMM myosin fragments, 200
Homologues, 123
Hooke's law, 16, 30–31
Huxley, A. F., 79, 265, 269–270
Huxley, H. E., 197, 199
Hydrodynamic beam equation, 108, 314
Hydrodynamic bending modes, 314–315
Hydrodynamics of slender rods, 314–316
Hydrolysis
  ATP, 229–244
  cytoskeleton polymerization, 152
  and duty ratios, 221–226
  free energy and, 236
  GTP, 185, 191–192
  motor domain distances, 245–246
  motor proteins, 205–207
  nucleotides, 190–193
  power stroke and, 279
  structural model for myosin, 237

I

Inertial forces, 18, 23, 288
Intermediate filament proteins
  commercial interests, 142
  families, 124–125
Intermediate filaments, 120
  diameter, 119
  dimers, 121–122
  structures, 132
  properties of, 121
  rigidity of in vitro, 145–146
Internal combustion engines, 226
Ion channels, 76, 79, 80–81, 82
Ion diffusion, 59
Isoforms, 123
Isometrically contracting muscles, 225, 237

364 INDEX

Isotropic, definition, 30
Isotropy, of proteins, 33

J
Johnson noise, 70

K
Keratins
  leucine zippers, 130
  localization, 125
  materials containing, 146
  polarity, 134
  rigidity of materials, 142
Keratocytes, 4, 167
KIF1A, 260–261
KIF3A, 204
Kinesin-related proteins, 211
Kinesins
  atomic structures, 201, 204–205
  ATPase cycles, 242–244
  ATPase rates, 242–243
  bead techniques, 252–253
  conformational changes, 211
  conventional, motility parameters, 267
  crossbridge model for, 280–282
  dimeric, 209
  force–velocity curve, 280–282
  gliding assay, 107
  hand-over-hand model, 335
  hydrolysis of ATP, 238–242
  mechanical lever, 212
  motility, 211
  motility assays, 218–219
  motor domain, 209
  processive motor, 219–221
  purification, 1
  single-molecule techniques, 252–255
  speeds, 213–216, 221
  spring content, 114
  stepwise motion, 251
  structural similarity with myosin, 206
  thermodynamic force, 255
  tubulin dimer repeat distance, 282
  working and path distances, 259
Kinetic energy, of molecules, 71
Kinetic equations, 309–310
Kramers rate theory, 85–86, 88, 89, 177, 307
Kramers time, 63. See also First-passage times
Kuhn lengths, 115

L
Laminar flow, 38
Langevin equation, 64, 113–114, 147, 298, 319

Laser tweezers. See Optical tweezers
Lattice rotation, definition, 127
Leading edge, 167
Lengths, of filaments, 132–133, 154–155, 155–156
Lennard–Jones potential, 34
Leucine zippers, 130. See also Coiled coils
Lever length, and working stroke, 258
Light, photon shot noise, 331–333
Light-chain binding domains, 207, 210, 258, 265–266, 282–283
Liposomes, microtubules inside of, 168
Listeria monocytogenes, 166, 175
LMM myosin fragments, 200
Loading, muscle contraction, 3
Lorenzian curve, 69
Lymn–Taylor model, 197
Lysozymes, 30

M
Macrophages, 44–45
Magnetic forces, single-molecule level, 11, 13
Marcus mechanism, 87
Mass Action, Law of, 77
Mass and dashpot model, motion under external forces, 16–18
Mass and spring model, 20–21, 22–24, 287–288
Mass, 14, 29–30
Maxwell element, motion of, 28
Mean position, 53
Mean-squared position, 53
Mean time to capture, 296
mec-7 gene, 123
MEC-7 tubulin, 123
Mechanical forces, motor proteins, 9–28
Mechanical levers, motor proteins, 212
Mechanical strain, generation of, 269–270
Methylcellulose, 220
Michaelis–Menten equation, 93–94, 310
Microfilaments. See Actin filaments
Microscopic Reversibility, Principle of, 95, 157
Microtubule-associated proteins (MAPs), 144–145
Microtubules
  Amos and Klug model, 129
  and ATP hydrolysis by kinesin, 239
  bending stiffness, 140–142
  buckling, 104–105, 109
  closure of, 128

crossbridges, 199
diameters, 119
doublet, 129
flexural rigidity, 104, 141, 144
gliding assays, 218
kinesin, 252–255
in liposomes, 168
motility gliding assay, 107
motor density, 221
polymerization, 167
relaxation, 109
rigidity of filaments, 143–145
single-molecule techniques, 252
structure, 127–130
Minus (–) ends, 127
Mitochondria, 30, 60
Molecular friction, 39
Molecular mass units, 30
Molecular ratchets, 92
Monotonic motion, 16, 22
Monomeric subunits, 152
Motility
  actin-polymerization driven, 174–175
  descriptions of, 263–264
  models, 263–283
  polymerization-driven, 166–168
  power stroke models, 265–268
  in vitro assays, 216–219
Motion, laws of, 11
Motor domains, 192, 218
Motor proteins. See also Specific proteins
  ATPase rates, 223–224
  crossbridges, 198–202
  definition, 1
  density, 221
  domain organization, 200–201
  evolution, 2
  families of, 202–205
  mechanical forces, 9–28
  overdamping of, 24
  speeds of, 213–227
  stereospecificity, 210–212
  structures of, 197–212
  tails, 202
Motor-related proteins, 204
Muscle contractions. See also Skeletal muscle
  crossbridge model, 269–273
  force–velocity curves, 273–276
  isometric, 225
  mechanical amplification, 283
  mechanical transients, 276–277
  relaxation, 3
  sarcomeres, 330
  speed of, 267
Myoglobin, 41
Myosin heads, and sarcomere compliance, 138
Myosin I, working stroke, 259–260

Myosin II
  cardiac muscle, 204
  duty ratios, 224, 227
  light-chain binding domain, 282–283
  motility parameters, 267
  nonprocessive motor, 219–220
  single-molecule assays, 255–259
Myosin V
  duty ratios, 222, 227
  motion along actin filaments, 260
  processivity, 282–283
Myosin VI, directionality, 205, 208
Myosins. *See also* Specific myosins
  actin-binding domain, 200
  association with actin, 257–258
  atomic structures, 205–209
  ATPase cycles, 242–244
  ATPase rates, 242–243
  attachment to filaments, 197
  binding site distances, 282
  contracting-spring model, 271
  crossbridges, 41, 199, 333–334
  detachment from actin, 90
  duty ratio, 243–244
  elasticity, 265–266
  energy storage in conformational changes, 26
  family of proteins, 202–203, 214–215
  fragments, 200
  genetic disruptions, 204
  hydrolysis cycle, 237
  hydrolysis of ATP, 232–238
  isolation, 1
  light-chain binding domain, 210
  overdamping, 24
  path distances, 259
  protein vibrations, 22
  rotating crossbridge model, 197–198
  speeds, 213–216
  spring content, 114
  structural similarity with kinesin, 206
  working distance, 246, 259
  working stroke, 208, 251

N

N sites, definition, 122
Navier–Stokes equation, 290
Nernst equation, and Boltzmann's law, 53
Net force, definition, 10
Neurospora, 260
Newtonian fluids, 38
Newton's First Law of Motion, 11
Newton's Second Law of Motion, 11
Nitella, 216
Nkin, 254, 260

Nuclear magnetic resonance (NMR), 13
Nucleation, equilibrium polymers, 184–185
Nucleotide-binding pockets, conservation of, 205
Nucleotides, 191–193, 197, 232. *See also* Specific nucleotides

O

Off-rates. *See* Dissociation rate constants
On-rates. *See* Association rate constants
Opsin proteins, 75
Optical forces, single-molecule level, 12–13
Optical techniques, protein mechanics, 5
Optical tweezers, 5, 225, 248–249, 251–259
Organelles, movement of, 61
Orthologues, 123
Oscillation, 16, 21
Osmotic pressure (π), 82
Overdamping, 41–44, 44–45, 287, 289–290

P

Paralogues, 123
Parseval's theorem, 299–300
Partition function, 50
Path distances, 246, 259, 277–279
Pawl and ratchet, 268
Persistence lengths, 110–112, 111, 145, 316–317
Photodetectors, sensitivity of, 250
Photodiode detectors, 250, 332–333
Photon shot noise, 331–333
Photons, momentum, 12
Piconewtons, 79
Pivotal springs, 320
Plus (+) end, 127
Polarity, 127, 133–134
Polarity assays, 218–219
Polymerization
  actin, 179–191
  actin filaments, 3
  association rate constants, 93
  coupling hydrolysis, 191–192
  cytoskeleton filaments, 151–163
  diffusion-limited, 173–174, 175
  dynamic instability, 185–186
  force generation by, 166–168
  growth of multistranded filaments, 156–158
  growth/shrinkage switching, 186–187
  nucleation, 184–185
  passive, 152–154

  reaction-limited, 173
  reaction transition states, 177
  reaction–diffusion equation, 325–328
  two-stranded filaments, 159
  in vitro, 169
  work during, 188–191
Polymers
  dimensionality, 161–162
  dynamic instability, 329–330
  mechanics, 99–116
  persistence lengths, 111
  semiflexible, 112
  single-stranded, 323–324
  subunits, 152
  two-stranded, 324–325
Polymorphism, genetic, 122
Potential energy, 24–25, 39, 304
Power spectra
  damped spring, 69
  Fourier analysis, 67–69
  Parseval's theorem, 299–300
  of signal, 67
  of thermal forces, 301–302
Power stroke
  crossbridge cycle, 221–222
  models, 197
  models of motility, 265–268
  myosin, 208
  per ATP hydrolysis cycle, 279
  role of thermal fluctuations, 268–269
Powerstroke distances
  and conformational change, 266
  definition, 245
  and muscle properties, 277–279
  and work, 274–276
Pre-exponential factor, definition, 85
Pressure, definition, 30
Probability flux, 58
Processive motors, 219–221
Processivity, 220, 243, 259–261
Protein filaments, damping ratios, 43
Proteins
  bonds between, 20
  composition, 29
  diffusion, 65–66
  diffusion rates, 62
  elongated, 291
  friction, 39
  global motions of, 41–44
  internal viscosity, 41
  limit to conformational changes, 306–307
  mass, 29–30
  mechanical components, 5
  persistence of motion, 18
  Reynolds numbers, 39
  segmental flexibility, 113

timescale of conformational changes, 19
van der Waals model of rigidity, 39
vibrational frequencies, 293–294
Young's modulus, 39

Q
Quinary structures, 29

R
Radius of curvature, 100
Random walk, 55–58, 294–295
Rate constants, 83–85, 272
Ray optics, 249
Reaction coordinate, 85
Reaction-limited rates, 173
Reaction–diffusion equation, 59, 325–328
Recovery strokes, 222
Relaxation, 3, 109, 140
Rescue rate, 186–187
Reversal force, 170
Reversibility, polymerization, 157
Reynolds numbers, 38, 39, 290
Ribosomal oscillation, 43
Rigidity
  and constituent bonds, 35
  filaments, 136–142, 142–147
  keratin-containing materials, 142
  ribosomes, 42
  van der Waals forces, 288–289
Rigor, 140, 218
RNA polymerase forces generated, 261
Rods, stiffness under tension:, 32
Root-mean-square displacement, 60
Rotating crossbridge model, 197
Rotating lever model, 271
Rotational symmetry, 122

S
Saccharomyces cerevisiae, 123, 124
Sarcomeres, 3, 136, 216, 330
Scanning confocal microscope, 251
Schizosaccharomyces pombe, 151
Second moment of inertia of the cross-section, 101
Semiflexible, definition, 112
Shear moduli, 47, 146
Shear thinning, 38, 40
Shigella flexneri, 166
Signals, 66–69, 299–300
Single-molecule techniques, 4, 5, 246–251. *See also* Specific techniques
Site-directed mutagenesis, 5
Skeletal muscle, 3, 136, 200. *See also* Muscle contractions

Slender filaments, shape of, 318
Slender rods. *See also* Polymers
  drag forces on, 105–107
  end-to-end length, 317–318
  ending of, 100
  fluctuation in shape, 110
  hydrodynamics, 108–109, 314–316
  thermal bending, 110
  time constant of force fibers, 109
  torsion, 116
Sliding speeds, duty ratios and, 273, 283
Species, genetic variations, 123
Speed of sound, 306
Speeds
  and ATPase rates, 243
  force, work and, 274–275
  motor density and, 221
  of motor proteins, 213–227
Sperm
  bending stiffness of microtubules, 140–142
  drag coefficients, 107
  filament sliding, 330–331
  flexural rigidity, 141
  motility, 167
  polymerization-driven motility, 176
  rigidity of, 144
Spheres, oscillating, 290–291
Spring and dashpot model, motion under external forces, 19
Spring and mass models, 20–21, 287–288
Springs
  cantilever, 32, 315–316
  cubic lattice, 35
  entropic, 112
  free energy, 305–306
  molecules attached to, 54
  motion under external forces, 14–16
  pivotal, 320
  power spectra of, 69
Stall forces, 253, 255
Standard free energy, 96
Steady state, definition, 52
Step size, 246. *See also* working distances
Stereocilia
  bending stiffness of actin, 138–140
  damping on hair bundles, 322–323
  hair cells, 45
  ion channels, 80–81
  stiffness, 321
Stereospecificity, 93, 210–212

Stiffness
  actin in stereocilia, 138–140
  bending, microtubules and sperm, 140–142
  chemical bonds, 21
  coil springs, 33
  and compliance, 27
  crossbridges, 263–264
  rods under tension, 32
  segmental proteins, 99
  single-molecule level, 12
  skeletal muscle, 137
  of springs, 16
  stereocilia, 80 879, 321
  and structural states, 79
Stokes' law, 15, 38
Straightness, of filaments, 133
Strength, of filaments, 133
Strongly bound states, 236
Structural states, 76, 78, 82
Subfragment 1 (S1), myosin, 200
Subunits, description, 152
Superfamilies, of proteins, 124
Supertwists, microtubules, 127–128
Swinging lever arm hypothesis, 197
Symmetry, helical, 130
Synaptic vesicle density, 30

T
Tadarida brasiliensis (bats), 322
Tails, motor proteins, 202
Tangent angle, 102
Tendons, relaxation, 3
"Tensegrity" model, 165
Tensile forces, 31, 312
Tensile strength, 31, 32
Terminal velocity, definition, 17
Thermal bending, 110–115, 316–320
Thermal energy, 49, 51
Thermal fluctuations, 144, 251, 268–269
Thermal forces
  autocorrelation of, 302–303
  definition, 49
  and diffusion, 49–73
  illustration of, 70
  magnitude of, 69–71
  noise, 70
  power spectra of, 301–302
  single-molecule level, 11, 12
Thermal motion, definition, 49
Thermal ratchets, 88, 89, 268. *See also* Kramers rate theory
Thermodynamic forces, single-molecule, 255
Thick filaments, 3, 136, 199, 216
Thin filaments, 3, 136–137, 199

Time-average force, crossbridges, 263
Tip links, stereocilia, 139
Titin, 39, 91, 147
Torsional rigidity, slender rods, 116
Total-internal-reflection microscope, 251
Transition states, 84–85, 90
Translational symmetry, 122
Treadmilling, of actin, 184
Tropomyosin, rigidity of, 143
Trypsin, density, 30
Tubulins
  αβ-tubulin heterodimers, 121–122
  classes of proteins, 124
  δ-tubulin localization, 124
  dimer repeat distance, 282
  ε-tubulin localization, 124
  family of proteins, 124
  force generation in vitro, 169
  γ-tubulin localization, 124
  MEC-7 isoform, 123
  movement of kinesin, 252
  nucleotide hydrolysis, 191–193
  polymerization, 93
  properties of, 121
  rigidity in vitro, 143–144
  structures, 119, 121

U
Ultracentrifuges, 13
Underdamping, 22, 287

V
van der Waals forces
  proteins, 39
  rigidity, 34, 288–289
  single-molecule level, 11
Velocity, 10, 14
Vibration, chemical bonds, 21
Vibrational energy, proteins, 293–294
Viruses, 30, 166
Viscoelastic, definition, 47
Viscoelastic materials, 19
Viscosity
  definition, 37
  of honey, 39
  molecular basis of, 39–41
  various liquids, 38
Viscous forces
  and inertial forces, 23
  mass and spring with damping, 288
  single-molecule level, 11, 12
Voigt elements, definition, 28

W
Water, viscosity, 38
Wave optics, 249
Weakly bound states, 236
Work
  definition, 204
  dependence on speed, 274–275
  depolymerization, 188–191
  and force, 24–25
  kinesin, 254
  polymerization, 188–191

Working distances, 222
  definition, 245
  and duty ratios, 224–225
  kinesin and myosin, 259
  myosin II, 257
Working stroke
  and lever length, 258
  and motor domains, 221–222
  myosin, 236–237
  myosin II, 257
Worm-like chains, 114–115

X
Xenopus, 4

Y
Young's moduli
  actin, 137
  bacterial flagellum, 146
  biological materials, 148
  covalent solids, 35
  definition, 31
  and flexural rigidity, 100–101
  keratin-containing materials, 142
  proteins, 32–33
  and shear modulus, 47
  single-motor techniques, 247
  and tensile strength, 31
  thin-filament proteins, 137
  van der Waals forces, 288–289

Z
Z-lines, 136

ABOUT THE AUTHOR

Jonathon Howard is currently a Professor in the Department of Physiology and Biophysics at the University of Washington. He is a Director of the Max Planck Institute for Molecular Cell Biology and Genetics in Dresden, where his research group will move in July, 2001. He earned a Ph.D. in Neurobiology at Australian National University, doing postdoctoral research there as well as at the University of Bristol and the University of California, San Francisco. Dr. Howard's current research interests include the mechanics of motor proteins and the cytoskeleton and mechanoelectrical transduction by sensory receptors. The recipient of several scholarships and fellowships, he most recently received a MERIT Award from the National Institutes of Arthritis and Musculoskeletal and Skin Diseases. The writing of *Mechanics of Motor Proteins and the Cytoskeleton* was inspired by Dr. Howard's teaching of a course on Cell Motility at the University of Washington.

ABOUT THE BOOK

Editor: Andrew D. Sinauer

Project Editor: Kerry Falvey

Production Manager: Christopher Small

Book Production: Janice Holabird and Joan Gemme

Graphic Art: networkgraphics

Book Design: Jefferson Johnson

Cover Design: Jefferson Johnson

Cover Manufacture: Henry N. Sawyer Co., Inc.

Book Manufacture: Courier Companies, Inc.

Table of Parameters

Symbol	Parameter	SI unit	Chapter(s)
$[\ldots]$	Concentration	M	4
$\langle\ldots\rangle$	Time average	—	4
a, a_c	Acceleration, centrifugal acceleration	m/s^2	2
a_n, b_n	Fourier coefficients	m	4
A	Area	m^2	3
$[A_n], [A_t]$	Concentration of n-mers, total subunits	M	9
b	Kuhn length	m	6
c	Concentration	m^{-3}	4
$c, c_\parallel, c_\perp$	Drag coefficient per unit length	N·s/m^2	6
d	Path distance	m	15
D	Diffusion coefficient	m^2/s	4
E	Young's modulus	N/m^2	3
$E(\ldots)$	Expectation	Various	4
EI	Flexural rigidity	N·m^2	6
f, ν	Frequency	Hz = s^{-1}	2
f	Force per unit length	N/m	6
f_{+-}, f_{-+}	Catastrophe rate, rescue rate	s^{-1}	11
$F, \langle F \rangle$	Force, average force	N	2, 16
F_{eq}	Equilibrium force	N	5, 15
G	Shear modulus	N/m^2	3
$G(G^0)$	Free energy (no force)	J	5
$G_x(f)$	Power spectrum	m^2/Hz	4
H	Enthalpy	J	5
i	Electric current	A	2
I	Second moment of inertia	m^4	6
j, J	Probability flux, flux	m$^{-2}\cdot$s^{-1}	4
k, k_1, k_{-1}	Rate constant, forward, reverse	s^{-1}	4
k_{cat}	Catalytic constant	s^{-1}	5
$K, K_{eq} (K_{eq}^0)$	Equilibrium constant (no force)	*	5
K_c	Critical concentration	M	9
K_d	Dissociation constant	M	5
K_M	Michaelis–Menten constant	M	5
L	Length	m	3
L	Langevin function	Dimensionless	6
L_p	Persistence length	m	6
m	Mass	kg	2
M	Bending moment	N·m	6
n_{av}	Number of polymers in a filament	Dimensionless	9

*Varies with the order of the reaction

Symbol	Parameter	SI unit	Chapter(s)
p	Probability	m^{-1}	4, 15
P	Pressure	$Pa = N/m$	2
P	Power	$W = J/s$	15
q	Charge	C	2
Q	Heat	J	2
Q_{10}	Q-ten	Dimensionless	5
r	Radius	m	2
r	Duty ratio	Dimensionless	13
R	End-to-end distance	m	6
R	Radius of curvature	m	6
$1/R$	Curvature	m^{-1}	6
Re	Reynolds number	Dimensionless	3
$R_x(\tau)$	Autocorrelation	m^2	4
s	Arc length	m	6
S	Entropy	J/K	5
t, T	Time	s	2
T	Temperature	$K = °C + 273.15$	2
$U, \langle U \rangle$	Potential energy, average energy	$J = N \cdot m$	2
v, V	Velocity	m/s	2
V	Volume	m^3	3
w, W	Work	J	2
$x, \langle x \rangle = \langle x \rangle_\infty$	Displacement, time-averaged displacement	m	2
$x_{rms} = \sqrt{\langle x^2 \rangle}$	Root-mean-square displacement	m	4
Z	Partition function	Dimensionless	4
γ	Drag coefficient	$N \cdot s/m$	2
δ	Incremental length, working distance	m	9, 13
Δ	Distance per ATP	m	13
ΔG	Change in free energy	J	5
$\Delta G_{S \to P}$	Free energy of reaction	J	5
$\Delta G_S, \Delta G_{12}$	Entropic energy, interaction energy	J	9
η	Viscosity	$N \cdot s/m^2$	2
θ	Shear angle, tangent angle	rad	3
κ, K	Stiffness	N/m	2
λ	Wavelength	m	4
ρ	Density	kg/m^3	2
σ	Poisson ratio	Dimensionless	3
σ^2	Variance	m^2	4
τ	Time constant, delay time	s	2, 4

Table of Physical Constants

Constant	Symbol	Value	SI Unit
Avogadro constant	N	6.022×10^{23}	molecules/mol
Boltzmann constant	k	1.381×10^{-23}	J/K
Electric constant	ε_0	8.854×10^{-12}	F/m
Elementary charge	e	1.602×10^{-19}	C
Faraday constant	$F = Ne$	9.649×10^{4}	J/mol
Molar gas constant	$R = Nk$	8.314	J/K·mol
Planck constant	h	6.626×10^{-34}	J·s
Speed of light	c	2.998×10^{8}	m/s

Note: For a comprehensive list of physical constants and their uncertainties, see http://physics.nist.gov/cuu/index.html

Table of SI Base Units

Quantity	SI unit	Symbol
Length	Meter	m
Mass	Kilogram	kg
Time	Second	s
Electric current	Ampere	A
Temperature	Kelvin	K
Amount	Mole	mol
Luminous intensity	Candela	cd

Note: For a discussion of the SI units, see http://physics.nist.gov/cuu/index.html

Table of Derived Units

Quantity	SI unit	Symbol	Composite symbol
Plane angle	Radian	rad	—
Frequency	Hertz	Hz	s^{-1}
Force	Newton	N	$kg \cdot m/s^2$
Pressure	Pascal	Pa	N/m^2
Energy	Joule	J	$N \cdot m$
Power	Watt	W	J/s
Charge	Coulomb	C	$A \cdot s$
Potential difference	Volt	V	W/A

Table of Prefixes

Factor	10^9	10^6	10^3	10^{-3}	10^{-6}	10^{-9}	10^{-12}	10^{-15}	10^{-18}
Name	giga	mega	kilo	milli	micro	nano	pico	femto	atto
Prefix	G	M	k	m	μ	n	p	f	a

Note: Exponents apply to the attached prefixes; example: 1 nm³ = 10^{-27} m³.